Nutritional and Physiological Functions of Amino Acids in Pigs

François Blachier • Guoyao Wu • Yulong Yin
Editors

Nutritional and Physiological Functions of Amino Acids in Pigs

Springer

Editors
François Blachier
Paris, France

Guoyao Wu
Texas A&M University
College Station
USA

Yulong Yin
Chinese Academy of Sciences
Institute of Subtropical Agriculture
Changsha, Hunan
China

ISBN 978-3-7091-1327-1 ISBN 978-3-7091-1328-8 (eBook)
DOI 10.1007/978-3-7091-1328-8
Springer Wien Heidelberg New York Dordrecht London

Library of Congress Control Number: 2013934530

Printed on acid-free paper

Springer is part of Springer Science+Business Media (www.springer.com)

Preface

We, as the three editors, are pleased to edit this book entitled "Nutritional and Physiological Functions of Amino Acids in Pigs," which provides a comprehensive overview of the nutritional and physiological functions of amino acids in swine during growth and development. The first two parts of the book are related to the rapid development of the gastrointestinal tract and digestive glands in piglets and to the related terminal digestion of peptides by the enterocytes before amino acid absorption and metabolism by these cells. In the second part of the book, a chapter is specifically devoted to the development of the gut microbiota in piglets. The physiological functions of amino acids and related metabolites in the body are then presented in the subsequent chapters. These articles take into account that, apart from the role of amino acids as building blocks for protein synthesis, these compounds also have numerous functions including hormone secretion, signaling role in intestine, and modulation of immune functions. Finally, given the similarities in anatomy, digestion, absorption, metabolism, physiology, and nutrition between swine and humans, a special chapter is dedicated to the use of pigs as an animal model to study amino acid-related human diseases, including obesity, diabetes, and vascular disease.

The last part of the book is devoted to methodological considerations for research on amino acids including the state of the art for performing experiments such as surgical techniques used in the field, measurement of protein digestibility in growing pigs, quantification of protein metabolism in pigs, and analysis of amino acids including those of endogenous origin.

This book was written not only from an academic point of view but also with a goal of translating basic scientific knowledge to practices in agriculture, nutrition, physiology, and medicine. Thus, we hope that this book will be useful for agronomists involved in pig breeding, for professionals involved in amino acid production, for nutritionists dealing with alimentary amino acid supplementation, for physiologists and pathophysiologists who may use the pig as a model for studying human medicine, and for students majoring in agricultural and veterinary sciences.

Lastly, we wish to thank warmly all the talented contributors who spent much of their time and efforts on writing this textbook and NSFC (#31110103909), the China Basic Research Program (#2013CB127301) and the companies of LuYe, WiHon Food Co., Ltd, Hunan Diyi Bio-tech Co., Ltd, Hunan Ground, Shenzhen Premix Inve Nutrition Co., Ltd, Jiuding Group, Twins group, Nufarm group, Aonong, Lucta, Xingjia Biotechnology, Singao, BFI, Insighter, Tianke, Sunhy, Evonik Degussa (China) Co., Ltd. for their support.

Paris, France — François Blachier
College Station, TX, USA — Guoyao Wu
Changsha, China — Yulong Yin

Contents

Part III Physiological Functions of Amino Acids in Pigs

Part IV Methodology for Research on Amino Acid

Part I
Anatomical Characteristics of the Gastrointestinal Tract and Digestive Glands of Pigs During Development

Chapter 1
Development of the Gastrointestinal Tract in Pigs

Kang Yao, Zhihong Sun, Zikui Liu, Zhefeng Li, and Yulong Yin

Contents

K. Yao • Y. Yin (✉)
Chinese Academy of Sciences, Institute of Subtropical Agriculture, Research Center for Healthy Breeding Livestock & Poultry, Hunan Engineering & Research Center of Animal & Poultry Science, Key Laboratory of Agro-ecological Processes in Subtropical Region, Scientific Observing and Experimental Station of Animal Nutrition and Feed Science in South-Central, Ministry of Agriculture, Changsha, Hunan 410125, People's Republic of China
e-mail: yinyulong@isa.ac.cn

Z. Sun
College of Animal Science and Technology, Southwest University, Chongqing 400715, China

Z. Liu
Veterinary Faculty, Hunan Agricultural University, Changsha, Hunan 410128, China

Z. Li
Hangzhou King Techina Technology Co., Hangzhou, China

F. Blachier et al. (eds.), *Nutritional and Physiological Functions of Amino Acids in Pigs*,
DOI 10.1007/978-3-7091-1328-8_1,

1.1 Introduction

Feeding intolerance and necrotizing enterocolitis are all two commonly encountered problems in neonatal animal that result in significant mortality and morbidity and significant costs of care. The extent to which gastrointestinal tract (GIT) dysfunction occurs in neonatal animal is related to the increased incidence of clinical complications such as feeding intolerance, poor growth, necrotizing enterocolitis, malabsorption syndromes, and diarrhea.

The characteristics of GIT development in mammals (including pigs) during prenatal and postnatal periods, the effects of early nutrients on GIT development, and the negative impacts of early malnutrition on GIT development are discussed in this chapter. One aim of this chapter is to bring a clear understanding of the characteristics and mechanisms of development pertaining to gastrointestinal morphology, digestion, absorption, immune and microbial community at all stages of prenatal and postnatal period.

1.2 Characteristics of the Development of the Gastrointestinal Tract in Mammals

1.2.1 Stomach

The patterns exhibited by the different species resemble each other, although the chronology and sequence of appearance of the specialized glandular cells may vary. In this paragraph, the sequence of development of the human stomach is briefly described. At about 4 weeks following fertilization, the foregut, caudal to the primitive lung buds, expands. At 6–9 weeks after fertilization, depressions, characterized by epithelial cells oriented in a radial fashion, appeared in the epithelium along the lesser curvature; and the stratified (or pseudo-stratified) epithelium is then replaced by a layer of cuboidal cells. At the age of 11 weeks the first differentiated glandular epithelial cells appears, with clear representation of the structure of parietal cells. Finally, the chief cells (or a "zymogenic cell") which release a precursor enzyme in the stomach appeared at 12–13 weeks of age (Deren 1971).

In the human infant at birth, well-defined parietal cells are visible, and mucous neck cells are clearly discernible. The surface parietal cells contain periodic acid-schiff-positive material similar to that found in the adult. Although the chief cells are sparsely distributed compared to the adult stomach, they do appear to be capable of secreting pepsin. However, the total thickness of both the mucosa and the other layers of the stomach are much thinner than that observed in the adult (Deren 1971). The subsequent development of the stomach is related to thickening of the glandular portion of the stomach, with the appearance of more mature zymogen cells (Salenius 1962).

1.2.2 *Intestine*

The intestinal development in the pig, like in the human, is initiated early in fetal development (Kelly and Coutts 2000). At 40 days of gestation, the intestinal morphogenesis has progressed to the point where recognizable villi are present. At that time, mRNA for enzymes and cytoskeletal proteins are readily detectable (Perozzi et al. 1993). As the GIT develops, tremendous growth occurs, with a doubling of intestinal length in the last trimester of pregnancy. However, the mucosal surface area increase is even more dramatic, largely because of the villus and microvillus growth during this period of development (Neu 2007a). In the weeks before parturition, the pig intestine grows more rapidly than the body as a whole and its relative weight increases 70–80 % over the last 3 weeks of gestation (Sangild et al. 2000). The intestine undergoes tremendous growth during fetal life. It elongates 1,000 folds from 5 to 40 weeks. The length doubles in the last 15 weeks of gestation reaching a mean length at birth of 275 cm (Neu 2007b). In the small intestine, fingerlike projections, the villi are already formed at 16 weeks gestation (Neu 2007b).

After birth, the mammalian intestine undergoes two main phases of development during the neonatal period (Kelly and Coutts 2000). The first phase involves the preparation for extrauterine life when maternal colostrum and milk will provide the sole nutrient source. The second phase of intestinal development is associated with a shift in the digestive capability of the epithelium from one which digests and absorbs a milk diet to one which efficiently utilizes complex solid feeds. With these both phases, it is recognized that extensive changes occur in the gross architecture, ultrastructure, growth and differentiation of the intestine (Kelly and Coutts 2000). During the immediate postnatal period a dramatic increase in the diameter, total weight and surface area of the intestine of pig occurs (Zhang et al. 1997). Total intestinal and mucosal weight of pig increased by 58 % and 80 %, respectively during the first 6 h of suckling, and the mucosal DNA content is 4.6-fold higher at 24 h after birth (Zhang et al. 1997). This increase in growth is driven by accelerated cell proliferation in the crypt stem cell compartment of the intestine (Kelly and Coutts 2000). Microvilli begin to cover the apical surface of the small intestinal epithelium up to the adulthood, and the intestinal surface provides the largest interface between the outside environment and the internal milieu (approximately 2,000,000 cm^2, which is about the size of a tennis court). The increase in surface area during development has significant implications in terms of nutrient absorptive capacity (Neu 2007b).

1.3 The Characteristics of Functional Development of Gastrointestinal Tract

1.3.1 *Digestion and Absorption*

The function of the GIT including the esophagus is, to a large degree, mechanical. Contents received from the esophagus and stomach are propelled further down the intestine and mixed with secreted fluids to digest and absorb the food constituents. The peristaltic transport of material in tract is a neuromuscular function affected by a number of factors (Arndorfer et al. 1997). Prenatally, GIT functional maturation of pig may be more dependent on internal signals (genetic, endocrine) than on luminal stimuli (enteral nutrients and bioactive factors) (Sangild et al. 2002). Postnatally, the digestive and absorptive capacity of neonatal pig is highly sensitive to change in the enteral dietary input (Jensen et al. 2001).

Absorption of nutrients, as measured by the uptake of monosaccharides and amino acids by the intestinal mucosa, is low during the first half of gestation of fetal pig but increases rapidly thereafter (Buddington and Malo 1996; Kong et al. 2012). During the final weeks of gestation of fetal pig, there is a particularly rapid increase in glucose uptake capacity (Sangild et al. 1993), while the ability of the mucosa to take up most amino acids, including lysine, remains unchanged (Sangild et al. 2000).

The primary factors in feeding premature infants are dependent on the development and maturation of digestion and absorption. The maturation of digestive and absorptive functions of carbohydrates, proteins, fats, minerals, and vitamins in neonate are determined in relation to availability of hydrolytic enzymes, such as lipases, proteases, amylases, glucosidases, and lactase (Lebenthal and Lebenthal 1999). Most digestive-absorptive processes are well developed in late preterm infants, and there should be very few limitations in terms of digestion and absorption of lactose, long-chain lipid, or protein (Neu 2006).

1.3.2 Protein

Protein digestion is initiated within the stomach, with the secretion of both acid and proteolytic enzymes (pepsin and papain). The protease zymogens are produced mainly by the chief cells of the fundic mucosa, and their activation and activity in the gastric lumen are stimulated by an acidic environment (pH 2.0–4.0) (Deren 1971). At the first 24–48 h after birth, the intragastric pH remains at about 5.5–7.0 (whereas the gastric pH in normal children and adults is 2.0–3.0) (Neu 2007b).

The enzymes in the small intestine (trypsin, chymotrypsin, and pancreatic proteases) perform the main function of digesting protein (Mayer 2003). The digestive cascade of pancreatic protease is catalyzed by food-stimulated secretion of enterokinase from the upper small intestinal epithelium. Enterokinase catalyzes the activation of trypsinogen to trypsin (Neu 2007b). The breakdown of large polypeptides into small dipeptides and tripeptides accomplishes two tasks: it allows for the process of digestion and absorption of nutrients to occur, and it renders potentially immunogenic proteins nonimmunogenic (peptides <8–10 amino acids in length are little immunogenic) (Mayer 2003). Active luminal enterokinase is detectable at 24 weeks gestation of human fetus but its activity level is relatively low and reaches only 25 % adult activity at term (Antonowicz and Lebenthal 1977). Similar to a lack of gastric acid, low enterokinase activities may limit protein digestion and may be responsible for an increased capability of larger antigens and/or microorganisms to pass into the intestine without breakdown by luminal enzymes. After luminal digestion, small peptides and amino acids are absorbed and transported to the villus capillaries. The proteases are usually present and fully active before 24 weeks gestation of human fetus and should not be limiting (Lebenthal et al. 1983). This suggests that hydrolyzed protein would be better tolerated.

Absorption of amino acids mainly occurs in the proximal region of the small intestine of pigs (Buraczewska 1981). The intestinal mucosal cells absorb amino acids via active transport, simple diffusion, and facilitated diffusion. There are at least four sodium-dependent amino acid transporters in the luminal plasma membrane of the intestinal mucosal cells responsible for transporting amino acids into the cytoplasm. The capacity for macromolecular absorption is very important in newborn pigs, which relies on passive immunity from colostral immunoglobulins (Lin 2006). The fetal type of enterocytes responsible for macromolecular uptake is present at birth.

There is little knowledge about the rates of amino acid absorption across the apical membrane during development. The rates of absorption for five amino acids using intact tissues from the proximal, mid-, and distal small intestines of pigs are ranged in age from 90 % of gestation to 42 days after birth (12 days after weaning) (Buddington et al. 2001). The results suggest that the rates of absorption (sum of carrier-mediated and apparent diffusion) were highest at birth (except for proline) and declined by an average of 30 % during the first 24 h of suckling.

1.3.3 Lipid

Fats are essential components of the diet and have a critical role in the growth and development of the neonate. The digestion of lipid can be split into several phases (Hamosh 1996). The luminal phase of lipid digestion involves lipase-catalyzed de-esterification of triglycerides to 3-monoglyerides and free fatty acids, and bile acid-mediated micellar solubilization. Bile acids synthesized in the liver are critical to efficient fat digestion and absorption (Neu 2007b). In the newborn and particularly in premature infants, pancreatic secretion of lipase and bile salt secretion are relatively inadequate (Carey and Hernell 1992). Briefly, the next phases, lipid absorption and assimilation, include the permeation of fatty acids and 2-monoglycerides from the lumen into the cell, intracellular re-esterification, chylomicron formation, and transport of the chylomicrons from the cell into the circulation (Neu 2007b).

Developmental deficiencies in luminal lipid digestion by pancreatic lipases and solubilization by bile acid micelles that result in "physiological" fat malabsorption in the newborn, as well as compensatory mechanisms such as gastric lipase digestion, have been well studied and are reviewed elsewhere (Black 2007).

1.3.4 Saccharide

During the final weeks of gestation of fetal pig, there is a particularly rapid increase in glucose uptake capacity (Sangild et al. 1993). Active glucose transport systems are operating in the intestine of 17-week-old human fetuses (Eggermont 1991). Maturation of mechanisms for carbohydrate absorption occurs in a defined sequence during human fetal development (Kien 1996). Disaccharidases are present by 8 gestational weeks of human fetuses, their concentrations increase around 14 weeks of gestation and then progress steadily to term values. Glucoamylase, a specific intestinal hydrolytic enzyme, develops very well by the 13 gestational weeks (Eggermont 1969). Trehalase appears by the 13 gestational weeks of human fetuses (Schmitz 1991).

The intestinal epithelial absorptive enzymes, such as lactase, sucrase, maltase, isomaltase, and glucoamylase, are at mature levels in the term newborn human neonate (Antonowicz and Lebenthal 1977). However, small intestinal lactase activity of human neonate is low relative to the other disaccharidases (Antonowicz and Lebenthal 1977). Despite this limitation, these infants appear to adapt rapidly to lactose feedings and also possess intestinal microflora that have the capability to ferment lactose into two and three carbon fragments that are absorbed in the distal intestine and utilized for ATP production (lactose salvage pathway) (Kien 1996).

Postnatal adaptive responses to ingested carbohydrates lead to efficient carbohydrate absorption. Inadequately absorbed carbohydrates are salvaged by colonic flora through fermentation of carbohydrates to hydrogen gas and short-chain fatty acids. The latter are readily absorbed through the colon epithelium. In this setting, carbohydrate tends to be absent from the stool. Noninvasive reflection of the status of carbohydrate absorption may be obtained from breath hydrogen testing, a technique of particular value in young infants (Mobassaleh et al. 1985).

1.3.5 Gastrointestinal Motility

There is a critical point of development at around 24–25 weeks of gestation when peristalsis is observed in all human fetuses (Sase et al. 1999). The small bowel motility patterns of human fetuses are poorly developed before 28 weeks gestation. The disorganized small intestine motility patterns of human fetuses between 27 and 30 weeks gestation, progressed to a more mature pattern so that the migrating myoelectric complexes are present at 33–34 weeks of gestation (Berseth 1996). Gastro-anal transit ranges from 8 to 96 h in premature human infants as compared with 4–12 h in adults (Neu 2007b). The motility of the small intestine is considerably less organized in premature human infants than in term infants (Berseth 1996). This is caused by an intrinsic immaturity of the enteric nervous system that delays transit, causing subsequent bacterial overgrowth and distension from gases that are the byproducts of fermentation (Neu 2007a). In preterm human infants, the motilin receptor is not present until 32 weeks gestation, and the cyclic release of motilin is not present (Neu 2007b).

1.3.6 Microbial Community

The resident microbiota confers many benefits to the gastrointestinal physiology of the host and is, therefore, an example of a truly symbiotic relationship (Hooper and Gordon 2001; Tang et al. 2013). Further benefits become visible in the period after birth, when the complexity of the intestinal environment increases considerably while changing from an exclusively milk-containing diet to an adult diet after weaning (Rumbo and Schiffrin 2005). During this period, the intestinal microbiota plays a crucial inductive role in the intestinal development. After birth, the intestinal microbiota takes some time before developing a stable community (Gaskins 2001). During the first few weeks after birth, microbial succession in the GIT of humans (Favier et al. 2002) and pigs (Moughan et al. 1992) is remarkably similar. After birth, the germ-free GIT is rapidly colonized by anaerobic and facultative anaerobic bacteria. Profound changes occur in the intestinal ecosystem when young mammals are weaned from their mother's milk (Neu 2007b). After weaning, clostridium, ruminococcus, enterococcus, and enterobacter spp. appeared, and microbial DGGE profiles became more complex and also more stable with increasing age (Bauer et al. 2006).

A cross-talk exists between microbes and the intestine whereby stimulation of the secretion of peptides by Paneth cells promotes angiogenesis, growth, and also an environment that prevents the growth of potentially pathogenic microorganisms (Hooper and Gordon 2001; Rumbo and Schiffrin 2005; Gaskins 2001; Favier et al. 2002; Moughan et al. 1992; Bauer et al. 2006; Neish 2002). Most very premature infants in the neonatal intensive care unit (NICU) are started on broad-spectrum antibiotic therapy shortly after birth during a "rule-out sepsis" workup. This can alter the normal flora with which the neonate would become colonized (Kosloske 1994). Rather than lactobacillus, bifidobacter, or other symbiotic microorganisms, resistant species indigenous to the NICU may colonize the infant's intestines. The commensal microflora may represent a key regulatory checkpoint for the intestinal inflammatory response. The intestinal epithelium partially relies on toll-like receptors (TLRs) to act as an interface between the luminal microflora and cellular signal transduction pathways. TLRs are cell surface receptors that recognize specific microbial ligands, from both pathogens and commensals, which enables the innate immune system to recognize non-self and activates both innate and adaptive immune responses (Takeda et al. 2003). Recent studies suggest that the epithelium and resident immune cells do not simply tolerate commensal microorganisms but are dependent on them (Rakoff-Nahoum et al. 2004). This is important for not only disease entities that we see in the NICU such as NEC, which has recently been shown to decrease with the use of probiotics (Vinderola et al. 2005), but also diseases that affect the infant later in life, such as allergy and atopy (Kalliomaki et al. 2001).

1.3.7 Immunology Function

The GIT is the largest organ in the immune system of the body, and as such is the location for the majority of lymphocytes and other immune effector cells. Most apparent are the functions of the Peyer's patches and other aggregated lymphoid follicles, which contain germinal centers where B and T lymphocytes proliferate and become antigen-specific lymphocytes, IgA secreting plasma cells, and memory cells (Koutsos and Arias 2006).

The immune system also acts to ensure tolerance to "self," to food and other environmental components, and to commensal bacteria. The development of tolerance is the result of active immune mechanisms requiring antigen-contact and acting in a T cell dependent fashion (Faria and Weiner 2005; Taylor et al. 2006). Both development and maintenance of tolerance are lifelong processes which start very early in life, even prenatally. Profound immunologic changes occur during pregnancy, involving a polarization of T helper (Th) cells towards a dominance of Th2 and regulatory T cell effector responses in both mother and fetus (Reinhard et al. 1998; Raghupathy 2001). Mature CD^{4+} and CD^{8+} T cells can be detected in the fetus as early as around week 20 in human pregnancy (Michaelsson et al. 2006). Therefore, during the third trimester of human pregnancy, fetal T cells are able to mount antigen-specific responses to environmental and food-derived antigens and antigen-specific T cells are detectable in cord blood in virtually all newborns indicating in utero sensitization (Szepfalusi et al. 2000). Although the fetus at term may be sensitized to certain antigens, it does lack a fully functional immune system and has a sterile GIT. Changes occur at, and soon after, birth in order to allow the immune system of the neonate to become competent and functional; and also to allow the gut to become colonized with bacteria. Pregnancy, the suckling period, and the periods during which formula and solid foods are introduced offer windows during which nutrition might affect the immunologic development of the fetus and young infants (Philip et al. 2006).

The mucosal immune system fulfills the primary function of defense against potential pathogens that may enter across vulnerable surface epithelia. However, a secondary function of the intestinal immune system is to discriminate between pathogen-associated and "harmless" antigens, expressing active responses against the former and tolerance to the latter. There are considerable evidences to indicate that the major functions of the mucosal immune systems are defective at birth (Adkins et al. 2004). Passive transfer of maternally derived antibody clearly has a protective effect and is likely to provide "cover" during a period when the neonatal immune system is insufficiently developed to respond actively (Bailey et al. 2005). Two important periods of maximum exposure to novel antigens occur in the young animal, immediately after birth and at weaning (Bailey et al. 2005). In both cases, the antigenic composition of the intestinal contents can shift suddenly, as a result of a novel diet and of colonization by novel strains of bacteria. Thus, birth and weaning are likely to represent hazard and critical control points in the development of appropriate responses to pathogens and harmless dietary and commensal antigens.

In the newborn animal, factors in maternal milk may regulate responses (TGFb) as well as providing passive effector molecules (Ig). At weaning these factors are removed and there is evidence that the balance may be disturbed. The magnitude and severity of this response are likely to be dependent on the extent to which the effector and regulatory arms of the immune system have been expanded during the preweaning period.

The intestinal barrier is critical in terms of preventing bacterial translocation and initiating the inflammatory response, which might affect the well-being of not only the intestine but distal organs such as the lung and central nervous systems as well (Liu et al. 2005; Speer 2004). The intestinal barrier consists of epithelial, immunologic, luminal, and mucosal factors that control antigen entry and the generation of immunologic phenomena in the gut (Lu and Walker 2001; Isolauri 2001). Probably the most significant barrier to antigen entry into the mucosa-associated lymphoid tissue is the presence of enzymes starting in the mouth and extending down to the stomach, small bowel, and colon (Mayer 2003). The next layer of the barrier is the epithelial cell (Mayer 2003). The microvillus membrane from newborn rabbit is inherently more disordered than microvillus membrane from adult animals. This difference in membrane organization might in part account for the increased attachment and penetration of macromolecules noted during the perinatal period (Pang et al. 1983).

The intrinsic barrier consists of the physical structures of the intestine. In particular, the tight junction between the epitheliums is a formidable barrier to entry of foreign material (Sanderson 2003). In inflamed states, as well as in the perinatal period, the tight junctions are less "tight," allowing for the passage of macromolecules into the underlying lamina propria (Mayer 2003). The components of the intestinal mucosal barrier are underdeveloped at birth and respond inappropriately to luminal stimuli (Walker 2002). As a result of this immature, inappropriate response to microbial stimuli, certain infectious diseases occur in greater frequency during infancy.

1.4 The Effects of Nutrients on Gastrointestinal Tract Development

The neonatal GIT requires nutrients to support their proliferation and differentiation as well as the secretion of enzymes, proteins and other materials (Koutsos and Arias 2006). The neonate is immunologically naive and is very susceptible to infection and damage from harmful antigens. Early nutrition plays an important role in protecting the developing intestine from harmful agents and in modulating immune responses following antigenic challenge (Kelly and Coutts 2000). Feeding represents a major change in the infant's internal environment, with the flow of swallowed amniotic fluid being replaced by colostrum, breast milk, or formula. Changes in dietary composition and intake undoubtedly have many highly significant effects on the growth and function of GIT.

Maternal colostrum and milk. Colostrum is characterized by a high concentration of IgG and relatively lower concentrations of IgA and IgM (Curtis and Bourne 1971; Klobasa and Butler 1987). The concentration of IgG in colostrum is several-fold higher than in sow plasma and declines rapidly during the first 24 h of secretion. Maternal colostrum and milk not only provide the newborn with nutrition but also confer passive immunity. For larger animals, such as horse, cow, pig, and sheep, colostrum is necessary for getting passive immunity. For animals such as rabbit and Guinea pig, transition of the antibodies are finished before birth. In addition, maternal milk might have the capacity to directly stimulate immune function (Kelly and Coutts 2000). Maternal milk provides a diverse range of substances that are developmentally delayed in the neonate and are thought to play a critical role in gastrointestinal defending. Maternal milk contained substances, including growth factors and cytokines, promotes the maturation of the intestine and the associated immune system (Kelly and Coutts 2000).

Breastfeeding is often considered as superior to infant formula feeding. When milk was infused directly into the rumen of calf, resulting in SCFA production, papillary growth was stimulated (Tamate et al. 1962). The most significant advantage of breast milk on the neonatal intestine might not be related to growth per se, but mucosal barrier and immune function (Burrin and Stoll 2002). This idea is based on the mounting evidence that breast milk is clearly superior to formula, with regard to, the incidence of infection and sepsis in neonatal infants (Iiboshi et al. 1994). Cytokines and lymphocytes present in breast milk can influence the development of the immune system. This suggests an immunoregulatory role for breast milk that is absent in infants consuming formula (Rees 2005). In addition, milk contains a large number of other components with antimicrobial activity including complex carbohydrates, glycoproteins, glycolipids, glycosaminoglycans, mucins, and oligosaccharides (Kelly and Coutts 2000).

1.4.1 Dietary Protein

Dietary protein is major macronutrient required for GIT growth of fetus and neonate. Dietary protein is also important for the maintenance of the immune system and influences the response to infection through several mechanisms (Millward 1999). One mechanism is the maintenance of gut barrier function, through the provision of threonine, cysteine, and other amino acids involved in the synthesis of mucus glycoproteins (Gil and Rueda 2002). Another is maintenance of general immunocompetence through the provision of specific amino acids for synthesis of cellular proteins of the immune system (Gil and Rueda 2002).

1.4.2 Individual Amino Acids

Individual amino acids have been proposed to act as signaling molecules that serve to regulate mRNA translation. Arginine is generally considered essential in neonates, because the metabolic pathways that synthesize this amino acid are not fully developed in neonates (Wu et al. 2004). Arginine is the most abound nitrogen carrier in tissue protein and may be responsible for maximal growth of young mammals (Flynn et al. 2002). Glutamine is a unique amino acid in that it functions as a key respiratory fuel for rapidly proliferating cells such as erythrocytes and lymphocytes and also as a precursor of nucleic acids, nucleotides, amino sugars, amino acids, and glutathione (Souba 1991). The amide nitrogen of glutamine is thought to be critical in the biosynthesis of nucleotides and hexosamines. Glutamine and nucleotides appear to act synergistically in the intestinal epithelial proliferation and differentiation (Huang et al. 2003). Dietary glutamine, through conversion to glutamate, probably contributes significantly to the intestinal glutathione synthesis (Duggan et al. 2002).

1.4.3 Short-Chain Fatty Acids

There is considerable evidence for the specific intestinal trophic effects of short-chain fatty acids (SCFA) on GIT development. Intraluminal infusions of SCFA have a stimulatory effect on colonic mucosal proliferation of neonatal rat. This effect is dose-dependent and vary among the SCFA's (butyrate > propionate > acetate) (Sakata 1987). For normal development of the ruminal epithelium to progress, a viable ruminal fermentation must be established, suggesting that there is a requirement for the presence of SCFA in the ruminal lumen to promote normal papillary development (Sander et al. 1959). Similarly, Lane and Jesse (1997) reported that infusing 50 % of the lamb's estimated net energy requirement in the form of SCFA at physiological concentrations resulted in increases in papillae length. Infusions of sodium propionate and sodium butyrate, but not sodium acetate, sodium chloride, or glucose, resulted in marked development of the ruminal papillae in calves (Tamate et al. 1962). A mixture of SCFA salts (propionate and butyrate included) fed as 10 % (wt/wt) of a concentrate starter ration resulted in increased incidence of ruminal parakeratosis, and, in all treated animals, a thickening of the stratum corneum was reported in calves (Gilliland et al. 1962) and lambs (Rickard and Ternouth 1965).

1.4.4 Long-Chain Fatty Acids

The study of the role of dietary long-chain fatty acids (LCFA) in pediatric nutrition is mainly focused on whether some of the n-3 long-chain, polyunsaturated fatty acids (LC-PUFA) improve the health and development of infants (Heird 2001; Li et al. 2011).

The n-3 LC-PUFA's or omega-3 fatty acids, particularly docosahexanoic acid (DHA), eicosapentaenoic acid (EPA), and arachidonic acid (AA), have pleiotropic biological effects on immune function, inflammation, hemodynamics, and bone metabolism. A series of studies have demonstrated that n-3 LC-PUFAs enhance intestinal adaptation after small-bowel resection and their effects are greater than those of less saturated oils (Vanderhoof et al. 1994). The mechanisms for these differences between fatty acids remain unclear, although studies have observed increased circulating PYY concentrations with fish oil feeding. In contrast to the evidence that LCFA's are beneficial to intestinal growth and well-being, studies with neonatal piglets indicated that the LCFA oleic acid can cause significant mucosal injury and increased permeability. The latter effect is more severe in newborn than in 1-month-old piglets (Velasquez et al. 1994).

The best described mediator of the responses to fatty acids is the peptide cholecystokinin (CCK), probably through enteroendocrine cells (EECs). However, it still remains to be proven that the sensing apparatus is intrinsic to EECs, which may function as conduits for transepithelial signals (McLaughlin 2007).

Indeed, the current basic model for nutrient sensing implicates EEC cells as transepithelial signal transduction units, detecting intraluminal nutrient signals apically and secreting regulatory peptides or biogenic amines basolaterally to affect appropriate physiological responses. This can include classical endocrine effects and paracrine effects on adjacent cells, especially afferent nerve terminals

1.5 The Negative Impacts of Early Malnutrition on Gastrointestinal Tract Development

The early nutrition including fetal and early postnatal periods is particularly important, and the lack of any nutrient consumption would delay the growth of the intestine and of the other tissues. The fetal hypothalamic–pituitary–adrenal (HPA) axis has numerous key roles in the development of organisms. The previous studies showed that maternal undernutrition could act as a "programming" factor to produce a reduction in responsiveness of the fetal HPA axis (Hawkins et al. 1999). A severe state of protein-energy malnutrition significantly diminished intestine length of rats when compared with control animals (Plagemann et al. 2000). Intestinal weight and total DNA were similarly diminished in the malnourished rats (Hatch et al. 1979). Length and weight of small intestine of rats feeding low protein diet were lower than those of rats feeding normal protein (Deng and Qiu 2003).

1.6 Summary and Future Perspectives

This chapter has described areas relating to the development of GIT. The interactions among the components of intestinal ecological system (that are host intestinal anatomy, microbial populations, immune systems, and nutrition of the animal) are indeed extremely complex. The immature intestine and immune system leave the neonate susceptible to viral and bacterial attack. Maternal milk appears to have a host of factors that can mature the function of neonatal GIT. The early nutrients also have important impacts on the morphological and functional development of neonate. We need a better understanding of the characteristics and mechanisms of development pertaining to gastrointestinal morphology, digestion, absorption, immune and microbial community at all stages of prenatal and postnatal period, in especial on cell proliferation and functional maturity in pigs. Furthermore, we need to divide the early stage of life into different time stage according to the characteristics and mechanisms of development of gastrointestinal morphology and function, and identify the pertinent regulatory factors for the development of gastrointestinal morphology and function.

References

Adkins B, Jones M, Bu YR et al (2004) Neonatal tolerance revisited again: specific CTL priming in mouse neonates exposed to small numbers of semi- or fully allogeneic spleen cells. Eur J Immunol 34(7):1901–1909

Antonowicz I, Lebenthal E (1977) Developmental pattern of small intestinal enterokinase and disaccharidase activities in the human fetus. Gastroenterology 72(6):1299–1303

Arndorfer RC, Stef JJ, Dodds WJ et al (1997) Improved infusion system for intraluminal esophageal manometry. Gastroenterology 73(1):23–27

Bailey M, Haverson K, Inman C et al (2005) The development of the mucosal immune system pre- and post-weaning: balancing regulatory and effector function. Proc Nutr Soc 64(4):451–457

Bauer E, Williams BA, Smidt H et al (2006) Influence of the gastrointestinal microbiota on development of the immune system in young animals. Curr Issues Intest Microbiol 7:35–52

Berseth CL (1996) Gastrointestinal motility in the neonate. Clin Perinatol 23(2):179–190

Black DD (2007) Development and physiological regulation of intestinal lipid absorption. I. Development of intestinal lipid absorption: cellular events in chylomicron assembly and secretion. Am J Gastrointest Liver Physiol 293(3):G519–G524

Buddington RK, Malo C (1996) Intestinal brush-border membrane enzyme activities and transport functions during theprenatal development of pigs. J Pediatr Gastroenterol Nutr 23:51–64

Buddington RK, Elnif J, Puchal-gardiner AA et al (2001) Intestinal apical amino acid absorption during development of the pig. Am J Physiol Regul Integr Comp Physiol 280(1):241–247

Buraczewska L (1981) Absorption of amino acids in different parts of the small intestine in growing pigs. I. Absorption of free amino acids and water. Acta Physiol Pol 32(4):419–428

Burrin DG, Stoll B (2002) Key nutrients and growth factors for the neonatal gastrointestinal tract. Clin Perinatol 29(1):65–96

Carey MC, Hernell O (1992) Digestion and absorption of fat. Semin Gastrointest Dis 3:189–208

Curtis J, Bourne FJ (1971) Immunoglobulin quantitation in sow (closure), serum, colostrum and milk and the serum of young pigs. Biochem Biophys Acta 236:319–332

Deng HY, Qiu XH (2003) Effects of early nutrition after birth on the development of small intestine and body growth in rats with intrauterine growth retardation. Chin J Contemp Pediatr 5:197–200

Deren JS (1971) Development of structure and function in the fetal and newborn stomach. Am J Clin Nutr 24:144–159

Duggan C, Gannon J, Walker WA (2002) Protective nutrients and functional foods for the gastrointestinal tract. Am J Clin Nutr 75:789–808

Eggermont E (1969) The hydrolysis of the naturally occurring alfa-glucosides by the human intestinal mucosa. Eur J Biochem 9:483–487
Eggermont E (1991) Problems of transfer of carbohydrates at the level of the intestinal mucosa. In: Schaub J, Van Hoof F, Vis HL (eds) Inborn errors of metabolism-Nestle. Raven, New York, pp 197–205
Faria AM, Weiner HL (2005) Oral tolerance. Immunol Rev 206(1):232–259
Favier CF, Vaughan EE, De Vos WM et al (2002) Molecular monitoring of succession of bacterial communities in human neonates. Appl Environ Microbiol 68(1):219–226
Flynn NE, Meininger CJ, Haynes TE et al (2002) The metabolic basis of arginine nutrition and pharmacotherapy. Biomed Pharmacother 56(9):427–438
Gaskins HR (2001) Intestinal bacteria and their influence on swine growth. In: Lewis AJ, Southern LL (eds) Swine nutrition. CRC, Boca Raton, FL, pp 585–608
Gil A, Rueda R (2002) Interaction of early diet and the development of the immune system. Nutr Res Rev 15(2):263–292
Gilliland RL, Bush LJ, Friend JD (1962) Relation of ration composition to rumen development in early-weaned dairy calves with observations on ruminal parakeratosis. J Dairy Sci 45 (10):1211–1217
Hamosh M (1996) Digestion in the newborn. Clin Perinatol 23(2):191–209
Hatch T, Lebenthal E, Branski D et al (1979) The effect of early postnatal acquired malnutrition on intestinal growth, disaccharidases and enterokinase. J Nutr 109(11):1874–1879
Hawkins P, Steyn C, McGarrigle HH et al (1999) Effect of maternal nutrient restriction in early gestation on development of the hypothalamic-pituitary-adrenal axis in fetal sheep at 0.8–0.9 of gestation. J Endocrinol 163(3):553–561
Heird WC (2001) The role of polyunsaturated fatty acids in term and preterm infants and breastfeeding mothers. Pediatr Clin North Am 48(1):173–188
Hooper LV, Gordon JI (2001) Commensal host–bacterial relationships in the gut. Science 292 (5519):1115–1118
Huang Y, Shao XM, Neu J (2003) Immunonutrients and neonates. Eur J Pediatr 162(3):122–128
Iiboshi Y, Nezu R, Kennedy M et al (1994) Total parenteral nutrition decreases luminal mucous gel and increases permeability of small intestine. J Parenter Enteral Nutr 18(4):346–350
Isolauri E (2001) Probiotics in human disease. Am J Clin Nutr 73(4):1142–1146
Jensen AR, Elnif J, Burrin DG et al (2001) Development of intestinal immunoglobulin absorption and enzyme activities in neonatal pigs is diet-dependent. J Nutr 131(12):3259–3265
Kalliomaki M, Salminen S, Arvilommi H et al (2001) Probiotics in primary prevention of atopic disease: a randomized placebo-controlled trial. Lancet 357(9262):1076–1079
Kelly DA, Coutts GP (2000) Development of digestive and immunological functions in neonates: role of early nutrition. Livest Prod Sci 66(2):161–167
Kien CL (1996) Digestion, absorption, and fermentation of carbohydrates in the newborn. Clin Perinatol 23(2):211–228
Klobasa F, Butler JE (1987) Absolute and relative concentrations of immunoglobulins G, M and A, and albumin in the lacteal secretions of sows of different lactation numbers. Am J Vet Res 48(2):176–182
Kosloske AM (1994) Epidemiology of necrotizing enterocolitis. Acta Paediatr 83(Suppl 396):2–7
Kong XF, Tan BE, Yin YL, Li XL, Jaeger LA, Bazer FW, Wu GY (2012) Arginine stimulates the mTOR signaling pathway and protein synthesis in porcine trophectoderm cells. J Nutr Biochem 23(9):1178–1183
Koutsos EA, Arias VJ (2006) Intestinal ecology: interactions among the gastrointestinal tract, nutrition, and the microflora. J Appl Poultry Res 15(1):161–173
Lane MA, Jesse BW (1997) Effect of volatile fatty acid infusion on development of the rumen epithelium in neonatal sheep. J Dairy Sci 80(4):740–746
Lebenthal A, Lebenthal E (1999) The ontogeny of the small intestinal epithelium. J Parenter Enteral Nutr 23(5 Suppl):3–6

Lebenthal E, Lee PC, Heitlinger LA (1983) Impact of development of the gastrointestinal tract on infant feeding. J Pediatr 102(1):1–9

Li FN, Li LL, Yang HS, Yuan XX, Zhang B, Geng MM, Xiao CW, Yin YL (2011) Regulation of soy isoflavones on weight gain and fat percentage: evaluation in a Chinese Guangxi minipig model. *Animal* 5:1903–1908

Lin C (2006) Amino acid nutrition for growth and health of neonatal pigs. Texas Teck University, Texas

Liu Z, Li N, Neu J (2005) Tight junctions, leaky intestines, and pediatric diseases. Acta Paediatr 94 (4):386–393

Lu L, Walker WA (2001) Pathologic and physiologic interactions of bacteria with the gastrointestinal epithelium. Am J Clin Nutr 73(Suppl):1124–1130

Mayer L (2003) Mucosal Immunity. Pediatrics 111:1595–1600

McLaughlin J (2007) Long-chain fatty acid sensing in the gastrointestinal tract. Biochem Soc Trans 35(5):1199–1202

Michaelsson J, Mold JE, McCune JM et al (2006) Regulation of T cell responses in the developing human fetus. J Immunol 176(10):5741–5748

Millward DJ (1999) Requirements and role in diet. In: Caballero B, Sadler MJ, Strain JJ (eds) Encyclopedia of human nutrition. Academic, London, pp 1661–1679

Mobassaleh M, Montgomery RK, Biller JA et al (1985) Development of carbohydrate absorption in the fetus and neonate. Pediatrics 75(1):160–166

Moughan PJ, Birtles MJ, Cranwell PJ et al (1992) The piglet as a model animal for studying aspects of digestion and absorption in milkfed human infants. World Rev Nutr Diet 67:40–113

Neish AS (2002) The gut microflora and intestinal epithelial cells: a continuing dialogue. Microbes Infect 4(3):309–317

Neu J (2006) Gastrointestinal maturation and feeding. Semin Perinatol 30(2):77–80

Neu J (2007a) Gastrointestinal development and meeting the nutritional needs of premature infants. Am J Clin Nutr 85:629–634

Neu J (2007b) Gastrointestinal maturation and implications for infant feeding. Early Hum Dev 83 (12):767–775

Pang KY, Bresson JL, Walker WA (1983) Development of the gastrointestinal mucosal barrier: evidence for structural differences in microvillus membranes from newborn and adult rabbits. Biochim Biophys Acta 727(1):201–208

Perozzi G, Barila D, Muriga C et al (1993) Expression of differentiated functions in the developing porcine small intestine. J Nutr Biochem 4(12):699–705

Philip CC, Susanne KE, Esther CJ et al (2006) Early nutrition and immunity-progress and perspectives. Br J Nutr 96(4):774–790

Plagemann A, Harder T, Rake A et al (2000) Hypothalamic nuclei are malformed in weanling offspring of low protein malnourished rat dams. J Nutr 130(10):2582–2589

Raghupathy R (2001) Pregnancy: success and failure within the Th1/Th2/Th3 paradigm. Semin Immunol 13(4):219–227

Rakoff-Nahoum S, Paglino J, Eslami-Varzaneh F et al (2004) Recognition of commensal microflora by toll-like receptors is required for intestinal homeostasis. Cell 118(2):229–241

Rees L (2005) Immunity and the infant gastrointestinal tract. S Afr Fam Pract 47:56–58

Reinhard G, Noll A, Schlebusch H et al (1998) Shifts in the TH1/TH2 balance during human pregnancy correlate with apoptotic changes. Biochem Biophys Res Commun 245(3):933–938

Rickard MD, Ternouth JH (1965) The effect of the increased dietary volatile fatty acids on the morphological and physiological development of lambs with particular reference to the rumen. J Agric Sci 65:371–382

Rumbo M, Schiffrin EJ (2005) Ontogeny of intestinal epithelium immune functions: developmental and environmental regulation. Cell Mol Life Sci 62(12):1288–1296

Sakata T (1987) Stimulatory effect of short-chain fatty acids on epithelial cell proliferation in the rat intestine: a possible explanation for trophic effects of fermentable fibre, gut microbes and luminal trophic factors. Br J Nutr 58(1):95–103

Salenius P (1962) On the ontogenesis of the human gastric epithelial cells. Acta Anat 50 (46 Suppl):1–76
Sander EG, Warner HN, Harrison HN et al (1959) The stimulatory effect of sodium butyrate and sodium propionate on the development of rumen mucosa in the young calf. J Dairy Sci 42(9):1600–1605
Sanderson IR (2003) The innate immune system of the gastrointestinal tract. Mol Immunol 40 (7):393–394
Sangild PT, Diernæs L, Christiansen IJ et al (1993) Intestinal transport of sodium, glucose and immunoglobulin in neonatal pigs: effect of glucocorticoids. Exp Physiol 78:485–497
Sangild PT, Fowden AL, Trahair JF (2000) How does the foetal gastrointestinal tract develop in preparation for enteral nutrition after birth? Livest Prod Sci 66(2):141–150
Sangild PT, Schmidt M, Elnif J et al (2002) Prenatal development of gastrointestinal function in the pig and the effects of fetal esophageal obstruction. Pediatr Res 52(3):416–424
Sase M, Tamura H, Ueda K et al (1999) Sonographic evaluation of antepartum development of fetal gastric motility. Ultrasound Obstet Gynecol 13(5):323–326
Schmitz J (1991) Digestive and absorptive function. In: Walker WA, Durie PR, Hamilton JR, Walker-Smith JA, Watkins JB, Decker BC (eds) Pediatric gastrointestinal disease. Decker, Philadelphia, PA, pp 266–280
Souba WW (1991) Glutamine: a key substrate for the splanchnic bed. Annu Rev Nutr 11:285–308
Speer CP (2004) Pre- and postnatal inflammatory mechanisms in chronic lung disease of preterm infants. Paediatr Resp Rev 5(Suppl A):241–244
Szepfalusi Z, Loibichler C, Pichler J et al (2000) Direct evidence for transplacental allergen transfer. Pediatr Res 48(3):404–407
Takeda K, Kaisho T, Akira S (2003) Toll-like receptors. Annu Rev Immunol 21:335–376
Tamate H, McGilliard AD, Jacobson NL et al (1962) Effect of various dietaries on the anatomical development of the stomach in the calf. J Dairy Sci 45(3):408–420
Tang X-S, Shao H, Li T-J, Tang Z-R, Huang RL, Wang S-P, Kong X-F, Wu X, Yin Y-L (2013) Dietary supplementation with bovine lactoferrampin-lactoferricin produced by pichia pastoris fed-batch fermentation affects intestinal microflora in weaned piglets. Appl Biochem Biotechnol 168:887–898. doi:10.1007/s12010-012-9827-0
Taylor A, Verhagen J, Blaser K et al (2006) Mechanisms of immune suppression by interleukin-10 and transforming growth factor-beta: the role of T regulatory cells. Immunology 117 (4):433–442
Vanderhoof JA, Park JH, Herrington MK et al (1994) Effects of dietary menhaden oil on mucosal adaptation after small bowel resection in rats. Gastroenterology 106(1):94–99
Velasquez O, Pace AR, Tso P et al (1994) Developing intestine is injured during absorption of oleic acid but not its ethyl ester. J Clin Invest 93(2):479–485
Vinderola G, Matar C, Perdigon G (2005) Role of intestinal epithelial cells in immune effects mediated by gram-positive probiotic bacteria: involvement of toll-like receptors. Clin Diagn Lab Immunol 12(9):1075–1084
Walker WA (2002) Development of the intestinal mucosal barrier. J Pediatr Gastroenterol Nutr 34:33–39
Wu G, Knabe DA, Kim SW (2004) Arginine nutrition in neonatal pigs. J Nutr 134(10):2783–2790
Zhang H, Malo C, Buddington RK (1997) Suckling induces rapid growth and changes in brush border digestive functions of newborn pigs. J Nutr 127(3):418–426

Chapter 2
Development of Digestive Glands in Pigs

Zhiru Tang, Zheng Ruan, and Yulong Yin

Contents

2.1 Introduction

The digestive glands include the pancreas, liver, and the gallbladder. The pancreas plays a central role in the digestive system by synthesizing and secreting specific digestive enzymes, and thus has an integrative role for the absorption and metabolism of nutrients in the organism. The liver performs numerous physiological and metabolic functions for the pig's whole body, but the most important one is maybe metabolism of

Z. Tang
College of Animal Science and Technology, Southwest University, Chongqing 400715, China

Z. Ruan
State Key Laboratory of Food Science and Technology, Nanchang University, Nanchang 330047, China

Y. Yin (✉)
Chinese Academy of Sciences, Institute of Subtropical Agriculture, Research Center for Healthy Breeding Livestock & Poultry, Hunan Engineering & Research Center of Animal & Poultry Science, Key Laboratory of Agro-ecological Processes in Subtropical Region, Scientific Observing and Experimental Station of Animal Nutrition and Feed Science in South-Central, Ministry of Agriculture, Changsha, Hunan 410125, People's Republic of China
e-mail: yinyulong@isa.ac.cn

F. Blachier et al. (eds.), *Nutritional and Physiological Functions of Amino Acids in Pigs*, DOI 10.1007/978-3-7091-1328-8_2,

nutrients and toxins arising from digestion and/or absorption in the intestines. The pig's liver also produces bile, involved in lipid digestion. Bile is secreted into the intestinal system via the gallbladder. In pig, as in other mammals, the liver is involved in protein synthesis, toxin and hormone breakdown, vitamin storage as well as blood cell and blood component processing. The gallbladder is a small sac involved in bile storage. Bile is concentrated in the gallbladder. The gallbladder is located at the back of the liver. It is connected with the liver via the cystic duct and the hepatic bile duct.

2.2 Development of the Pancreas in Pigs

The pancreas is both an exocrine and an endocrine organ. Digestive enzymes are conducted via a duct system to the duodenum, and the endocrine products enter the blood directly. The bulk of the tissue in the pancreas is exocrine in nature.

The exocrine pancreas produces pancreatic juice, which contains enzymes, nonenzymatic proteins, and salts. These substances reach the duodenum via a single pancreatic duct in the pig (van den Borne et al. 2007). Enzymes able to digest proteins, carbohydrates, lipids, and nucleic acids, respectively, are produced in the acinar cells of the exocrine pancreas (Malagelada 1981; Pitchumoni et al. 1986). Electrolytes capable of neutralizing the acidic stomach contents and establishing the proper pH for enzymatic action are produced mainly in the intra-acinar and ductal cells (Schulz 1980). The islet cells of the endocrine pancreas produce hormones (e.g., insulin, glucagon) that regulate the utilization of the absorbed digestion products (Sitbon and Mialhe 1978; Christensen and Just 1988; Motyl et al. 1986; Malmlef et al. 1990). In the embryo, the pancreas develops as an outgrowth of the endoderm; the cell layer that delineates the primitive gut. Pancreatic progenitors derived from the endoderm form two buds (dorsal and ventral) which later fuse to form a single organ. Within these buds, the progenitor cells give rise, through a stepwise process, to endocrine, acinar, and duct cells.

The neonatal period is a very important developmental stage in the life of pig. The pancreas, an important digestive organ responsible for production of various hydrolytic enzymes, also shows a marked tissue mass gain (up to 80 %) during the first 2 postnatal days in newborn pigs (Widdowson and Crabb 1976; Tarvid et al. 1994).

In naturally suckled piglets, both body weight and pancreas weight increased with age. The pancreatic weight increased much faster than the body weight, particularly during the first postnatal day. The ratio of pancreas weight to body weight is significantly greater in 1-day and 3-day-old piglets than in newborn piglets. Mubira (1998) demonstrates that the pancreas growth is disproportionally faster than the body taken as a whole, the protein concentration in the pancreatic tissue is lower in 1- and 3-day-old than in the newborn piglets, The concentration of DNA in the pancreatic tissue remains unchanged with age and the total contents of protein and DNA increase progressively with age. Mubira (1998) also demonstrates that pancreatic lipase activity expressed either as per gram tissue weight, per milligram protein or per milligram DNA is significantly lower in 1- and 3-day-old piglets when compared with newborn piglets, the enlargement of the acini is apparently due to an increase in the cell numbers although the difference among the three groups of animals in the number of cells per acinus did not reach a significant level and the size of the acinar cells did not change significantly with age.

However, the mechanisms responsible for this rapid pancreatic tissue growth are far from being completely understood, but are apparently related to colostrum ingestion since starved, water-fed, or formula-fed newborns show little changes (Tungthanathanich et al. 1992; Burrin et al. 1992). Further studies have shown that porcine colostrum contains a number of hormones and growth-promoting peptides, such as insulin, cortisol, epidermal growth factor (EGF), bombesin, and insulin-like growth factors-I and-II (IGF-I and -II) (Jaeger et al. 1990) and epithelial-like cells (Cera et al. 1987). The size of acinar cells decreased significantly in piglets fed 5 % lactose solution while it increased significantly in piglets fed porcine colostrum. The acinar cell mitotic index estimated from the incorporation rate of Brdu increased significantly in piglets fed either porcine colostrum or trypsinized porcine colostrums, the index being even greater in the trypsinized-colostrum group than in the colostrum group (Mubira 1998).

The rapid pancreatic tissue growth in naturally suckling piglets is apparently related to ingestion of colostrum. Widdowson and Crabb (1976) found that rapid pancreatic growth would not occur if piglets were prevented from sucking and fed water only. Mubira (1998) study showed that pancreatic weight increased significantly in piglets bottle-fed porcine colostrum both in absolute and relative terms, but no increment was observed in the piglets bottle-fed lactose solution. All these observations indicate that colostrum may contain some trophic factors capable of stimulating pancreatic growth in newborn animals.

A number of growth-promoting peptides and hormones, such as EGF, IGF-I, IGF-II, bombesin, and insulin, have been found in the mammary gland secretions in pigs and in several other species (Jaeger et al. 1987; Donovan et al. 1994; Simmen et al. 1988; Read et al. 1984; Westrom et al. 1987). The concentration of these peptides is generally high in colostrum and often several fold greater than that in the maternal circulation (Donovan et al. 1994; Simmen et al. 1988; Read et al. 1984). There is evidence showing that these colostrum growth-promoting peptides are stable in the gastrointestinal lumen in suckling pigs (Shen and Xu 1996) and in suckling rats (Philipps et al. 1997) as it can be absorbed intact into the blood circulation in newborn pigs (Xu and Wang 1996) and in suckling rats (Thornburg et al. 1984). It has also been found in laboratory rodents that pancreatic cells possess specific receptors for EGF, IGF-I, and IGF-II (Potau et al. 1984; Korc et al. 1983; Williams et al. 1984). It was observed in an earlier study that newborn piglets fed infant milk formula supplemented with IGF-I or IGF-II had heavier pancreas and greater pancreatic RNA and DNA contents than did the control animals (Xu et al. 1994). Subcutaneous administration of bombesin and EGF stimulated pancreatic growth in rats (Papp et al. 1987; Dembinski et al. 1982) and EGF and insulin stimulated mouse pancreatic acinar cell growth in vitro (Logsdon 1986).

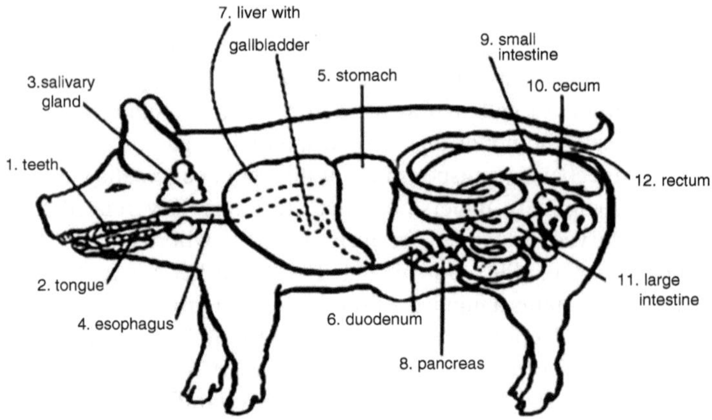

Fig. 2.1 The location of pig's liver and gallbladder

2.3 The Structure and Function of the Liver in Pigs

The liver is the largest internal organ, and the body' second largest, after the skin (Baruch 2000). The liver and its functions are referred to as the hepatic system, from the Greek word meaning liver. Schematically, liver metabolism can be partly view as the readjustment of molecules into a form usable by the pig's body tissues. In the case of toxins, liver metabolism converts poisonous molecules into less harmful molecules, finally eliminated into the urine.

2.3.1 The Structure of the Liver

As shown in Fig. 2.1, the liver in pigs is located behind the diaphragm in the abdominal cavity, connected to the digestive system through the bile ducts, gallbladder, and duodenum. Blood supply to the liver is approximately 75 % from the venous system, directly from the intestines. Pig livers have four main lobes for filtering toxins and nutrients from the bloodstream: the right, left, caudate and quadrate. Bile ducts connect the lobes to the gallbladder, where excess bile is stored until needed.

Four lobes are surrounded by a capsule of fibrous connective tissue called Glisson's capsule. The Glisson's capsule in turn is covered by the visceral peritoneum (tunica serosa), except where the liver adheres directly to the abdominal wall or other organs. The parenchyma of the liver is divided into lobules, which are incompletely partitioned by septa from Glisson's capsule (Ekataksin and Wake 1991). The parenchyma within the lobules is supported only by fine reticular fibers (which are discernible only with special preparations).

The blood vessels supplying the liver (portal vein and hepatic artery) enter at the hilum (or porta hepatis), from which the common bile duct (carrying bile secreted by the liver) and lymphatic vessels also leave (Fig. 2.2).

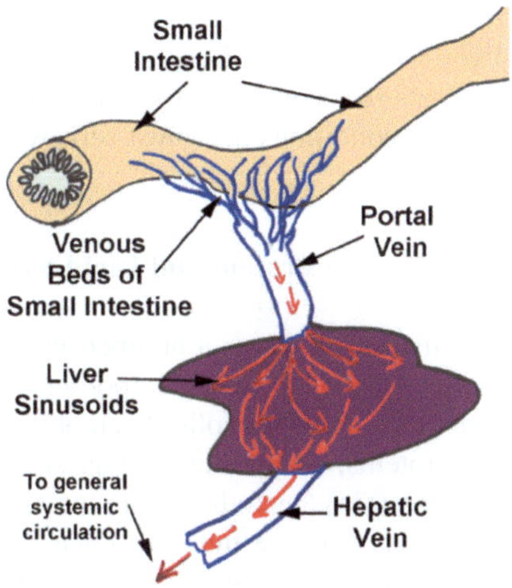

Fig. 2.2 The liver's structure and anatomical placement

Within the liver sinusoids, the oxygen-poor but nutrient rich blood from the portal vein mixes with the oxygenated blood from the hepatic artery. From the sinusoids, the blood enters a system of veins which converge to form the hepatic veins. The hepatic veins follow a course independent of the portal vessels and enter the inferior vena cava. Liver cells, or hepatocytes, are large polygonal cells, usually tetraploid and often binucleate in the adult (Sen and Jalan 2005). Each nucleus has two or more nucleoli. The average life span of liver cells is 5 months. They contain abundant rough endoplasmic reticulum and mitochondria, large deposits of glycogen and lipid droplets of various sizes, and several small elaborate Golgi complexes. They also contain many peroxisomes, a variable amount of smooth endoplasmic reticulum and lysosomes (Vega 2004). In standard histological preparations, liver cells usually appear vacuolated because the glycogen and lipids are removed during processing. Liver cells are capable of considerable regeneration when liver substance is lost (Harada et al. 2009).

2.3.2 *The Function of the Liver*

The liver is extremely versatile, having both endocrine and exocrine functions. Liver is involved in numerous metabolic activities and acts as a storage depot. The list below is an attempt to categorize its numerous and diverse activities.

2.3.2.1 Blood Supply

As mentioned above, the liver receives a dual blood supply. The hepatic portal vein provides the majority of the incoming blood, the rest coming from the hepatic artery. The blood from the portal vein has already supplied the small intestine, pancreas, and spleen, and is largely deoxygenated. It contains nutrients and noxious materials absorbed in the intestine. The blood from the hepatic artery supplies the liver with oxygen. Because the blood from the two sources intermingles as it perfuses the liver cells, these cells must carry out their many activities under low oxygen conditions that most cells would not be able to tolerate.

2.3.2.2 The Protein and Lipid Synthesis

The liver synthesizes a number of proteins, as suggested by its abundant rER and Golgi. One of the most important of these is albumins (responsible for several functions including colloid osmotic pressure), the protein parts of several kinds of lipoproteins, nonimmune alpha and beta globulins, prothrombin, and numerous glycoproteins, including fibronectin. These proteins are considered as endocrine secretions because they are released directly from the liver cells into the blood supply. The liver also synthesizes cholesterol and the lipid portion of lipoproteins. The enzymes for these activities are found in the sER.

2.3.2.3 Metabolic Functions

The liver is involved in a wide range of metabolic activities. Among its major metabolic functions are gluconeogenesis (the formation of glucose from non-carbohydrate precursors), and the deamination of amino acids. In addition to the proteins listed above, the liver also releases a number of other products directly into the blood as endocrine secretions. These include products of carbohydrate metabolism, e.g., glucose released from stored glycogen, and modified secretions from other organs, e.g., triiodothyronine (T3).

2.3.2.4 Storage of Metabolites

Liver cells store large deposits of glucose in the form of glycogen, which can be stained by the periodic acid-Schiff or PAS procedure. They also store lipids in the form of droplets of varying sizes. The latter can be revealed with Sudan staining after tissue fixation. The removal of these materials in routine preparations gives liver cells a foamy or vacuolated appearance. Lysosomes within hepatocytes may be a normal storage site for iron, in a form of a ferritin complex. Excessive amounts of iron may accumulate in cells as an unusable yellowish-brown pigment called hemosiderin, a partly denatured form of ferritin. Liver lysosomes also contain pigment granules (lipofuscin) and partially digested cytoplasmic organelles. The liver also stores vitamins, especially vitamin A, which is transported from the liver to the retina to be used in the formation of visual pigments. Most of the vitamin A is not stored in the hepatocytes, however, but in special fat-storing cells, called Ito cells, within the sinusoids. Ito cells also store the other three fat-soluble vitamins, D, E, and K, as well as vitamin B12.

2.3.2.5 Detoxification and Inactivation of Noxious Substances

The liver is, after the gastrointestinal tract, the first organ to receive metabolic substrates and nutrients from the intestine, but it is also the first to receive any toxic substances, for instance carcinogens or drugs that have been ingested. The sER of liver cells contains enzymes involved in the degradation and conjugation of toxins (in addition to those responsible for synthesizing cholesterol and the lipid portion of lipoproteins). Under conditions of challenge by drugs, toxins or metabolic stimulants, the sER becomes the predominant organelle. Stimulation by one drug, e.g., alcohol, enhances its ability to detoxify some other compounds. On the other hand, some toxins can get metabolized by the liver to even more damaging compounds.

2.4 The Structure and Function of the Gallbladder in Pigs

The gallbladder is one of the most dispensable organs in mammal. The gallbladder is a small sac and is the storage depot for bile. Bile is only concentrated in the gallbladder. The gallbladder is located at the back part surface of the liver (Fig. 2.1). It is connected with the liver via the cystic duct and the hepatic bile duct. Together they form the common bile duct, which leads into the small intestine.

2.4.1 The Structure of the Gallbladder in Pigs

The gallbladder is a tubular organ. Its tunica mucosa is a very regular simple columnar epithelium, with no goblet cells or glands. The deep folds in the bottom of the mucosa, which are often cut in cross section, sometimes look like "glands," but they have no secretory activity. There is a scanty lamina propria, and a thin tunica muscularis. There is no muscularis mucosae, and this is one of the "landmarks" for this organ. The gallbladder is only partially covered by a fold of peritoneum; its opposite side is seated in a pocket in the surface of the liver. Bile ducts carry bile to the gall bladder for temporary storage.

The low magnification image above shows the walls of the gallbladder (Fig. 2.3). They vary in height depending on how full the bladder is. At higher magnification, the nature of the gallbladder epithelium is revealed: a simple columnar form, containing no goblet cells. Though these cells do have a few microvilli on their free surfaces, they cannot be considered as real brush borders. There is no muscularis mucosae in these structures, and the tunica muscularis is stringy and scanty with collagen fibers. Since it has only the function of squirting bile into the duct that carries it to the duodenum, it appears little developed.

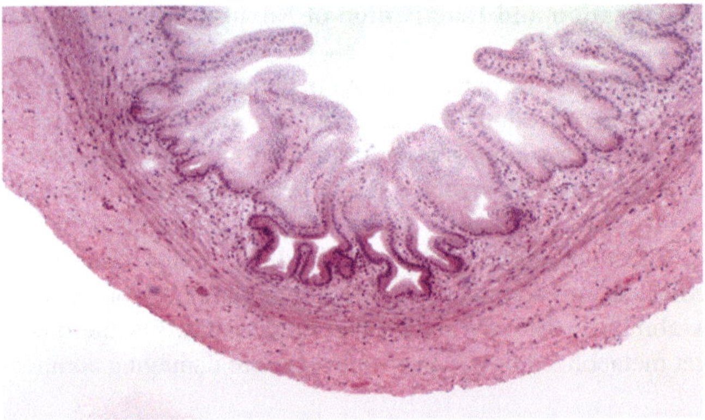

Fig. 2.3 The wall of the gallbladder at the low magnification image

2.4.2 The Function of the Gallbladder in Pigs

Bile is continuously formed in bile canaliculi in the liver hepatocytes, were it is collected and secreted into the hepatic bile duct. Bile is a complex fluid containing water, electrolytes, and a battery of organic molecules including bile acids (water-soluble derivatives of cholesterol), cholesterol, phospholipids, and bilirubin (Mcmaster and Elman 1926). The rate of bile fluid secretion is dependent on the circadian rhythm of the animal.

Bile is the exocrine secretion of the liver, released into ducts which eventually form the common hepatic duct. It, in turn, unites with the cystic duct from the gallbladder to form the common bile duct which enters the duodenum.

Bile contains bile salts, also called bile acids, which are important in emulsifying the lipids of the digestive tract, thereby promoting easier digestion by lipases and absorption. Bile salts are largely recycled between the liver and gut.

Bile also contains cholesterol and phospholipids, and another function of bile salts is to keep them in solution. Cholesterol and phospholipids serve as metabolic substrates for other cells and as precursors for components of cell membranes and steroid hormones. They are largely reabsorbed in the gut and recycled.

Bile also contains pigments, which detoxify bilirubin, the breakdown product of hemoglobin from aging and of abnormal red blood cells destroyed in the spleen. The detoxified bilirubin, mainly in the form of bilirubin glucuronide, is eliminated through the feces (and gives them their color). Failure of the liver to absorb bilirubin, or to conjugate it to its glucuronide, or to secrete the bilirubin glucuronide, may result in jaundice. Finally, bile contains electrolytes (largely reabsorbed in the gut). The latter maintain bile as an isotonic fluid.

References

Baruch Y (2000) The liver: a large endocrine gland. J Hepatol 32(3):505–507

Burrin DG, Shulman RJ, Reeds PJ, Davis TA, Gravitt KR (1992) Porcine colostrum and milk stimulate visceral organ and skeletal muscle protein synthesis in neonatal piglets. J Nutr 122:1205–1213

Cera K, Mahan DC, Simmen FA (1987) In vitro growth-promoting activity of porcine mammary secretions: initial characterization and relationship to known peptide growth factors. J Anim Sci 65:1149–1159

Christensen K, Just A (1988) Interactive effects of live weight, basal diet and fat on essential fatty acid status and blood concentrations of glucose, insulin, and thyroxine measured postprandially in pigs. Comp Biochem Physiol 91A:279–291

Dembinski A, Gregory H, Konturek SJ, Polanski M (1982) Trophic action of epidermal growth factor on the pancreas and gastroduodenal mucosa in rats. J Physiol 325:34–42

Donovan SM, McNeil LK, Jimenez-Flores R, Odle J (1994) Insulin-like growth factors and insulin-like growth factors binding proteins in porcine serum and milk throughout lactation. Pediatr Res 36:159–168

Ekataksin W, Wake K (1991) Liver units in three dimensions: I. Organization of argyrophilic connective tissue skeleton in porcine liver with particular reference to the "compound hepatic lobule". Am J Anat 191(2):113–153

Harada T, Nakagawa Y, Wadkins RM, Potter PM, Wheelock CE (2009) Comparison of benzil and trifluoromethyl ketone (TFK)-mediated carboxylesterase inhibition using classical and 3D-quantitative structure–activity relationship analysis. Bioorg Med Chem 17(1):149–164

Jaeger LA, Lamar CH, Bottoms GT, Cline TR (1987) Growth-stimulating substances in porcine milk. Am J Vet Res 48:1531–1533

Jaeger LA, Lamar CH, Cline TR, Cardona CJ (1990) Effect of orally administered epidermal growth factor on the jejunal mucosa of weaned pigs. Am J Vet Res 51:47

Korc M, Matrisian LM, Planck SR, Magun BE (1983) Binding of epidermal growth factor in rat pancreatic acini. Biochem Biophys Res Commun 111:1066–1073

Logsdon CD (1986) Stimulation of pancreatic acinar cell growth by CCK, epidermal growth factor and insulin in vitro. Am J Physiol 251:G487–G494

Malagelada JR (1981) Gastric, pancreatic, and biliary responses to a meal. In: Johnson LR (ed) Physiology of the gastrointestinal tract. Raven, New York, pp 172–196

Malmlef K, Orberg J, Helberg S, Cortova Z, Bjurkgren S (1990) The diurnal influence on utilisation of dietary protein in the growing pig. J Anim Physiol Anim Nutr 63:180–187

Mcmaster PD, Elman R (1926) On the expulsion of bile by the gall bladder; and a reciprocal relationship with the sphincter activity. J Exp Med 44(2):173–198

Motyl T, Debski B, Kukulska W, Ostaszewski P, Krupa P (1986) The influence of fasting on plasma amino acid indices and urinary nitrogen compounds excretion in wool and meat breeds rabbits. J Anim Physiol Anim Nutr 55:110–117

Mubira JM (1998) Growth and development of the pancreas and stomach in neonatal pigs. Thesis for the degree of doctor of philosophy, University of Hong Kong

Papp M, Dobronyl L, Vaxga G, Scaipignato C (1987) Bombesin promotes pancreatic growth in suckling rats. Experientia 43:201–202

Philipps AF, Rao R, Anderson GG, McCracken DM, Lake M, Koldovsky O (1997) Fate of insulin-like growth factors-I and -II administered orogastrically to sucking rats. Pediatr Res 37:586–592

Pitchumoni CS, Sheele G, Lee PC, Lebenthal E (1986) Effects of nutrition on the exocrine pancreas. In: Go VLW, Gardner JD, Brooks FP (eds) The exocrine pancreas: biology, pathobiology, and diseases. Raven, New York, pp 387–406

Potau N, LA MosnerJ W, Goldfine LD (1984) Cholocystokinin and insulin-like growth factor-binding to pancreatic receptors: evidence of a role for intracellular calcium. Biochem Biophys Res Commun 119:359–364

Read LC, Upton FM, Francis GL, Wallace JC, Dahlenberg GW, Ballard FJ (1984) Changes in the growth-promoting activity of human milk during lactation. Pediatr Res 18:133–139
Schulz I (1980) Bicarbonate transport in the exocrine pancreas. Ann NY Acad Sci 341:191–209
Sen S, Jalan R (2005) Ideal hepatocyte: quest for the Holy Grail. J Gastroenterol Hepatol 20(1):1–4
Shen WH, Xu RJ (1996) Stability of epidermal growth factor in the gastrointestinal lumen of suckling and weaned pigs. Life Sci 59:197–208
Simmen FA, Simmen RCM, Reinhart G (1988) Maternal and neonatal somatomedin C/insulin-like growth factor-I (IGF-I) and IGF binding proteins during early lactation in the pig. Dev Biol 130:16–27
Sitbon G, Mialhe P (1978) Pancreatic hormones and plasma glucose: regulation mechanisms in the goose under physiological conditions. Horm Metab Res 10:12–16
Tarvid I, Cranwell PD, Ma L, Vavala R (1994) The early postnatal development of protein digestion in pigs. L Pancreatic enzymes. In: Souffrant WB (eds). Proceedings of the Vf 1 international symposium on digestive physiology in pigs. Edited by Rostock, Germany. pp 199–202
Thornburg W, Matrisian L, Magun B, Koldovsky O (1984) Gastrointestinal absorption of epidermal growth factor in suckling rats. Am J Physiol 246:G80–G85
Tungthanathanich P, Xu RJ, Reynolds GW, Simpson HV, Mellor DJ (1992) The effects of milk diets on small intestinal growth in newborn piglets. Proc Nutr Soc NZ 17:51–55
van den Borne JJ, Weström BR, Kruszewska D, Botermans JA, Svendsen J, Wolinski J, Pierzynowski SG (2007) Exocrine pancreatic secretion in pigs fed sow's milk and milk replacer, and its relationship to growth performance. J Anim Sci 85:404–412
Vega GL (2004) Obesity and the metabolic syndrome. Minerva Endocrinol 29(2):47–54
Westrom BR, Ekman R, Svendsen L, Svendsen J, Karlsson BW (1987) Levels of immunoreactive insulin, neurotensin and bombesin in porcine colostrums and milk. J Pediatr Gastroenterol Nutr 6:460–465
Widdowson EM, Crabb DE (1976) Changes in the organs of pigs in response to feeding for the first 24 h after birth I. The internal organs and muscles. Biol Neonate 28:261–271
Williams JA, Bailey A, Humbel R, Goldfine LD (1984) Insulin like growth factors bind to specific receptors in isolated pancreatic acini. Am J Physiol 246:G96–G99
Xu RJ, Wang T (1996) Gastrointestinal absorption of insulin-like growth factor-I in neonal pigs. J Pediatr Gastroenterol Nutr 23:430–437
Xu RJ, Mellor DJ, Birtles MF, Breier BH, Gluckman PD (1994) Effects of oral IGF-I or IGF-II on digestive organ growth in newborn piglets. Biol Neonate 66:280–287

Chapter 3
Development and Renewal of Intestinal Villi in Pigs

Huansheng Yang, Xia Xiong, and Yulong Yin

Contents

Abbreviations

HNF-1α	Hepatocyte nuclear factor-1α
Cdx1	Caudal-related homeobox protein 1
Cdx2	Caudal-related homeobox protein 2
VFE	Vacuolated fetal-type enterocytes
SEM	Scanning electron microscopy
CVA	Crypt–villus axis
TCF	T-cell factor

H. Yang • X. Xiong • Y. Yin (✉)
Chinese Academy of Sciences, Institute of Subtropical Agriculture, Research Center for Healthy Breeding Livestock & Poultry, Hunan Engineering & Research Center of Animal & Poultry Science, Key Laboratory of Agro-ecological Processes in Subtropical Region, Scientific Observing and Experimental Station of Animal Nutrition and Feed Science in South-Central, Ministry of Agriculture, Changsha, Hunan 410125, People's Republic of China
e-mail: yinyulong@isa.ac.cn

F. Blachier et al. (eds.), *Nutritional and Physiological Functions of Amino Acids in Pigs*,
DOI 10.1007/978-3-7091-1328-8_3,

BMP	Bone morphogenetic protein
TGF-β	Transforming growth factor beta
GATA 4	GATA-binding transcription factor 4
GATA 5	GATA-binding transcription factor 5
GATA 6	GATA-binding transcription factor 6
HES1	Hairy/enhancer of split homolog 1
Atoh1	Atonal homolog 1
EAAC1	Excitatory amino acid carrier 1
SGLT1	Sodium/glucose cotransporter 1
SLC6A19	Solute carrier family 6, member 19
ASCT2	Alanine-serine-cysteine transporter 2
PP	Polish landrace/Pietrain
DHW	Duroc/Hampshire/wild boar
GLUT5	Glucose transporter member 5
ENS	Enteric nervous systems
IGF-I	Insulin-like growth factor I
IGF-II	Insulin-like growth factor II
EGF	Epidermal growth factor

3.1 Introduction

The gastrointestinal mucosa of pig not only acts as an organ for diet digestion and absorption but also plays an important role in fighting against pathogenic bacteria and toxic substances presented in intestinal lumen. The development of gastrointestinal mucosa begins in the early embryo and self-renews during the whole life. Its homeostasis is preserved via regulating the proliferation, growth arrest, differentiation, and apoptosis of epithelial cell. Moreover, the developmental and renewal processes of gastrointestinal mucosa are regulated by various factors, such as genetic, neural, hormonal and dietary influences and disturbed by weaning stress. Pig gastrointestinal mucosa is not a very common experimental model, and therefore, some information discussed in this chapter is obtained from other species.

3.2 Structural Development

3.2.1 Ontogeny of the Intestinal Mucosa

The growth and ontogenetic development of the mammalian small intestine is a topologically and temporally highly organized process, which decides about the formation of specialized intestinal epithelium that is defined by a set of genes expressed in different epithelial cells. These specialized cells fulfill various important physiological roles, including digestive and absorptive functions, certain endocrine and immunological roles, and secretion of water, electrolytes, and mucus (Pácha 2000). The early ontogenetic development starts during gestation but the most dynamic changes occur before and after birth. This developmental process can be divided into five phases: (1) morphogenesis, (2) cytodifferentiation and fetal development including preparation of the epithelium for processing colostrum and milk, (3) birth and colostrum suckling period, i.e., the shift from an intrauterine to an extrauterine environment, (4) suckling period, and (5) weaning transition of the offspring from mother's milk to a solid diet (Henning 1981; Menard and Calvert 1991; Pácha 2000). The first two phases occur during prenatal life and prepare the intestine for the postnatal life.

3.2.1.1 Morphogenesis

The mammalian small intestine develops from a simple tubular structure in the early embryo to a mature, highly specialized and organized organ system. In the early embryonic development, the anterior and posterior invaginations of visceral endoderm form a primitive gut tube which is lined by an undifferentiated, stratified epithelium. The primitive gut tube can be divided into three parts: foregut, midgut, and hindgut. Among the three parts, the midgut is destined to become the small intestine. During the developmental process, the midgut endoderm undergoes rapid remodeling converting the stratified epithelium into a simple columnar epithelium. At the same time, the underlining mesenchyme grows upward and forms finger-like projections (nascent villi). The formation of nascent villi and microvilli occurs simultaneously. Then the cellular proliferation along the nascent villi and in the intervillus region forms a proliferating compartment known as the intervillus epithelium. Then, this intervillus epithelium goes through reshaping to form crypts by penetrating into the underlying mesenchyme. In addition to crypts and villus, the proximal colon of the human fetus or neonatal rodents has transient villus-like structures (Ménard et al. 1994). Although they are not true villi, these villus-like transient structures are characterized by properties similar to enterocytes of the small intestine. They are not characterized only by their capability of endocytosis (Ono 1977), but also by the fact that they are equipped with brush-border enzymes (Lacroix et al. 1984; Zweibaum et al. 1984; Foltzer-Jourdainne et al. 1989) and transporters. Moreover, results from chimeric mice revealed that neonatal crypts and adult crypts do not have similar cellular organization. In the neonatal period, the crypts contain cells of a mixed genotype. In contrast, the crypts of mature intestine are composed with cells of one genotype which are derived from a single progenitor (Schmidt et al. 1988).

3.2.1.2 Cytodifferentiation and Fetal Development

Cellular phenotypes are defined by the expression of specific sets of genes in individual cells. The conversion of the stratified epithelium into a simple columnar epithelium in a proximal-to-distal wave is called cytodifferentiation, and the expression of certain intestine-specific genes is first detected during this period (Davis et al. 2004). The expression of individual genes in the intestine is regulated by specific transcription factors, such as HNF-1α, Cdx1, and Cdx2 (Boudreau et al. 2002). The cytodifferentiation involves cytoplasmic membranes which belong to two domains: the apical and basolateral membranes with biochemical, morphological, and functional differences (Amerongen et al. 1989). This process determines the polarity of epithelial cells because only the basolateral membrane seems to adhere to the basement membrane of the extracellular matrix and the apical membrane is exposed to the luminal gastrointestinal content. The apical and basolateral membranes are separated by different complex protein systems including tight junctions which form the paracellular apical junctions of intestinal epithelia. Tight junctions are composed of specific proteins such as occludin (Furuse et al. 1993), claudins (Itoh et al. 1999), and zonula occludens (Stevenson et al. 1986), which interact with the actin cytoskeleton to stabilize the tight junctions at the apical end of the cell. This conjunctive intestinal epithelium provides the largest mucosal barrier between the internal host and the external environment.

At the same time than the formation of crypts and villi, four epithelial lineages differentiate from the immature primitive cells: absorptive enterocytes, goblet cells, endocrine cells, and Paneth cells (Cheng and Leblond 1974a, b). Absorptive enterocytes, constituting up to 90 % of epithelial cells in the crypt and more than 95 % of villus cells, are absorptive cells with an apical microvillus membrane that contains transporters, receptors, and membrane-anchored hydrolases (Cheng and Leblond 1974a, b; Traber et al. 1991). Goblet cells secrete protective mucus which is essential for the maintenance of the gut epithelial integrity (Forstner 1978). Endocrine cells are involved in local and systemic metabolism by secreting various growth factors and hormones (Xu and Zhang 2003). The Paneth cells are involved in the modulation of the whole body metabolism through the secretion of cytokines (Shanahan 1997). The developing intestine has the capacity to establish and maintain the functional differences of each cell lineage in the differentiation programs, and this capacity of the intestinal epithelium is already apparent in the late fetal life (Roth et al. 1991; Simon and Gordon 1995). It needs to be emphasized that enterocytes in fetal and newborn mammals are not the same as those of adult mammals. Indeed, enterocytes of fetal and newborn mammals have an apical canalicular system leading to production of large vacuoles which are important for colostrum macromolecules uptake. The vacuolated fetal-type enterocytes (VFE), located at the upper part of villi, are firstly observed in the small intestine proximal region of the pig and lamb fetuses, then in the second trimester of pregnancy, and subsequently in the middle and distal regions. After birth this fetal-type enterocytes are gradually replaced by enterocytes lacking apical canalicular system (Skrzypek et al. 2007a).

3.2.1.3 Birth and Colostrum Suckling Period

The pig gut is exposed to various stress factors particularly in the early development period. In response to the stress, the gut goes through profound changes resulting in accelerated tissue growth and functional maturation (Le Dividich and Seve 2000; Trahair and Sangild 2004; Xu et al. 2000). After birth, changes are induced by dramatic shift from parenteral to enteral nutrition and a vast number of bioactive substances present in the colostrum and milk, and the morphology of the small intestine of piglets experiences great changes during the first 3 days after birth. Xu et al. (1992) reported a 72 % increase in small intestinal weight, a 115 % increase in mucosa weight, a 24 % increase in small intestinal length, a 15 % increase in small intestinal diameter, a 33–90 % increase in villus height and a 14–51 % increase in villus diameter during the first day of piglets. It was also demonstrated that the cellular population in the small intestinal mucosa, as indicated by the DNA content, increased progressively with age, and increased from 84 to 154 % during the first 3 days. Such dramatic changes result from the increase of local intestine blood flow in parallel with the reduction of the basal vascular resistance, the accumulation of proteins of the colostrum in enterocytes, and changes in epithelial cell turnover with increasing mitosis and inhibition of apoptosis (Xu et al. 1992). The types of enterocytes also changed rapidly during this period, especially in the duodenum and in the proximal jejunum. The VFE in the duodenum are only seen at birth. In the proximal jejunum, the VFE are present in the upper part of the villi, and are replaced by mature enterocytes until the day 3 of life. VFE in the mid and distal jejunum and in ileum diminishes gradually after birth and disappears until the day 21 of life (Skrzypek et al. 2007a).

Recently, scanning electron microscopy (SEM) has been used as a powerful tool to observe the surface of the intestine in three dimensions, enabling observations up to single microvillus. At birth, the surface of the mucosa of duodenum is folded and most of it is covered by finger-like villi. Only few villi are branched or incompletely divided (Skrzypek et al. 2005). The villi surface is not smooth, and many transversal furrows are apparent. Transverse incisions on the villi surface, with at least one quarter of villi circumference, can be observed. The enterocyte brush border is well developed, and the microvilli at the edge of enterocyte are shorter than in the central apical area. On the third day of life, the villi of duodenum are finger-like, and the surface of the villi corpus is smoother and the transversal furrows are thinner than at the day of birth (Skrzypek et al. 2005, 2010; Fig. 3.1). In jejunum, the villi of unsuckling neonatal piglets are thin with finger-like shape, but are not uniform in length, with shorter villi in between the taller ones. Because of numerous deep transversal furrows, the villi surface is rough and the apical enterocyte membrane is increased. Compared to the duodenum, the jejunum has relatively more goblet cells. On the third day after birth, there are abundant villi of various lengths in jejunum. The transversal furrows on villi surface are shallower than at birth but still numerous, and no well-defined extrusion zone can be observed (Skrzypek et al. 2005, 2007b, 2010; Fig. 3.2). In ileum, the mucosa is covered by the villi of various sizes at birth. The villi are of flat finger-like shape, as if they are not completely divided. In contrast to duodenum and jejunum, the enterocyte in ileum are of various sizes and the enterocyte seems to be with large lysosomal vacuoles. The goblet cells are more abundant than upper parts of the gut. At the day 3 after birth, the ileal villi are still of flat finger-like shape, but many of them are partly-divided.

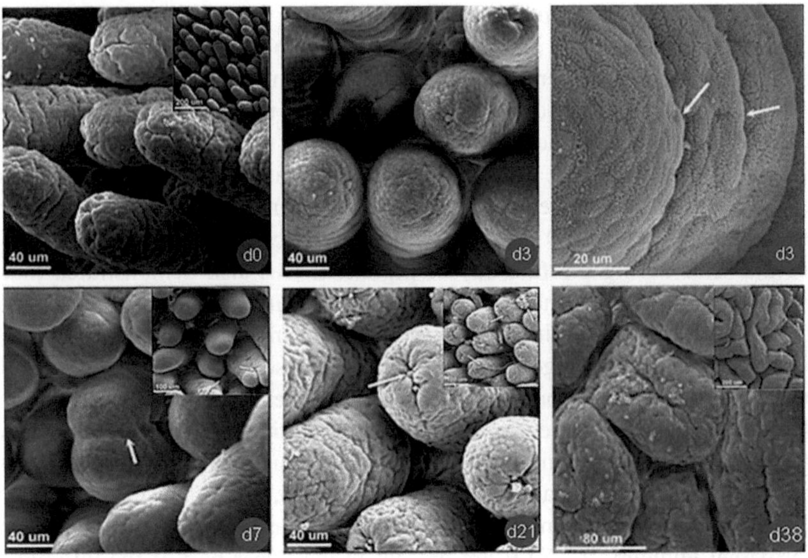

Fig. 3.1 SEM micrograph of duodenal mucosa on day 0 (at birth), day 3, day 7, day 21, and day 38 after birth in neonatal piglets (original images from J Physiol Pharmacol. 2005, 56 Suppl 3:71–87)

Fig. 3.2 SEM micrograph of jejunum mucosa on day 0 (at birth), day 3, day 7, day 21, and day 38 after birth in neonatal piglets (original images from J Physiol Pharmacol. 2005, 56 Suppl 3:71–87)

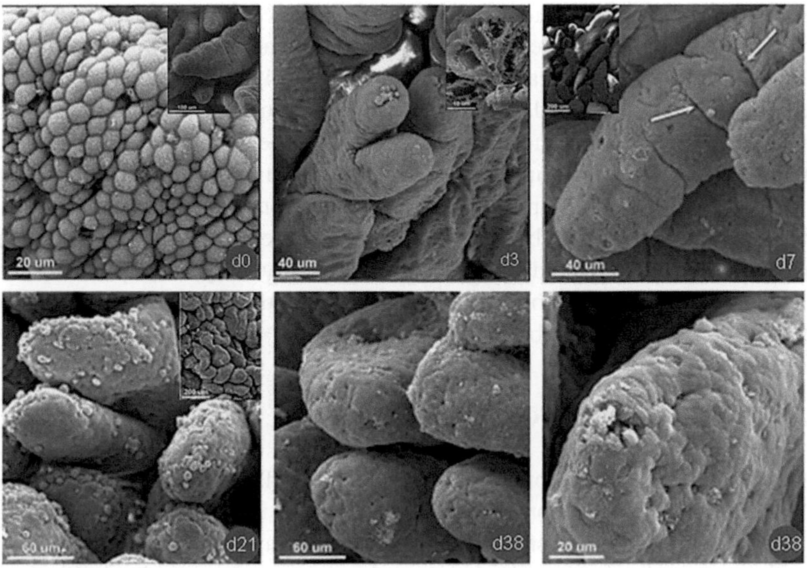

Fig. 3.3 SEM micrograph of ileum mucosa on day 0 (at birth), day 3, day 7, day 21, and day 38 after birth in neonatal piglets (original images from J Physiol Pharmacol. 2005, 56 Suppl 3:71–87)

Like duodenum, the surface of villi is relatively smoother, but some large enterocytes still exist on the top of villi (Skrzypek et al. 2005, 2007b, 2010; Fig. 3.3).

3.2.1.4 Suckling Period

The weaning of piglets usually takes place between 3 and 4 weeks of life (Boudry et al. 2004; Zabielski et al. 2008; Zhang et al. 2012a). During the sucking period, there is an intense rebuilding of small intestine, and the most intensive processes are noted in the epithelium (Boudry et al. 2004; Zabielski et al. 2008). In duodenum, although the architecture of villi on day 7 after birth is similar to that at birth, the transversal furrows are much shallower and less numerous than at day 0. Moreover, the density of villi decreases by 58 % and 8 % compared with birth and day 3 of life, respectively. At day 14, villi are of irregular shape and the villi density is significantly decreased (Skrzypek et al. 2005, 2010). At the day 21 after birth, although most of villi are finger-shaped, a few twin-shaped villi or villi with longitudinal indention are present. Corresponding to the increase in villi length, the number of transversal furrows is further reduced on the surface of villi. In addition, the apex of duodenal villi is not smooth any more since the presence of deep, knife incision-like, furrows and "cell packets" are apparent. These "cell packets" are made of groups of enterocytes that underwent apoptosis (Skrzypek et al. 2005, 2007b, 2010; Boudry et al. 2004; Fig. 3.1). In the jejunum, the shape of villi at day 7 change from finger-like to flat and tongue-like with the transversal furrows being located on the entire villi. Moreover, the thickness of mucosa significantly decreased and the crypt depth increased by 81 % in comparison with that measured at birth. At day 14, the depth of crypt further increased and achieved the highest value (Skrzypek et al. 2010; Fig. 3.2). On the 21st day of life, the shape of most of the jejunal villi changed to wide tongue-like; and the depth and total numbers of transversal furrows is further reduced. The villus tips are still containing numerous goblet cells. In ileum, the villi on the day 7 are of finger-like, leaf-like and tongue-like shapes, and the villi surface is relatively smooth but still with some transversal furrows. On day 21 after birth, the mucosa is covered by villi with a great diversity in shape, and the villi surface is shed by numerous cells including a large number of goblet cells (Skrzypek et al. 2005, 2007b, 2010; Fig. 3.3).

3.2.1.5 Weaning Transition

During weaning transition, piglets undergo nutritional, environmental, and psychological stresses, with concomitant marked changes in the structure of the small intestine. The most marked structural changes of the small intestine during weaning transition are a reduction in villus height and an increase in crypt depth (Pluske et al. 1997; Fan et al. 2004). Following weaning at day 21, the villus height of piglets reduces to 75 % of the preweaning value at 1 day after weaning, and further declines to approximately 50 % of the preweaning value at 5 days after weaning (Hampson 1986). Crypt depth of duodenum, jejunum, and ileum also increases with age after weaning (Gu et al. 2002; Kelly et al. 1991; van Beers-Schreurs et al. 1998). Associated with the reduction in villus height and the increase in crypt depth, the morphology of the villi are also changed. In duodenum of weaned piglets (38 day old), the villi change from finger-like to leaf-like, with a wide irregular base with no transversal furrows. At that time, few micrometer spaces, on the top of villi, between the epithelial cells as well as partly broken cells are observed (Skrzypek et al. 2005; Fig. 3.1). In jejunum, following weaning on day 38 after birth, the mucosa become thinner, and the villi are of various shapes: tongue-like and fold-like with predominant incomplete division projections. In addition, no transversal furrows are observed, and there are numerous epithelial cells shed along the villi surface (Fig. 3.2). In the ileum of weaned piglets, the villi are dominated by tongue-like shapes. However, single finger-like and leaf-like villi as well as incompletely divided villi are still present. On the surface of the villi, the transversal furrows are not observed, and numerous shed cells are observed in the apical region (Skrzypek et al. 2005; Fig. 3.3). In addition, the structural changes in the small intestine are affected by the age of weaning. The changes in the piglets weaned at 14 days of age are more conspicuous than those weaned at 21 days of age (Pluske et al. 1997).

3.2.2 Epithelial Cells Differentiation Along the Crypt–Villus Axis

The maturation of the intestinal epithelium is completed at the beginning of adulthood, which forms a complex equilibrium system with multiple cell types

(Cheng and Leblond 1974a, b; Mutoh et al. 2002). This equilibrium system undergoes continual renewal that involves highly coordinated processes of cellular proliferation, lineage-specific differentiation, migration, and apoptosis along the crypt–villus axis (CVA) (Gordon and Hermiston 1994). The epithelial cells differentiation along the CVA is observed during gestation. In rodent intestine, it is firstly apparent between day 17 and day 18 of gestation, when the endoderm is converted to an epithelial monolayer overlying nascent villi (Clatworthy and Subramanian 2001; Mathan et al. 1976). Stem cells located near the base of crypts give rise to progenitor cells, which expand through rapid proliferation before undergoing cell cycle arrest and ultimately differentiation into four principal cell lineages: absorptive enterocytes, goblet cells, endocrine cells, and Paneth cells (Clatworthy and Subramanian 2001). The fate of epithelial cells depends on the direction of their migration. After division, some cells go down to the bottom of the crypts and are transformed into Paneth cells with a life-time averaging 20 days. Most of cells migrate up the villus and differentiate into enterocytes, goblet cells, and endocrine cells. These cells are ultimately shed into the intestinal lumen every 3–5 days (Karam 1999). Microarrays results also establish that markers of enterocyte and goblet cell differentiation are maximally expressed in villus cells, whereas Paneth cell markers are up-regulated in crypt cells (Mariadason et al. 2005). Many signaling pathways and transcription factors with regulating functions upon small intestine cell maturation are identified. The known signaling pathways that are implicated in the regulation of cell fate determination and lineage specification in the intestine include Wnt–beta-catenin–TCF (Korinek et al. 1998; Mariadason et al. 2001; van de Wetering et al. 2002), BMP-TGF-beta-SMAD (Batts et al. 2006; He et al. 2004), Notch and its downstream factors HES1 and Atoh1 (Yang et al. 2001), and hedgehog signaling (van den Brink et al. 2004). Many transcription factors, including cdx-1 and cdx-2, kruppel-like factor 4, GATA4, 5, and 6 (Gao et al. 1998), and several forkhead family members (Burgess 1998; Clatworthy and Subramanian 2001; Shie et al. 2000), have also been suggested to be involved in the regulation of the intestinal cell maturation. Some of these transcription factors represent downstream targets of the different signaling pathways. Moreover, integrin-mediated cell-substratum and E-cadherin-mediated cell–cell adhesion, chemotactic gradients, as well as a lot of cytokines, hormones, and growth factors, have also been involved in the regulation of intestinal cell maturation (Burgess 1998; Kedinger et al. 1998). Lastly, microarrays and proteomic analysis suggest that intestinal cell maturation is primarily regulated at the transcriptional level according to the significant correlation between proteomic changes and corresponding gene expression changes along CVA (Chang et al. 2008).

3.3 Functional Development

The ontogeny of the intestinal mucosa function is discussed in Chap. 1 (Development of digestive glands in pigs) and Chap. 4 (Terminal digestion of polypeptides and amino acid absorption by the pig intestine epithelial cells during development).

This section mainly focuses on the functional development of mucosa along CVA. Epithelial cells differentiation CVA is accompanied by its functional specialization. There are four approaches in studying the differentiation-dependent expression of enterocyte function in vivo, including serial sectioning technique, quantitative immunohistochemical analysis, quantitative cytochemical analysis, and sequential cell isolation in combination with biochemical and biomolecular analysis (Smith 1985). Among the four approaches, the first three approaches are difficult to quantify and limited by availability of specific antibodies. Sequential cell isolation in combination with biochemical and biomolecular assays is a useful approach to analyze digestive enzymes and nutrient transporters activities in differentiating enterocytes under various conditions. With this approach, Raul et al. (1977) analyzed the activities of alkaline phosphatase, enterokinase, aminopeptidase, sucrase, amylase in villus and crypt cells of normal rats from 5 days after birth until 8 weeks. These authors found out that the activities of enterokinase and alkaline phosphatase were always located in the upper villus during postnatal development; whereas aminopeptidase and sucrase activities appeared in the crypt cells after birth and then rose to villus during the fourth week of life. The activity of amylase was located along the entire CVA during the first 5 days of life, and then reached its maximum activity in crypt. However, after the fourth week, the maximum activity was detected in the upper villus. Furthermore, Rowling and Sepúlveda (1984), by using sequential cell isolation in combination with biochemical analysis, found a 2–3-fold increase in the number of Na^+-pumping sites accompanying cell differentiation in rabbit jejunal epithelium. In neonatal pigs, Fan et al. (2001) examined the activities of alkaline phosphatase, aminopeptidase N, sucrase, lactase, and Na^+/K^+-ATPase along the crypt–villus using sequential cell isolation in combination with biochemical analysis. The activity of alkaline phosphatase increased quadratically during the enterocyte differentiation along the CVA in both the proximal and the distal small intestine. In addition, aminopeptidase N and sucrase activities showed a linear pattern of increase accompanying enterocyte differentiation along the CVA in both the proximal and the distal segments. Moreover, lactase activity increased cubically during the enterocyte differentiation along the CVA. In enterocytes, total Na^+-ATPase activity includes two components: a ouabain-sensitive Na^+/K^+-ATPase activity and a ouabain-insensitive Na^+-ATPase activity. Both ouabain-sensitive and ouabain-insensitive Na^+-ATPase activities are increased when enterocytes differentiated along the CVA in the small intestine. The ability of nutrient absorption is also altered when enterocytes differentiate along the CVA in the small intestine. The maximal transport activity of L-glutamate was increased during the enterocyte differentiation along the CVA of neonatal porcine small intestine, but the transporter affinity of L-glutamate was decreased during enterocyte differentiation (Fan et al. 2004). At the same time than the L-glutamate uptake alteration, the expression of EAAC-1, the major glutamate transporter, was increased with neonatal porcine enterocyte differentiation along the CVA; and its expression was regulated both at the transcription and translation levels (Fan et al. 2004). As opposed to EAAC1, there is a high level of maximal

SGLT1 uptake activity along the CVA of neonatal porcine small intestine. Although the mRNA abundance of SGLT1 is increased during enterocyte differentiation, there is no significant difference in SGLT1 protein abundance between crypt and villus (Yang et al. 2011). Similar to SGLT1, the SLC6A19 mRNA abundance is increased during enterocyte differentiation along the CVA. However, the B^0AT1 protein is evenly expressed in the epithelium along the CVA. In addition, apical maximal Na^+-Gln uptake activity, which is largely modulated by B^0AT1, is expressed along the entire jejunal CVA (Yang 2011). In contrast, the mRNA abundance of ASCT2, an intestinal AA exchanger, is decreased with neonatal porcine enterocyte differentiation along the CVA, but no difference in ASCT2 protein expression is observed (Buddington 1992).

3.4 Factors Influencing Intestinal Mucosa Development

The survival of animals and humans require physiological regulation of the intestinal mucosa operating as a functional unit. This contributes to the maintenance of epithelial homeostasis by forming a selective barrier to the harsh environment of intestinal lumen. The formation of this functional unit begins in the early embryo and completes at the beginning of adulthood. Moreover, the intestinal epithelium undergoes continual renewal all along life. These developmental and renewal processes are influenced by various factors, such as genetic, neural, hormonal, and dietary factors and weaning stress.

3.4.1 Genetic Influence

Intensive growth of the piglet small intestine is faster than growth of the whole organism due to intensive remodeling of the epithelium. This remodeling is regulated by genetic influence. Skrzypek et al. (2007b) compared the postnatal development of small intestinal mucosa architecture in Polish landrace/Pietrain (PP) and Duroc/Hampshire/wild boar (DHW) crossbreed piglets by scanning electron microscopy. They found differences in villi shape modification, in transversal furrows disappearance, in extrusion zone formation and in the presence of apoptotic cell packets, reflecting differences of PP and DHW piglets in mucosa structure development and renewal. With the age-related alteration in mucosa architecture, marked changes in nutrient transporters and enzymes were also observed during the suckling and weaning periods. Detailed studies of digestive enzymes and nutrient transport showed that the age-related changes in enzymes and nutrient transporters are genetically programmed and little affected by diet or hormones (Henning 1980; Leeper and Henning 1990; Nanthakumar and Henning 1993; Toloza and Diamond 1992). Moreover, molecular biology analysis showed that the ontogenic mechanisms involved in intestine apical fructose transporter GLUT5 expression and function are independent of dietary signals (Davidson et al. 1992; Shu et al. 1997). Although the development and renewal of mucosa can be reprogrammed by interactions of genetic determinants with other factors, it is ultimately controlled by transcription regulation via multiple transcriptional elements with activatory or repressive roles (Traber and Silberg 1996).

3.4.2 Neural Influence

The enteric nervous systems (ENS), a large network of neurons and glial cells located along the gastrointestinal tract, provide an intricate network for the reflex control of intestinal mucosa (Pácha 2000). In prenatal animals and humans, the nutrients and biologically active substances involved in intestine development are mainly transferred from the mother via placenta. The contribution of ENS is presumably small but increases with fetal development (Zabielski 2007). However, in adults, almost all main gut processes such as secretion, absorption, immune responses, blood flow and complex motility patterns (such as mixing, peristalsis and migrating motor complexes) are regulated or controlled by the ENS (Burzynski et al. 2009). In comparison with adults, virtually nothing is known about the possible involvement of the nervous system in mediating mucosa structure and function development. Nevertheless, it has been suggested that the ENS function was different between early postnatal development and adulthood. Evidences suggest that ENS is involved in the regulation of the development of intestinal motility, and maybe also other function of mucosa (Zabielski 2007; Burns et al. 2009).

3.4.3 Hormonal Influence

The role of hormones in intestinal mucosa development was studied more intensively than that of ENS. A large number of hormones and cytokines have been shown to affect intestinal mucosa development and nutrients transport (Zabielski 2007). The hormones of the IGF family, including insulin, insulin-like growth factor I (IGF-I) and insulin-like growth factor II (IGF-II), showed positive effects on intestinal mucosa development and intestinal adaptation (Ben et al. 2010; Lund 1998; Lemmey et al. 1991). Treatment with oral insulin significantly increased enterocyte proliferation, and decreased cell apoptosis, in rats (Ben et al. 2010). IGF-I was also reported as able to enhance crypt cell migration and cell proliferation (Liao and Lönnerdal 2010; Chen et al. 1999). Moreover, transforming growth factor alpha, epidermal growth factor, and hepatocyte growth factor are also able to promote crypt cell proliferation as measured by the ^{3}H-thymidine incorporation assay (Nishimura et al. 1998; Sheng et al. 2006). In recent years, many newly discovered hormones or regulatory peptides have been also reported as able to regulate intestinal mucosa development. Leptin, whose receptor is widely distributed in the small intestine mucosa, has been shown to enhance small intestinal length and mitotic index, and to reduce the percentage of vacuolated enterocytes in the small intestine of neonatal piglets (Słupecka et al. 2005). Ghrelin, a growth-hormone-releasing acylated peptide, is able to delay piglet intestinal mucosa development by reducing the length of intestinal villi and increasing crypt depth (Kotunia et al. 2006). Glucagon-like peptide-2 also involves in the stimulation of stem cell proliferation and in the simultaneous reduction of the programmed cell death in neonatal piglets, thus acting as an important regulator of growth and maturation of the small intestinal mucosa (Burrin et al. 2005).

3.4.4 Dietary Influence

The compounds in diet not only play a nutritional role but also directly stimulate the growth and development of the intestinal mucosa (Buts et al. 1993; Yao et al. 2011; Kong et al. 2012b). The alterations in dietary input (amniotic fluid, maternal milk, postweaning diet) during the ontogeny of the intestinal mucosa impose different functional demands in relationship with its structure (Buddington 1994). Similarly, amniotic fluid, maternal colostrum and milk contain many biologically active substances that stimulate the development of the intestinal mucosa (Weaver et al. 1988; Xu 1996). Although the family of biologically active substances in milk is continuously increasing, EGF, IGF-I, glucocorticoids, and insulin were confirmed as compounds with important roles in stimulating epithelial cells proliferation and differentiation (Burrin et al. 1996; Buts et al. 1997; Houle et al. 1997; Xu 1996; Yeh et al. 1991). The importance of nutrients in amniotic fluid for mucosa development, and the ability of the fetal intestine to absorb exogenous nutrients were confirmed by infusing galactose into amniotic fluid of rabbits. This infusion increases fetal mucosal weights and total intestinal capacities to transport aldohexoses (Buchmiller et al. 1992). Although milk is the principal source of nutrients for most suckling mammals, the composition of milk is different between species (Jenness and Sloan 1970). In addition, the regulatory role of maternal milk in intestinal mucosa development is not only dependent on the content of nutrients and non-nutrient components, but also depends on the gradual alteration of the milk composition (Keen et al. 1981). The milk for rat pups exceeds the demand during the first week of life, whereas it becomes the limiting factor during the second week. The transition from milk to a solid diet results in significantly shorter villi and deeper crypts in the small intestinal mucosa of swine. A variety of nutrient deficiencies can impair animal growth after weaning and some of these deficiencies preferentially target intestinal mucosa development (Williams et al. 1996). The villi heights of rat pups are smaller in all segments of the intestine when there dams were subjected to protein deficiency (Subramoniam 1979).

3.4.5 Weaning Stress

Weaning is one of the most significant event in the pig life, as it corresponds to a transition from milk, which contains highly digestible protein, fat, and lactose, to dry and less-digestible starch-based diet. This transition results in reduced energy intake for the maintenance of the epithelial structure (Kim et al. 2012). Moreover, weaning of piglets coincides with, or is preceded by, the appearance of adults hydrolytic and transport characteristic of adults, and is accompanied by increases in the circulating concentrations of hormones and cytokines which are involved in the intestinal development (Buddington 1994). In response to weaning, the small intestinal mucosa of piglets undergoes major changes in structure and function (Pluske et al. 1997; Xu et al. 1996). The most obvious structural changes are the reduction in villus height and increase in crypt depth, changes which are suspected to result from the increased rate of cell loss and the increase in crypt cell proliferation (Pluske et al. 1997). It has been shown that villus height is reduced to 75 % of the preweaning value one day after weaning; and then further declines to approximately 50 % of the preweaning value 5 days after weaning (Hampson 1986). Along with the reduction in villus height and increase in crypt depth, the villi morphology also changes from long finger-like projections to leaf- or tongue-like structures (Cera et al. 1988).

References

Amerongen HM, Mack JA, Wilson JM, Neutra MR (1989) Membrane domains of intestinal epithelial cells: distribution of Na+, K+-ATPase and the membrane skeleton in adult rat intestine during fetal development and after epithelial isolation. J Cell Biol 109:2129–2138

Batts LE, Polk DB, Dubois RN, Kulessa H (2006) Bmp signaling is required for intestinal growth and morphogenesis. Dev Dyn 235:1563–1570

Ben Lulu S, Coran AG, Mogilner JG, Shaoul R, Shamir R, Shehadeh N, Sukhotnik I (2010) Oral insulin stimulates intestinal epithelial cell turnover in correlation with insulin-receptor expression along the villus-crypt axis in a rat model of short bowel syndrome. Pediatr Surg Int 26:37–44

Boudreau F, Rings EH, van Wering HM, Kim RK, Swain GP, Krasinski SD, Moffett J, Grand RJ, Suh ER, Traber PG (2002) Hepatocyte nuclear factor-1α, GATA-4, and caudal related homeodomain protein Cdx2 interact functionally to modulate intestinal gene transcription. J Biol Chem 277:31909–31917

Boudry G, Péron V, Le Huërou-Luron I, Lallès JP, Sève B (2004) Weaning induces both transient and long-lasting modifications of absorptive, secretory, and barrier properties of piglet intestine. J Nutr 134:2256–2262

Buchmiller TL, Fonkalsrud EW, Kim CS, Chopourian HL, Shaw KS, Lam MM, Diamond JM (1992) Upregulation of nutrient transport in fetal rabbit intestine by transamniotic substrate administration. J Surg Res 52:443–447

Buddington RK (1992) Intestinal nutrient transport during ontogeny of vertebrates. Am J Physiol 263:R503–R509

Buddington RK (1994) Nutrition and ontogenetic development of the intestine. Can J Physiol Pharmacol 72:251–259

Burgess AW (1998) Growth control mechanisms in normal and transformed intestinal cells. Philos Trans R Soc Lond B Biol Sci 353:903–909

Burns AJ, Roberts RR, Bornstein JC, Young HM (2009) Development of the enteric nervous system and its role in intestinal motility during fetal and early postnatal stages. Semin Pediatr Surg 18:196–205

Burrin DG, Stoll B, Guan X, Cui L, Chang X, Holst JJ (2005) Glucagon-like peptide 2 dose-dependently activates intestinal cell survival and proliferation in neonatal piglets. Endocrinology 146:22–32

Burrin DG, Wester TJ, Davis TA, Amick S, Heath JP (1996) Orally administered IGF-I increases intestinal mucosal growth in formula-fed neonatal pigs. Am J Physiol 270:R1085–R1091

Burzynski G, Shepherd IT, Enomoto H (2009) Genetic model system studies of the development of the enteric nervous system, gut motility and Hirschsprung's disease. Neurogastroenterol Motil 21:113–127

Buts JP, De Keyser N, Kolanowski J, Sokal E, Van Hoof F (1993) Maturation of villus and crypt cell functions in rat small intestine. Role of dietary polyamines. Dig Dis Sci 38:1091–1098

Buts JP, De Keyser N, Marandi S, Maernoudt AS, Sokal EM, Rahier J, Hermans D (1997) Expression of insulin receptors and of 60-kDa receptor substrate in rat mature and immature enterocytes. Am J Physiol 273:G217–G226

Cera KR, Mahan DC, Cross RF, Reinhart GA, Whitmoyer RE (1988) Effect of age, weaning and postweaning diet on small intestinal growth and jejunal morphology in young swine. J Anim Sci 66:574–584

Chang J, Chance MR, Nicholas C, Ahmed N, Guilmeau S, Flandez M, Wang D, Byun DS, Nasser S, Albanese JM, Corner GA, Heerdt BG, Wilson AJ, Augenlicht LH, Mariadason JM (2008) Proteomic changes during intestinal cell maturation in vivo. J Proteomics 71:530–546

Chen K, Nezu R, Wasa M, Sando K, Kamata S, Takagi Y, Okada A (1999) Insulin-like growth factor-1 modulation of intestinal epithelial cell restitution. JPEN J Parenter Enteral Nutr 23: S89–S92

Cheng H, Leblond CP (1974a) Origin, differentiation and renewal of the four main epithelial cell types in the mouse small intestine I. Columnar cell. Am J Anat 141:461–479

Cheng H, Leblond CP (1974b) Origin, differentiation and renewal of the four main epithelial cell types in the mouse small intestine V. Unitarian theory of the origin of the four epithelial cell types. Am J Anat 141:537–561

Clatworthy JP, Subramanian V (2001) Stem cells and the regulation of proliferation, differentiation and patterning in the intestinal epithelium: emerging insights from gene expression patterns, transgenic and gene ablation studies. Mech Dev 101:3–9

Davidson NO, Hausman AM, Ifkovits CA, Buse JB, Gould GW, Burant CF, Bell GI (1992) Human intestinal glucose transporter expression and localization of GLUT5. Am J Physiol 262:C795–C800

Davis TA, Bush JA, Vann RC, Suryawan A, Kimball SR, Burrin DG (2004) Somatotropin regulation of protein metabolism in pigs. J Anim Sci 82(E-Suppl):E207–E213

Fan MZ, Matthews JC, Etienne NM, Stoll B, Lackeyram D, Burrin DG (2004) Expression of apical membrane L-glutamate transporters in neonatal porcine epithelial cells along the small intestinal crypt-villus axis. Am J Physiol Gastrointest Liver Physiol 287:G385–G398

Fan MZ, Stoll B, Jiang R, Burrin DG (2001) Enterocyte digestive enzyme activity along the crypt-villus and longitudinal axes in the neonatal pig small intestine. J Anim Sci 79:371–381

Foltzer-Jourdainne C, Kedinger M, Raul F (1989) Perinatal expression of brush-border hydrolases in rat colon: hormonal and tissue regulations. Am J Physiol 257:G496–G503

Forstner JF (1978) Intestinal mucins in health and disease. Digestion 17:234–263

Furuse M, Hirase T, Itoh M, Nagafuchi A, Yonemura S, Tsukita S, Tsukita S (1993) Occludin: a novel integral membrane protein localizing at tight junctions. J Cell Biol 123:1777–1788

Gao X, Sedgwick T, Shi YB, Evans T (1998) Distinct functions are implicated for the GATA-4, -5, and -6 transcription factors in the regulation of intestine epithelial cell differentiation. Mol Cell Biol 18:2901–2911

Gordon JI, Hermiston ML (1994) Differentiation and self-renewal in the mouse gastrointestinal epithelium. Curr Opin Cell Biol 6:795–803

Gu X, Li D, She R (2002) Effect of weaning on small intestinal structure and function in the piglet. Arch Tierernahr 56:275–286

Hampson DJ (1986) Alterations in piglet small intestinal structure at weaning. Res Vet Sci 40:32–40

He XC, Zhang J, Tong WG, Tawfik O, Ross J, Scoville DH, Tian Q, Zeng X, He X, Wiedemann LM, Mishina Y, Li L (2004) BMP signaling inhibits intestinal stem cell self-renewal through suppression of Wnt-beta-catenin signaling. Nat Genet 36:1117–1121

Henning SJ (1980) Postnatal development of jejunal sucrase: independence from cyclic AMP. Biol Neonate 37:246–253

Henning SJ (1981) Postnatal development: coordination of feeding, digestion, and metabolism. Am J Physiol 241:G199–G214

Houle VM, Schroeder EA, Odle J, Donovan SM (1997) Small intestinal disaccharidase activity and ileal villus height are increased in piglets consuming formula containing recombinant human insulin-like growth factor-I. Pediatr Res 42:78–86

Itoh M, Furuse M, Morita K, Kubota K, Saitou M, Tsukita S (1999) Direct binding of three tight junction-associated MAGUKs, ZO-1, ZO-2, and ZO-3, with the COOH termini of claudins. J Cell Biol 147:1351–1363

Jenness R, Sloan RE (1970) The composition of milks of various species: a review. Dairy Sci Abstr 32:599–612

Karam SM (1999) Lineage commitment and maturation of epithelial cells in the gut. Front Biosci 4:D286–D298

Kedinger M, Lefebvre O, Duluc I, Freund JN, Simon-Assmann P (1998) Cellular and molecular partners involved in gut morphogenesis and differentiation. Philos Trans R Soc Lond B Biol Sci 353:847–856

Keen CL, Lönnerdal B, Clegg M, Hurley LS (1981) Developmental changes in composition of rat milk: trace elements, minerals, protein, carbohydrate and fat. J Nutr 111:226–236

Kelly D, Smyth JA, McCracken KJ (1991) Digestive development of the early-weaned pig. I. Effect on continuous nutrient supply on the development of the digestive tract and on changes in digestive enzyme activity during the first week post-weaning. Br J Nutr 65:169–180

Kim JC, Hansen CF, Mullan BP, Pluske JR (2012) Nutrition and pathology of weaner pigs: nutritional strategies to support barrier function in the gastrointestinal tract. Anim Feed Sci Technol 173:3–16

Kong XF, Tan BE, Yin YL, Li XL, Jaeger LA, Bazer FW, Wu GY (2012b) Arginine stimulates the mTOR signaling pathway and protein synthesis in porcine trophectoderm cells. J Nutr Biochem 23(9):1178–1183

Korinek V, Barker N, Moerer P, van Donselaar E, Huls G, Peters PJ, Clevers H (1998) Depletion of epithelial stem-cell compartments in the small intestine of mice lacking Tcf-4. Nat Genet 19:379–383

Kotunia A, Woliński J, Słupecka M, Dolman D, Kato I, Kuwahara A, Zabielski R (2006) Exogenous ghrelin retards the development of the small intestine in pig neonates fed with artificial milk formula. Book of abstracts, digestive physiology in pigs. 25–27 May 2006, Vejle, Denmark. pp 82

Lacroix B, Kedinger M, Simon-Assmann P, Rousset M, Zweibaum A, Haffen K (1984) Developmental pattern of brush border enzymes in the human fetal colon. Correlation with some morphogenetic events. Early Hum Dev 9:95–103

Le Dividich J, Seve B (2000) Effects of underfeeding during the weaning period on growth, metabolism and hormonal adjustments in the piglet. Dom Anim Endocrinol 19:63–74

Leeper LL, Henning SJ (1990) Development and tissue distribution of sucrase-isomaltase mRNA in rats. Am J Physiol 258:G52–G58

Lemmey AB, Martin AA, Read LC, Tomas FM, Owens PC, Ballard FJ (1991) IGF-I and the truncated analogue des-(1-3) IGF-I enhance growth in rats after gut resection. Am J Physiol 260:E213–E219

Liao Y, Lönnerdal B (2010) Global microRNA characterization reveals that miR-103 is involved in IGF-1 stimulated mouse intestinal cell proliferation. PLoS One 5(9):e12976

Lund PK (1998) Molecular basis of intestinal adaptation: the role of the insulin-like growth factor system. Ann N Y Acad Sci 859:18–36

Mariadason JM, Bordonaro M, Aslam F, Shi L, Kuraguchi M, Velcich A, Augenlicht LH (2001) Down-regulation of β-catenin TCF signaling is linked to colonic epithelial cell differentiation. Cancer Res 61:3465–3471

Mariadason JM, Nicholas C, L'Italien KE, Zhuang M, Smartt HJ, Heerdt BG, Yang W, Corner GA, Wilson AJ, Klampfer L, Arango D, Augenlicht LH (2005) Gene expression profiling of intestinal epithelial cell maturation along the crypt-villus axis. Gastroenterology 128:1081–1088

Mathan M, Moxey PC, Trier JS (1976) Morphogenesis of fetal rat duodenal villi. Am J Anat 146:73–92

Menard D, Calvert R (1991) Fetal and postnatal development of the small and large intestine: patterns and regulation. In: Morisset J, Solomon TE (eds) Growth of the gastrointestinal tract: gastrointestinal hormones and growth factors. CRC, Boca Raton, FL, pp 159–174

Ménard D, Dagenais P, Calvert R (1994) Morphological changes and cellular proliferation in mouse colon during fetal and postnatal development. Anat Rec 238:349–359

Mutoh H, Hakamata Y, Sato K, Eda A, Yanaka I, Honda S, Osawa H, Kaneko Y, Sugano K (2002) Conversion of gastric mucosa to intestinal metaplasia in Cdx2-expressing transgenic mice. Biochem Biophys Res Commun 294:470–479

Nanthakumar NN, Henning SJ (1993) Ontogeny of sucrase-isomaltase gene expression in rat intestine: responsiveness to glucocorticoids. Am J Physiol 264:G306–G311

Nishimura S, Takahashi M, Ota S, Hirano M, Hiraishi H (1998) Hepatocyte growth factor accelerates restitution of intestinal epithelial cells. J Gastroenterol 33:172–178

Ono K (1977) Absorption of horseradish peroxidase by the principal cells of the large intestines of postnatal developing rats. Anat Embryol (Berl) 151:53–62

Pácha J (2000) Development of intestinal transport function in mammals. Physiol Rev 80:1633–1667

Pluske JR, Hampson DJ, Williams IH (1997) Factors influencing the structure and function of the small intestine in weaned pig: a review. Livest Prod Sci 51:215–236

Raul F, Simon P, Kedinger M, Haffen K (1977) Intestinal enzymes activities in isolated villus and crypt cells during postnatal development of the rat. Cell Tissue Res 176:167–178

Roth KA, Rubin DC, Birkenmeier EH, Gordon JI (1991) Expression of liver fatty acid-binding protein/human growth hormone fusion genes within the enterocyte and enteroendocrine cell populations of fetal transgenic mice. J Biol Chem 266:5949–5954

Rowling PJ, Sepúlveda FV (1984) The distribution of (Na^+/K^+)-ATPase along the villus crypt-axis in the rabbit small intestine. Biochim Biophys Acta 771:35–41

Schmidt GH, Winton DJ, Ponder BA (1988) Development of the pattern of cell renewal in the crypt-villus unit of chimaeric mouse small intestine. Development 103:785–790

Shanahan F (1997) A gut reaction—lymphoepithelial communication in the intestine. Science 275:1897–1898

Sheng G, Bernabe KQ, Guo J, Warner BW (2006) Epidermal growth factor receptor-mediated proliferation of enterocytes requires p21waf1/cip1 expression. Gastroenterology 131:153–164

Shie JL, Chen ZY, O'Brien MJ, Pestell RG, Lee ME, Tseng CC (2000) Role of gut-enriched Kruppel-like factor in colonic cell growth and differentiation. Am J Physiol Gastrointest Liver Physiol 279:G806–G814

Shu R, David ES, Ferraris RP (1997) Dietary fructose enhances intestinal fructose transport and GLUT5 expression in weaning rats. Am J Physiol 272:G446–G453

Simon TC, Gordon JI (1995) Intestinal epithelial cell differentiation: new insights from mice, flies and nematodes. Curr Opin Genet Dev 5:577–586

Skrzypek T, Valverde Piedra JL, Skrzypek H, Kazimierczak W, Biernat M, Zabielski R (2007a) Gradual disappearance of vacuolated enterocytes in the small intestine of neonatal piglets. J Physiol Pharmacol 58(Suppl 3):87–95

Skrzypek T, Valverde Piedrab JL, Skrzypeka H, Kazimierczaka W, Szymańczykb S, Pawłowskab M, Zabielskic R (2007b) Intestinal villi structure during the development of pig and wild boar crossbreed neonates. Livest Sci 109:38–41

Skrzypek T, Valverde Piedra JL, Skrzypek H, Woliński J, Kazimierczak W, Szymańczyk S, Pawłowska M, Zabielski R (2005) Light and scanning electron microscopy evaluation of the postnatal small intestinal mucosa development in pigs. J Physiol Pharmacol 56(Suppl 3):71–87

Skrzypek T, Valverde Piedrab JL, Skrzypeka H, Kazimierczaka W, Szymanczykb SE, Zabielskic R (2010) Changes in pig small intestinal absorptive area during the first 14 days. Livest Sci 133:53–56

Słupecka M, Woliński J, Zabielski R (2005) Leptin affects the p53 expression in the gut mucosa of neonatal piglets. Proceedings of the 13th Euro-conference on apoptosis, Budapest, Hungary, 2005 October 1–4, Abstract. pp 207.

Smith MW (1985) Expression of digestive and absorptive function in differentiating enterocytes. Annu Rev Physiol 47:247–260

Stevenson BR, Siliciano JD, Mooseker MS, Goodenough DA (1986) Identification of ZO-1: a high molecular weight polypeptide associated with the tight junction (zonula occludens) in a variety of epithelia. J Cell Biol 103:755–766

Subramoniam A (1979) Rat small intestinal morphology with special reference to villi: effects of maternal protein deficiency and hydrocortisone. Acta Anat (Basel) 104:439–450

Toloza EM, Diamond J (1992) Ontogenetic development of nutrient transporters in rat intestine. Am J Physiol 263:G593–G604

Traber PG, Gumucio DL, Wang W (1991) Isolation of intestinal epithelial cells for the study of differential gene expression along the crypt-villus axis. Am J Physiol 260:G895–G903

Traber PG, Silberg DG (1996) Intestine-specific gene transcription. Annu Rev Physiol 58:275–297

Trahair JF, Sangild PT (2004) Studying the development of the small intestine: philosophical and anatomical perspectives. In: Zabielski R, Gregory PC, Weström B (eds) Biology of the intestine in growing animals. Elsevier, Amsterdam, pp 1–54

van Beers-Schreurs HM, Nabuurs MJ, Vellenga L, Kalsbeek-van der Valk HJ, Wensing T, Breukink HJ (1998) Weaning and the weanling diet influence the villous height and crypt depth in the small intestine of pigs and alter the concentrations of short-chain fatty acids in the large intestine and blood. J Nutr 128:947–953

van de Wetering M, Sancho E, Verweij C, de Lau W, Oving I, Hurlstone A, van der Horn K, Batlle E, Coudreuse D, Haramis AP, Tjon-Pon-Fong M, Moerer P, van den Born M, Soete G, Pals S, Eilers M, Medema R, Clevers H (2002) The β-catenin/TCF-4 complex imposes a crypt progenitor phenotype on colorectal cancer cells. Cell 111:241–250

van den Brink GR, Bleuming SA, Hardwick JC, Schepman BL, Offerhaus GJ, Keller JJ, Nielsen C, Gaffield W, van Deventer SJ, Roberts DJ, Peppelenbosch MP (2004) Indian Hedgehog is an antagonist of Wnt signaling in colonic epithelial cell differentiation. Nat Genet 36:277–282

Weaver LT, Freiberg E, Israel EJ, Walker WA (1988) Epidermal growth factor in human amniotic fluid. Gastroenterology 95:1436

Williams EA, Rumsey RD, Powers HJ (1996) An investigation into the reversibility of the morphological and cytokinetic changes seen in the small intestine of riboflavin deficient rats. Gut 39:220–225

Xu RJ, Zhang YQ (2003) Gastrointestinal regulatory peptides. In: Xu RJ, Granwell PD (eds) Neonatal pig gastrointestinal physiology and nutrition. Nottingham University Press, Nottingham, pp 61–93

Xu RJ, Mao YL, Tso MY (1996) Stability of gastrin in the gastrointestinal lumen of suckling, weanling and adult pigs. Biol Neonate 70:60–68

Xu RJ, Mellor DJ, Tungthanathanich P, Birtles MJ, Reynolds GW, Simpson HV (1992) Growth and morphological changes in the small and the large intestine in piglets during the first three days after birth. J Dev Physiol 18:161–172

Xu RJ, Wang F, Zhang SH (2000) Postnatal adaptation of the gastrointestinal tract in neonatal pigs: a possible role of milk-borne growth factors. Livestock Prod Sci 66:95–107

Xu RJ (1996) Development of the newborn GI tract and its relation to colostrum/milk intake: a review. Reprod Fertil Dev 8:35–48

Yang C, Albin DM, Wang Z, Stoll B, Lackeyram D, Swanson KC, Yin Y, Tappenden KA, Mine Y, Yada RY, Burrin DG, Fan MZ (2011) Apical Na^+-D-glucose cotransporter 1 (SGLT1) activity and protein abundance are expressed along the jejunal crypt-villus axis in the neonatal pig. Am J Physiol Gastrointest Liver Physiol 300:G60–G70

Yang C (2011) Expression of porcine intestinal nutrient transporters along crypt-villus axis and during postnatal development. Doctoral dissertation, University of Guelph, Guelph, Ontario, Canada

Yang Q, Bermingham NA, Finegold MJ, Zoghbi HY (2001) Requirement of Math1 for secretory cell lineage commitment in the mouse intestine. Science 294:2155–2158

Yao K, Wang L, Ding BY, Fu DB, Liu YL, Zhu HL, Liu J, Li YT, Kang P, Yin YL, Wu GY, Hou YQ (2012a) Alpha-ketoglutarate inhibits glutamine degradation and enhances protein synthesis in intestinal porcine epithelial cells. Amino Acids 42(6):2491–2500

Yeh KY, Yeh M, Montgomery RK, Grand RJ, Holt PR (1991) Cortisone and thyroxine modulate intestinal lactase and sucrase mRNA levels and activities in the suckling rat. Biochem Biophys Res Commun 180:174–180

Zabielski R, Godlewski MM, Guilloteau P (2008) Control of development of gastrointestinal system in neonates. J Physiol Pharmacol 59(Suppl 1):35–54

Zabielski R (2007) Hormonal and neural regulation of intestinal function in pigs. Livest Sci 108:32–40

Zhang Y-G, Yin Y-L, Fang J, Wang Q (2012a) Pig production in subtropical agriculture. J Sci Food Agric 92:1016–1024

Zweibaum A, Hauri HP, Sterchi E, Chantret I, Haffen K, Bamat J, Sordat B (1984) Immunohistological evidence, obtained with monoclonal antibodies, of small intestinal brush border hydrolases in human colon cancers and foetal colons. Int J Cancer 34:591–598

Part II
Terminal Digestion, Amino Acid Absorption/Metabolism, and Microbiota in the Pig Intestine During Development

Chapter 4
Terminal Digestion of Polypeptides and Amino Acid Absorption by the Pig Intestine Epithelial Cells During Development

François Blachier, Yulong Yin, Guoyao Wu, and Mireille Andriamihaja

Contents

4.1 Introduction

Terminal digestion of alimentary and endogenous polypeptides followed by absorption of oligopeptides and amino acids are performed by the absorptive intestinal epithelial cells of the small intestine mucosa. The intestinal transport of amino acids is critical for the supply of amino acids to all tissues (Bröer 2008). Although intestinal amino acid absorption in the pig has been the subject of numerous reports, there is a paucity of available data regarding the terminal digestion of peptides in pig enterocytes.

F. Blachier (✉) • M. Andriamihaja
UMR 914 Nutrition Physiology and Ingestive Behavior, INRA-CRNH-IdF-AgroParisTech, Paris, France
e-mail: francois.blachier@agroparistech.fr

Y. Yin
Institute of Subtropical Agriculture, Chinese Academy of Sciences, Changsha, China

G. Wu
Department of Animal Science and Faculty of Nutrition, Texas A&M University, College Station, TX, USA

F. Blachier et al. (eds.), *Nutritional and Physiological Functions of Amino Acids in Pigs*,
DOI 10.1007/978-3-7091-1328-8_4, © Springer-Verlag Wien 2013

4.2 Neonatal Intestinal Growth in Piglets, Terminal Digestion, and Intestinal Absorption of Amino Acid and Oligopeptides

The major site for the digestion and absorption of N-containing compounds in diets containing wheat and milk protein concentrates in young pigs is in the first 25–50 % of the total small intestine from the pylorus, while for other dietary sources (that are a mixture of wheat and protein concentrates from soya bean and lupins), the major site of digestion and absorption is in the 50–75 % of the small intestine (Leibholz 1985). However, for instance, all villus enterocytes from the small intestine of newborn pigs have been found to be able to transport alanine (Smith 1981). Using neonatal piglets between the ages of 14 and 18 days, it was found that in the small intestine, the activity of aminopeptidase N is higher in the distal than in the proximal segment (Fan et al. 2001a, b). From histological observations, it was concluded that the dipeptidases are predominantly localized in the cytoplasm of enterocytes, and as expected, in the villus enterocytes but not in the crypts (Noren et al. 1977). Dipeptides taken up by villus enterocytes from the lumen of the small intestine are rapidly hydrolyzed to release the constituent amino acids.

In pigs, the very rapid postnatal growth of the small intestine and its mucosal protein as observed by Widdowson et al. (1976) has been attributed to mucosal hyperplasia, mucosal protein biosynthesis, and endocytosis of ingested immunoglobulins (Burrin et al. 1992; Simmen et al. 1990; Werhahn et al. 1981). These rapid changes after birth coincide with changes in intestinal morphology (Xu et al. 1992) and enterocyte ultrastructure (Komuves et al. 1993). In the intestine of neonatal pigs, it has been found that the speed of epithelial cell migration increases from birth until 3 weeks (Klein 1989). Indeed, in piglets, one day after birth, the enterocyte replacement is estimated to take between 9 and 10 days in the proximal region of the small intestine and between 7 and 10 days in the distal region. However, 3 weeks after birth, the cell turnover is considerably shortened to 2–4 days (Moon 1971). From more recent work (Zhang et al. 1997), it has been demonstrated that most of the spectacular changes in intestinal structure and functions occur within the first 6 h of suckling and that the rapid neonatal intestinal growth is due to both hyperplasia and hypertrophy. Measurement of intestinal brush border membrane-associated aminooligopeptidases in neonatal pigs reveals an increase in their enzymatic activities after onset of suckling and suggest that these enzymes are important for hydrolyzing the peptides recovered from milk proteins during the suckling period (Zhang et al. 1997).

The absorption of amino acids and oligopeptides from the intestinal lumen to the portal bloodstream in pigs has been the subject of several studies. In order to move from the luminal content to the bloodstream, amino acids have to firstly enter the intestinal absorptive epithelial cells through transporters present in the brush border membranes at the apical side, and then to be exported from the enterocyte to the bloodstream through transporters present in the baso-lateral membranes. Amino acid absorption has been studied during development in pigs with rates of absorption being measured for five amino acids that are substrates for the acidic (aspartate), basic (lysine), neutral (leucine and methionine), and imino (proline) amino acid carriers (Buddington et al. 2001). The rates of absorption (which were calculated as the sum of carrier-mediated and apparent diffusion) were highest at birth for all the amino acids considered (except for proline), and then declined after 24 h of suckling. Thereafter, there were declines for leucine, methionine, and proline (but not for aspartate and lysine) in rates of absorption up to 42 days after birth (that is the longest period of time tested, Buddington et al. 2001). Using isolated loops of the proximal, middle, and terminal small intestine of pigs, it has been determined that the absorption capacity of nitrogen from a mixture of proteins, peptides, and amino acids is lower in the proximal small intestine when compared with other regions of the small intestine (Buraczewska 1981). Using the same experimental model, it has been shown that the peptide fraction of the mixture disappeared more rapidly from the intestinal lumen than did the free amino acids, leading the authors to propose that peptidases in the intestinal mucosa play an important role in the process of amino acid absorption (Buraczewska 1981). By measuring the net portal amino acid absorption in pigs (calculated by multiplying the porto-arterial plasma amino acid concentration difference by the portal vein flow rate) after ingestion of different mixtures of proteins and amino acids, it has been determined that free amino acids like lysine and threonine are more rapidly absorbed than protein-bound lysine and threonine (Yen et al. 2004). By measuring amino acid influx across the brush border membrane of the pig ileal epithelium, it has been demonstrated that bipolar amino acids have similar affinities for transport by systems B and ASC (Munck et al. 2000). Using pig jejunal brush border membrane vesicles, Maenz and Patience (Maenz and Patience 1992) have shown that the transport system for L-threonine (system B) is functionally distinct from other amino acid transport systems. In Tibetan suckling piglets, it has been reported that the expression pattern of the amino acid transporter system b(0,+), which mediates apical uptake of cationic amino acids and cysteine, is similar in duodenum and proximal jejunum and the abundance of the transporter mRNA is increased during the suckling period (Wang et al. 2009). Lastly, the Na^+-dependent high affinity X_{AG}^- system and/or the low affinity B° system have been shown to be involved in the transport of L-glutamate from the lumen into the intestinal epithelial cells (Fan et al. 2004).

In a recent work with multi-catheterized minipigs, Remond et al. (2011) have demonstrated that small intestine has the capacity to sequester large amounts of dietary cysteine in the process of absorption of this amino acid, thus limiting its release into the bloodstream. The cysteine may also be utilized for the synthesis of glutathione in enterocytes.

Dipeptide transport in the pig jejunum has been documented (Winckler et al. 1999) as well as the presence of a PepT1-like transporter which is likely to contribute to the overall amino acid absorption from the small intestine. The functional characterization of the pig intestinal peptide transporter PepT1 was studied, and the results obtained have demonstrated that this peptide transporter is able to transport a variety of di- and tripeptides but not tetrapeptides (Klang et al. 2005). Figure 4.1 is shown to illustrate how L-glutamate (taken as an example of amino acid uptake) is transported from the intestinal lumen to the bloodstream.

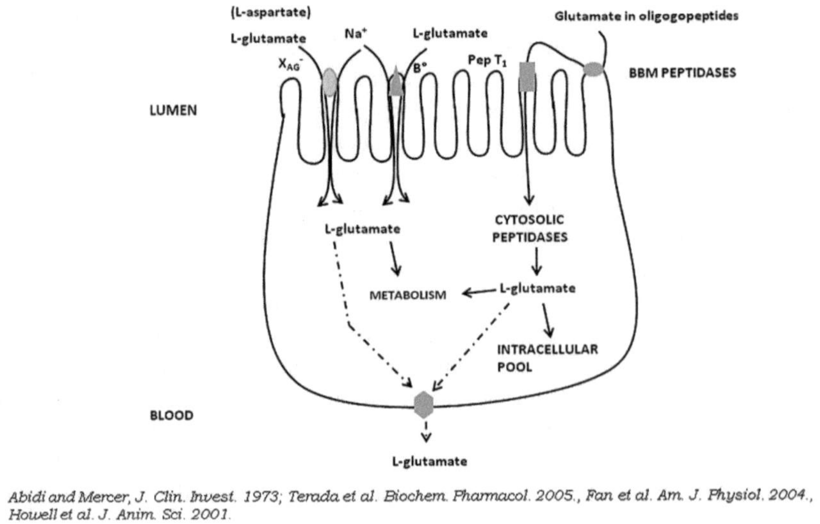

Fig. 4.1 Schematic view of the absorption of amino acids (L-glutamate being taken as an example) from the lumen of the small intestine to the portal bloodstream. L-glutamate can be transported inside enterocytes through the Na^+-dependent high affinity X_{AG}^- system and/or the low affinity B° system present in the brush border membranes (BBM). Oligopeptides (containing, for instance, glutamate) has to be degraded to dipeptides and tripeptides by brush border membranes before being transported inside enterocytes through PepT1 transporters. Then, L-glutamate is released from di and tripeptides by cytosolic peptidases. Intracellular glutamate which is not used for metabolism is released into the portal bloodstream through baso-lateral membranes using transport systems

Amino acid absorption by pig small intestine can be modified by different factors. For instance, the oral administration of insulin-like growth factor-I (IGF-I) to piglets is able to stimulate the Na^+-dependent glutamine absorption by the jejunum without modifying the jejunal mucosal mass (Alexander and Carey 2002). Conversely, phytate, in addition to interfering with protein digestion, may inhibit amino acid absorption (Selle et al. 2012).

Supplementing the diet with alpha-ketoglutarate has been suggested as a mean to spare dispensable amino acids in pigs (Kristensen et al. 2002). Although transporters for di- and tricarboxylic acids have been detected in the pig small intestine, only a small percentage of alpha-ketoglutarate administered lumenally to pigs appears in the portal circulation, a result that is attributed to mucosal metabolism, and possibly to limited absorption by the small intestine (Lambert et al. 2002). Results of experiments involving pigs have suggested that small intestine absorption is probably also affected by low basolateral transfer from the enterocytes to the bloodstream as proposed by Buddington et al. (2004).

4.3 Amino Acid Absorption by the Piglet Colon

There is some evidence which suggest that the colonic epithelial cells of the piglets possess the capacity for the transport of some amino acid from the lumen. For instance, Smith and James (1980) have shown that it is possible to measure unidirectional fluxes of different amino acids across the newborn pig colon. Long-chain amino acids appeared to enter the proximal colon of newborn pigs more readily than did short- chain amino acids (Sepulveda and Smith 1979). The proximal colon of newborn pigs is able to transport methionine from the luminal side, but this capacity markedly decreases in 4-day-old piglets and remains undetectable in 10-day-old piglets (James and Smith 1976). The evolution of the capacity of colonocytes for amino acid transport in the piglets after birth may be related to marked ultrastructural changes of absorptive cells like the reduction of the height of the microvilli within few days following birth (Wooding et al. 1978). However, the nutritionally relevance of amino acid transport by the piglet colon has been questioned (Darragh et al. 1994).

References

Alexander AN, Carey HV (2002) Insulin-like growth factor-I stimulates Na^+-dependent glutamine absorption in piglet enterocytes. Dig Dis Sci 47(5):1129–1134

Bröer S (2008) Amino acid transport across mammalian intestinal and renal epithelia. Physiol Rev 88(1):249–286

Buddington RK, Elnif J, Puchal-Gardiner AA, Sangild PT (2001) Intestinal apical amino acid absorption during development of the pig. Am J Physiol Regul Integr Comp Physiol 280(1): R241–R247

Buddington RK, Pajor A, Buddington KK, Pierzynowski S (2004) Absorption of alpha-ketoglutarate by the gastrointestinal tract of pigs. Comp Biochem Physiol A Mol Integr Physiol 138(2):215–220

Buraczewska L (1981) Absorption of amino acids in different parts of the small intestine in growing pigs. III. Absorption of constituents of protein hydrolysates. Acta Physiol Pol 32 (5):569–584

Burrin DG, Shulman RJ, Reeds PJ, Davis TA, Gravitt KR (1992) Porcine colostrum and milk stimulate visceral organ and skeletal muscle protein synthesis in neonatal piglets. J Nutr 122 (6):1205–1213

Darragh AJ, Cranwell PD, Moughan PJ (1994) Absorption of lysine and methionine from the proximal colon of the piglet. Br J Nutr 71(5):739–752

Fan MZ, Stoll B, Jiang R, Burrin DG (2001a) Enterocyte digestive enzyme activity along the crypt-villus and longitudinal axes in the neonatal pig small intestine. J Anim Sci 79(2):371–381

Fan MZ, Archbold T, Sauer WC, Lackeyram D, Rideout T, Gao Y, de Lange CF, Hacker RR (2001b) Novel methodology allows simultaneous measurement of true phosphorus digestibility and the gastrointestinal endogenous phosphorus outputs in studies with pigs. J Nutr 131(9):2388–2396

Fan MZ, Matthews JC, Etienne NM, Stoll B, Lackeyram D, Burrin DG (2004) Expression of apical membrane L-glutamate transporters in neonatal porcine epithelial cells along the small intestinal crypt-villus axis. Am J Physiol Gastrointest Liver Physiol 287(2):G385–G398

James PS, Smith MW (1976) Methionine transport by pig colonic mucosa measured during early post-natal development. J Physiol 262(1):151–168

Klang JE, Burnworth LA, Pan YX, Webb KE Jr, Wong EA (2005) Functional characterization of a cloned pig intestinal peptide transporter (pPepT1). J Anim Sci 83(1):172–181

Klein RM (1989) Small intestinal cell proliferation during development. In: Lebenthal E (ed) Human gastrointestinal development. Raven, New York

Komuves LG, Nicols BL, Hutchens TW, Heath JP (1993) Formation of crystalloid inclusions in the small intestine of neonatal pigs: an immunocytochemical study using colloidal gold. Histochem J 25(1):19–29

Kristensen NB, Jungvid H, Fernandez JA, Pierzynowski SG (2002) Absorption and metabolism of alpha-ketoglutarate in growing pigs. J Anim Physiol Anim Nutr (Berl) 86(7–8):239–245

Lambert BD, Stoll B, Niinikoski H, Pierzynowski S, Burrin DG (2002) Net portal absorption of enterally fed alpha-ketoglutarate is limited in young pigs. J Nutr 132(11):3383–3386

Leibholz J (1985) The digestion of protein in young pigs and the utilization of dietary methionine. Br J Nutr 53(1):137–147

Maenz DD, Patience JF (1992) L-Threonine transport in pig jejunal brush border membrane vesicles. Functional characterization of the unique system B in the intestinal epithelium. J Biol Chem 267(31):22079–22086

Moon HW (1971) Epithelial cell migration in the alimentary mucosa of the suckling pig. Proc Soc Exp Biol Med 137(1):151–154

Munck LK, Grondahl ML, Thorboll JE, Skadhauge E, Munck BG (2000) Transport of neutral, cationic and anionic amino acids by systems B, b(o,+), X(AG), and ASC in swine small intestine. Comp Biochem Physiol A Mol Integr Physiol 126(4):527–537

Noren O, Dabelsteen E, Sjostrom H, Josefsson L (1977) Histological localization of two dipeptidases in the pig small intestine and liver, using immunofluorescence. Gastroenterology 72(1):87–92

Remond D, Buffiere C, Pouyet C, Papet I, Dardevet D, Savary-Auzeloux I, Williamson G, Faure M, Breuille D (2011) Cysteine fluxes across the portal-drained viscera of enterally fed minipigs: effect of an acute intestinal inflammation. Amino Acids 40(2):543–552

Selle PH, Cowieson AJ, Cowieson NP, Ravindran V (2012) Protein-phytate interactions in pig and poultry nutrition: a reappraisal. Nutr Res Rev 25(1):1–17

Sepulveda FV, Smith MW (1979) Different mechanisms for neutral amino acid uptake by new-born pig colon. J Physiol 286:479–490

Simmen FA, Cera KR, Mahan DC (1990) Stimulation by colostrum or mature milk of gastrointestinal tissue development in newborn pigs. J Anim Sci 68(11):3596–3603

Smith MW (1981) Autoradiographic analysis of alanine uptake by newborn pig intestine. Experentia 37(8):868–870

Smith MW, James PS (1980) Amino acid transport by small and large intestine of newborn pig. Experentia 36(3):319–320

Wang W, Gu W, Tang X, Geng M, Fan M, Li T, Chu W, Shi C, Huang R, Zhang H, Yin Y (2009) Molecular cloning, tissue distribution and ontogenetic expression of the amino acid transporter b(0,+) cDNA in the small intestine of Tibetan suckling piglets. Comp Biochem Physiol B Biochem Mol Biol 154(1):157–164

Werhahn E, Klobasa F, Butler JE (1981) Investigation of some factors which influence the absorption of IgG by the neonatal piglet. Vet Immunol Immunopathol 2(1):35–51

Widdowson EM, Colombo VE, Artavanis CA (1976) Changes in the organs of pigs in response to feeding for the first 24 h after birth. Biol Neonate 28:272–281

Winckler C, Breves G, Boll M, Daniel H (1999) Characteristics of dipeptide transport in pig jejunum in vitro. J Comp Physiol B 169(7):495–500

Wooding FB, Smith MW, Craig H (1978) The ultrastructure of the neonatal pig colon. Am J Anat 152(3):269–285

Xu RJ, Mellor DJ, Tungthanathanich P, Birtles MJ, Reynolds GW, Simpson HV (1992) Growth and morphological changes in the small and the large intestine in piglets during the first three days after birth. J Dev Physiol 18(4):161–172
Yen JT, Kerr BJ, Easter RA, Parkhurst AM (2004) Difference in rates of net portal absorption between crystalline and protein-bound lysine and threonine in growing pigs fed once daily. J Anim Sci 82(4):1079–1090
Zhang H, Malo C, Buddington RK (1997) Suckling induces rapid intestinal growth and changes in brush border digestive functions of newborn pigs. J Nutr 127:418–426

Chapter 5
Developmental Amino Acid Metabolism in the Pig Small and Large Intestine Epithelial Cells

François Blachier, Guoyao Wu, Yulong Yin, Yongqing Hou, and Mireille Andriamihaja

Contents

5.1 Introduction

From experimental works performed on pigs, it appears clearly that a significant part of several dispensable and indispensable amino acids present in the small intestine content are metabolized during their transcellular journey through the small intestine epithelial cells (enterocytes) when moving from the luminal side of the intestine to

F. Blachier (✉) • M. Andriamihaja
UMR 914 Nutrition Physiology and Ingestive Behavior, INRA-CRNH-IdF-AgroParisTech, Paris, France
e-mail: francois.blachier@agroparistech.fr

G. Wu
Department of Animal Science and Faculty of Nutrition, Texas A&M University, College Station, TX, USA

Y. Yin
Institute of Subtropical Agriculture, Chinese Academy of Sciences, Changsha, China

Y. Hou
Laboratory of Animal Nutrition and Feed Science, Wuhan Polytechnic University, Wuhan, China

F. Blachier et al. (eds.), *Nutritional and Physiological Functions of Amino Acids in Pigs*, DOI 10.1007/978-3-7091-1328-8_5, © Springer-Verlag Wien 2013

the portal bloodstream. Importantly, this metabolism in enterocytes is different according to the pig developmental stages, and we will see in this chapter that this has important metabolic and physiological consequences. It has been determined that some amino acids are catabolized in enterocytes and some other amino acids are generated by the small intestine epithelium. These results are mostly derived from both in vivo and in vitro experiments. A major in vivo experimental design used to estimate the apparent amino acid intestinal absorption is the porto-arterial balance sheet which consists of measuring the amino acid concentrations at different time following a meal in both the arterial and portal blood as well as measuring continuously the blood flow in the portal vein. This allows to have an estimation of the apparent intestinal absorption of amino acids as well as to presume if a given amino acid is globally degraded or generated during its transport through the intestinal mucosa. Although very informative, these in vivo methods are limited by the fact that the portal vein does not exclusively drain amino acids from the intestine but also from other visceral tissues (notably stomach, pancreas, and spleen). Numerous teams in the world have also used alimentary proteins labelled with stable isotope to estimate the metabolic fate of amino acids during their intestinal absorption and further on in the body.

The situation is further complicated by the fact that a part of both endogenous and alimentary proteins present in the small intestine luminal content can be transferred to the large intestine through the ileocaecal junction (Darragh and Hodgkinson 2000). In the large intestine, microbiota proteases intensively degrade these proteins in peptides and amino acids. Amino acids are not believed to be absorbed through the large intestine epithelium of pigs except in a short period following birth (Darragh et al. 1994). In contrast, amino acids are further converted to numerous bacterial metabolites, some of them being then absorbed through the colonic mucosa or used further by the microbiota (Blachier et al. 2007). Some of these metabolites can also be absorbed through the colonic epithelium and interfere with host metabolism.

The in vitro study of amino acid metabolism in the small and large intestine epithelial cells generally uses isolated living enterocytes and colonocytes for measuring their metabolic capacity towards different amino acids. Although this in vitro design has proven much useful in understanding the metabolic fate of amino acids in intestinal cells, it is only fair to say that the results obtained do not necessarily reflect what is happening in vivo in the intestinal mucosa. Indeed, the intestinal mucosa does not exclusively contain epithelial cells but also numerous other cell types including smooth muscular and immune cells. As a matter of fact, it is tempting to say that the best way to see clearer in this complicated situation of amino acid metabolism in the intestinal mucosa is to compare the data obtained from in vivo and in vitro studies.

In this chapter, we examine the metabolic fate of several amino acids in the intestine epithelium according to the stage of pig development, and we will see that this intestinal metabolism is related to energy supply to enterocytes and colonocytes, to the synthesis of metabolites with important local and peripheral physiological functions, and to the synthesis of other amino acids.

5.2 Metabolic Fate of Glutamate, Glutamine (and Related Amino Acids: Aspartate, Alanine, Proline, Citrulline, and Ornithine) in Pig Enterocytes and Colonocytes

The Na^+-dependent high-affinity X_{AG}^- system and/or the low-affinity B° system have been shown to be involved in the transport of L-glutamate from the lumen into the pig small intestine epithelial cells (Fan et al. 2004). Although Na^+-independent carriers (system L and B° transporters) are present in the brush border membranes of porcine jejunal enterocytes (Fan et al. 1998), the Na^+-dependent B° transport system appears to represent the major route of luminal glutamine absorption (Fan et al. 1998). In piglet small intestine, the oral administration of insulin-like growth factor-I (IGF-I) stimulates Na^+-dependent glutamine absorption (Alexander and Carey 2002).

L-glutamate and L-glutamine are extensively metabolized in pig enterocytes. L-glutamate can be used in enterocytes for protein synthesis (Bos et al. 2005), can enter an intracellular pool (Blachier et al. 2009), or can be extensively metabolized in other pathways including those involved in enterocyte ATP production (Fig. 5.1).

A number of studies have shown that a large proportion of L-glutamate is metabolized during its transcellular journey through pig enterocytes. Indeed, in pigs, virtually all the enteral L-glutamate is metabolized by the gut during absorption (Reeds et al. 1996). In a 7-kg piglet, an increase in L-glutamate concentrations was observed in portal and arterial blood plasma when the basal milk formula, administered enterally at a rate of 510 μmol kg/h, was supplemented with a relatively high dose of monosodium glutamate, i.e., 1,250 μmol/kg/h (Janeczko et al. 2007). Similarly, in larger (60 kg) pigs, transient portal and arterial hyperglutamatemia was observed when the diet (i.e., 800 g meal containing 142 g casein) was supplemented with 10 g monosodium glutamate (Blachier et al. 1999), indicating that only very large doses of L-glutamate may exceed the intestinal capacity to catabolize this amino acid.

It has been demonstrated that dietary L-glutamate is the most important contributor among substrates for mucosal oxidative energy metabolism in pigs (Stoll et al. 1999). ATP production and utilization are very active in pig enterocytes. Indeed, although the gastrointestinal tract represents only approximately 5 % of body weight, it is responsible for around 20 % of whole-body oxygen consumption (Vaugelade et al. 1994). The pig intestinal epithelial cells have notably high energy demand (Madej et al. 2002; Wu and Knabe 1995) due to the rapid renewal of the epithelium within few days (Wiese et al. 2003), thus requiring intense anabolic metabolism. Sodium extrusion at the basolateral membranes through the activity of Na/K ATPase following nutrient absorption may also represent a major ATP-consuming process in enterocytes (Buttgereit and Brand 1995).

The metabolic steps involved in L-glutamate utilization in pig enterocytes involve firstly transamination with oxaloacetate to produce alpha-ketoglutarate and L-aspartate (Blachier et al. 2009) (Fig. 5.1). L-glutamate can also be transaminated in the presence of pyruvate to produce L-alanine and alpha-ketoglutarate. Transamination

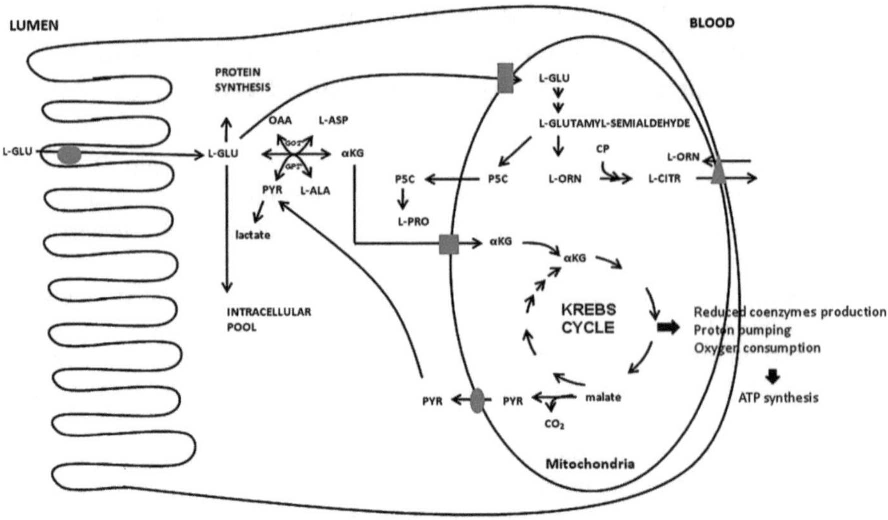

Riby et al. Ped. Res. 1990., Wu and Morris Biochem. J. 1998, Henslee and Jones Arch. Biochem. Biophys. 1982

Fig. 5.1 Schematic view of the metabolism of L-glutamate (L-GLU) in small intestine absorptive epithelial cells. This schema is related to the involvement of L-glutamate in the production of other amino acids as well as conversion of L-glutamate into α-ketoglutarate (α-KG) for entry into the Krebs cycle. OAA is oxaloacetate, L-ASP is L-aspartate, PYR is pyruvate, L-ALA is L-alanine, L-ORN is L-ornithine, L-CITR is L-citrulline, CP is carbamoylphosphate, P5C is pyrroline-5-carboxylate, PRO is L-proline

appears to be the principal route by which L-glutamate is converted to alpha-ketoglutarate in pig enterocytes, because these cells have little capacity for the conversion of L-glutamate into alpha-ketoglutarate (and ammonia) through glutamate dehydrogenase (Madej et al. 2002). L-glutamate and L-glutamine are similarly oxidized by pig enterocytes (Blachier et al. 1999). However, for L-glutamine, this amino acid has to firstly enter the mitochondria in order to be degraded to L-glutamate and ammonia by the phosphate-dependent glutaminase activity (Wu et al. 1994, 1995a; Fig. 5.2).

L-glutamate that arises within the mitochondria can be either metabolized locally or exported into the cytosol, where the amino acid is metabolized into alpha-ketoglutarate by transamination before reentering the mitochondria and the tricarboxylic acid cycle (Duée et al. 1995). When both L-glutamate and L-glutamine are simultaneously presented to isolated pig enterocytes, L-glutamate can inhibit L-glutamine utilization and oxidation by inhibiting mitochondrial glutaminase activity (Blachier et al. 1999). Presumably, the sparing effect of L-glutamate over L-glutamine is dependent on the relative concentration of both amino acids in mitochondria.

L-glutamate and/or L-glutamine can be used by pig enterocytes to produce other amino acids including L-aspartate (Blachier et al. 1999), L-alanine (Wu et al. 1995a), L-proline (Wu and Knabe 1994), L-ornithine (Blachier et al. 1992), and L-citrulline (Wu et al. 1994) (Fig. 5.1). Note that neither ornithine nor citrulline is present in

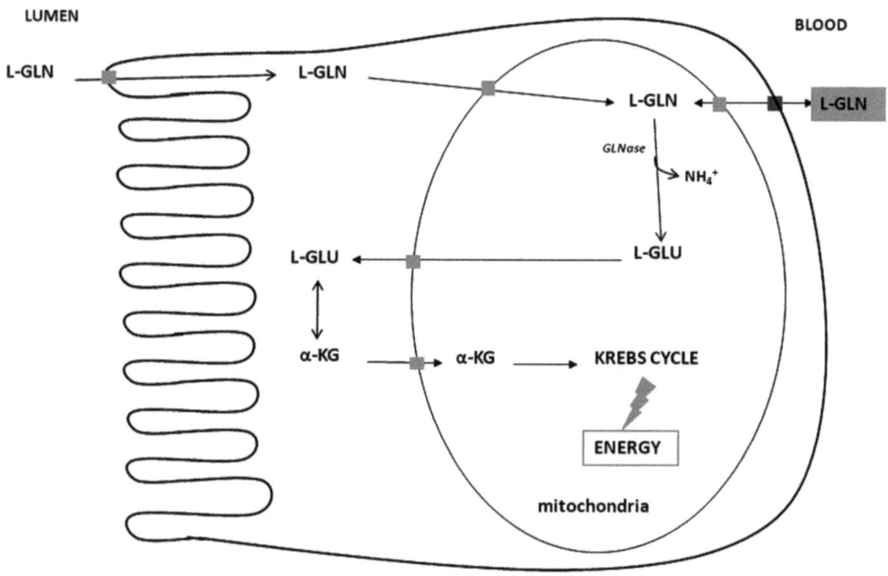

Fig. 5.2 Schematic view of the metabolism of L-glutamine (L-GLN) in small intestine absorptive epithelial cells. This schema is related to the metabolism of L-GLN to α-ketoglutarate (α-KG) for entry into the Krebs cycle. GLNase is glutaminase

proteins. L-aspartate and L-alanine are thus produced in the course of glutamate catabolism in pig enterocytes. L-aspartate, either produced endogenously from glutamate/glutamine or originating from the exogenous luminal content, can enter mitochondria and is then oxidized in the TCA cycle (Windmueller and Spaeth 1976), thus representing another oxidative fuel for enterocytes. The stepwise conversion of L-glutamate to L-ornithine is performed in mitochondria. Then, L-ornithine can serve as a precursor for citrulline production (Blachier et al. 1991b). L-citrulline is then released into the portal vein and passes through the liver without major uptake and is then presumably used for de novo synthesis of L-arginine in kidneys (Flynn and Wu 1996).

Together with L-cysteine and glycine, L-glutamate is the precursor for the synthesis of glutathione in pig enterocytes (Fig. 5.3).

Studies in pigs suggest extensive utilization of dietary cysteine by the intestine (Stoll et al. 1998). It has been reported that in fed piglets, mucosal glutathione is derived largely from the direct metabolism of enteral L-glutamate (Reeds et al. 1997). The ratio of reduced to oxidized glutathione is an important measure for both the determination of the intracellular redox status and for the cell's capacity to control intracellular concentrations of both oxygen-reactive and nitrogen-reactive species (Chakravarthi et al. 2006; Kemp et al. 2008).

Metabolism of L-glutamate and L-glutamine in enterocytes has been studied in growing piglets. Enterocytes isolated from piglet small intestine at birth display a high capacity for L-glutamine oxidation (Darcy-Vrillon et al. 1994), indicating that this amino acid can be used as a fuel at this stage of development when the growth

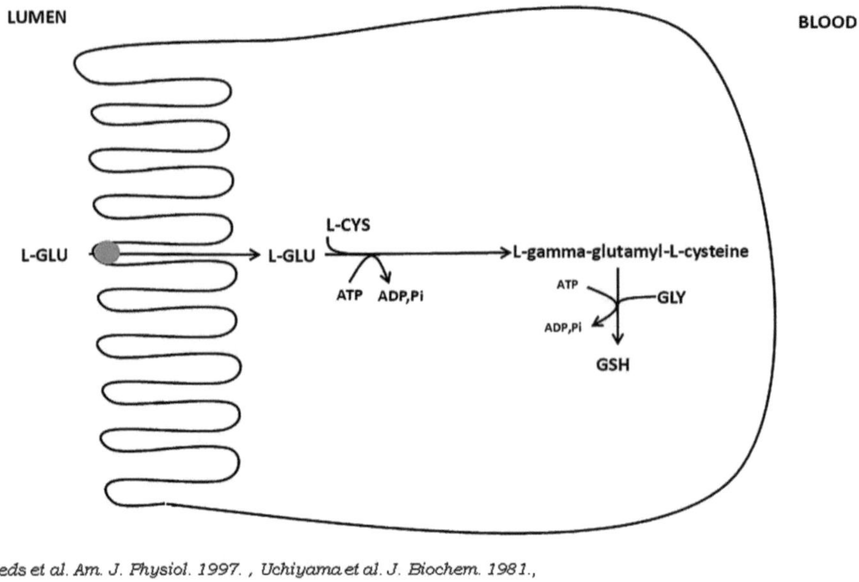

Fig. 5.3 Schematic view of the conversion of L-glutamate (L-GLU) into glutathione (GSH) in small intestine absorptive epithelial cells. L-CYS is L-cysteine, GLY is glycine

of the intestine mucosa is very intense (Widdowson et al. 1976; Klein and McKenzie 1983a, b). Furthermore, enterocytes isolated from newborn pigs are able to convert L-glutamine into L-citrulline and then L-arginine (Blachier et al. 1993; Wu et al. 1994). This metabolic capacity of enterocytes may reflect the high content in sow's milk proteins of glutamate and glutamine and the relatively low content of arginine (Wu and Knabe 1994). In suckling piglets, it has been observed that, soon after birth, the net enterocyte's capacity to convert L-glutamine to L-citrulline is severely decreased compared with piglets at birth (Wu et al. 1995b). In contrast, L-glutamine still represents a major oxidative energy substrate for enterocytes in these animals. Indeed, in suckling piglet enterocytes, L-glutamine is approximately eight times more rapidly oxidized than in enterocytes isolated from weaned pigs (Darcy-Vrillon et al. 1994). In addition, conversion of enteral glutamate to proline in neonatal piglet gastrointestinal tract has been measured by Murphy (1996). Some interactions have been reported between glutamine and carbohydrate metabolism in piglet enterocytes. For instance, in these cells, the metabolism of D-glucose in the pentose cycle which allows NADPH synthesis appears to increase the net conversion of L-glutamine into L-proline (Wu 1996).

Colonic epithelial cells are also able to metabolize glutamate and glutamine. An important characteristic of the colonic epithelium, as said above, is that except for a short period after birth, there is little or no transfer of amino acids from the lumen to the portal blood (Darragh et al. 1994). Then, amino acids (including L-glutamate and L-glutamine) must be taken from arterial blood. Pig colonocytes can use L-glutamine as oxidative substrates (Darcy-Vrillon et al. 1994). Like in enterocytes, L-glutamine is firstly converted into L-glutamate and ammonia by the mitochondrial glutaminase, and then into alpha-ketoglutarate, mainly by transamination, followed by entry into the tricarboxylic cycle.

Colonocytes can also use luminal organic acids generated from the microbial activity, including short-chain fatty acids, as oxidative substrates (Darcy-Vrillon et al. 1993). Dietary substrates for short-chain fatty acid production are mainly dietary fibers, resistant starch, and some amino acids including L-glutamate (Bindelle et al. 2011; Htoo et al. 2007). Regarding amino acid metabolism by the microbiota, although alimentary protein digestion followed by amino acid and oligopeptide absorption by the small intestine is efficient in pigs (Stein et al. 1990), substantial amounts of nitrogenous compounds of both exogenous (alimentary) and endogenous origin may enter the pig large intestine through the ileocaecal junction. The first event in the large intestine protein degradation is hydrolysis of proteins and polypeptides by microbiota and residual pancreatic proteases and peptidases, which results in peptide and amino acid release, followed by the production of numerous bacterial metabolites. In the large intestinal lumen, L-glutamate released from proteins and peptides is the precursor for acetate and butyrate production (Blachier et al. 2007), but the relative contributions of L-glutamate and alimentary indigestible polysaccharide to acetate and butyrate production have not been determined. Glutamine synthetase, in terms of protein expression and catalytic activity, has been measured in pig colonocytes (Eklou-Lawson et al. 2009). Because the L-glutamine-degrading enzyme glutaminase is also highly expressed in colonocytes, this raises the open question of the physiologic meaning of the expression within the same cells of L-glutamine-synthesizing and -degrading enzymes. Because ammonia at concentrations that can be found in the pig colonic lumen inhibits short-chain fatty acid oxidation in colonocytes (Darcy-Vrillon et al. 1996), it has been proposed that cytosolic glutamine synthetase activity, which converts L-glutamate and ammonia into L-glutamine, may represent a way to reduce the intracellular concentration of ammonia during its transfer from the lumen to the bloodstream (Eklou-Lawson et al. 2009), thus avoiding metabolic disturbances.

Utilization of L-glutamate in colonocytes is not restricted to energy metabolism. Indeed, L-glutamine metabolism in pig isolated colonocytes leads to a net production of L-aspartate, L-alanine, and lactate (Darcy-Vrillon et al. 1993). However, unlike what is observed in enterocytes, pig colonocytes produce more L-aspartate than L-alanine from L-glutamine.

5.3 Metabolic Fate of Arginine (and Related Amino Acids, i.e., Ornithine, Citrulline, and Proline) in Pig Enterocytes and Colonocytes

In pig enterocytes isolated from postweaning animals, arginine is degraded mainly into urea, ornithine, and citrulline and to a much lesser extent to nitric oxide (NO) (Wu et al. 1995a; Blachier et al. 1993; M'Rabet-Touil et al. 1993, 1995).

The synthesis of ornithine and urea from arginine is made from the catalytic activity of arginase. Although urea represents a metabolic end-product, a part of ornithine derived from arginine can be used as a substrate together with carbamoylphosphate to allow citrulline synthesis through the activity of carbamoylphosphate synthetase isoform I (CPS I). Citrulline can also be produced together with NO through the activity of nitric oxide synthase (NOS) which catalyzes the conversion of arginine into these two metabolites. However, the amount of citrulline produced through the NOS pathway appears lower than the amount of citrulline produced through the arginase/CPS I pathways (Blachier et al. 1991a).

Metabolism of arginine in pig enterocytes has been studied in growing piglets and has revealed that the metabolism of arginine is orientated towards synthesis in neonatal and suckling piglets, but then is orientated towards degradation up to the weaning period (Wu and Knabe 1995; Blachier et al. 1993). Indeed, at birth, several amino acids are used by enterocytes as precursors for arginine synthesis. Thus glutamine, proline, and citrulline can be used as precursors for the net synthesis of arginine in neonatal enterocytes (Wu et al. 1994; Wu 1997; Blachier et al. 1993; Flynn and Wu 1996; Wu and Knabe 1995). Interestingly, the arginine content in sow's colostrum and milk proteins (and as free arginine) is relatively low when compared with the other amino acids, with glutamine and proline being very abundant in porcine milk (Wu and Knabe 1994; Davis et al. 1994). Thus, it has been proposed that the metabolic capacity of the absorptive epithelial intestinal cells to synthesize arginine would compensate for the relatively low content of this amino acid in milk protein (Wu and Knabe 1995), and would thus contribute to the endogenous supply of this amino acid in a context of piglet rapid growth.

Accordingly, the arginase activity in enterocytes is very low but increases slightly in the suckling period to a relatively high level in weaned animals (M'Rabet-Touil et al. 1996; Wu 1995). Also, the NOS activity is very low in enterocytes of neonatal piglets, but increases progressively up to the period of weanling (M'Rabet-Touil et al. 1993). The production of NO from arginine in enterocytes isolated from weaned pigs can be increased by the production of NADPH (a cofactor used for NOS catalytic activity) from D-glucose in the pentose phosphate pathway (Blachier et al. 1991a). In addition, although the activity of ornithine decarboxylase, the enzyme which allows the synthesis of putrescine and CO_2 from ornithine, is present at a low level in enterocytes at birth, this activity falls down to a value close to the limit of detection in suckling and weaned animals (Blachier et al. 1992).

What can be the physiological consequences of these metabolic changes during piglet development? Firstly, this low arginine catabolism in enterocytes will spare a conditionally essential amino acids in a context of limited alimentary supply and de novo synthesis. Then, the increase in NO biosynthesis capacity by enterocytes from birth to weaning, although representing a low arginine utilization, is likely related to the emergence in the suckling period of different NO-dependent physiological processes including the protection of the gastrointestinal mucosa (Stark and Szurszewski 1992; Miller et al. 1993; Quintero and Guth 1992; Konturek et al. 1992; MacKendrick et al. 1993); the regulation of the intestinal motility (Calignano et al. 1992; Hata et al. 1990); and the modulation of the intestinal epithelial permeability (Kubes 1992, 1993). In newborn piglets, the low ornithine decarboxylase activity would contribute to the endogenous pool of putrescine produced from ornithine in enterocytes. However, from measurement of the capacity of enterocytes isolated from neonatal piglets for polyamine (putrescine, spermidine, and spermine) uptake, it appears that most of the intracellular polyamines likely originate from extracellular sources (Blachier et al. 1992). These sources include the polyamines present in the intestinal luminal content. Interestingly, since sow's milk contains relatively high concentration of spermidine (Kelly et al. 1991) and spermine (Motyl et al. 1995), these polyamines may represent a significant source for piglet's enterocyte intracellular content. In the piglets, growth of the intestinal mucosa is very intense during the first days after birth (Widdowson et al. 1976; Klein and McKenzie 1983b) and several experimental works have established that polyamines are central for the intestinal mucosal growth (Wang et al. 1991; Ginty et al. 1989). Finally, the synthesis of arginine from ornithine and the synthesis of ornithine from arginine demonstrate the existence of an intracellular arginine–ornithine cycle in suckling piglets (Blachier et al. 1993; Wu 1995).

The results obtained in vitro using enterocytes isolated from neonatal piglets which showed that these cells are orientated towards arginine production from various precursors including glutamine were complemented by in vivo experiments using suckling piglets treated with gabaculine, i.e., an inhibitor of ornithine aminotransferase. This inhibitor decreases the sequential intestinal conversion of glutamine to arginine. Using this experimental design, it was possible to demonstrate that the intestinal production of citrulline plays an important role in the endogenous synthesis of arginine and thus in its homeostasis in neonatal piglets (Flynn and Wu 1996). In these animals, proline is able to ameliorate arginine deficiency during enteral but not parenteral feeding, pointing out that the gut likely acts as an important player for the whole-body arginine supply in piglets by using proline as a substrate for arginine biosynthesis (Brunton et al. 1999; Wu 1997). The synthesis of arginine from proline may be mitigated by the plasma concentration of lactate (Dillon et al. 1999). Indeed, in enterocytes isolated from suckling piglets, lactate is able to inhibit citrulline and arginine synthesis from proline via an inhibition of proline oxidase, i.e., the enzyme responsible for the conversion of proline into delta1-pyrroline-5-carboxylate (Dillon et al. 1999). Although the factors responsible for the developmental changes in arginine metabolism in enterocytes remain unclear, it has been demonstrated that glucocorticoids play an important role in mediating the enhanced catabolism of arginine during weaning (Flynn and Wu 1997).

In pig colonic epithelial cells, much less studies have been performed regarding arginine metabolism. Amino acids from the luminal origin are not believed to be absorbed to any significant extent through the pig colonic epithelium except for a short period after birth (Smith and James 1976). Then, in contrast to the situation found in the small intestine enterocytes, which import amino acids from both luminal and blood origins, colonic epithelial cells import amino acids almost exclusively from arterial blood for metabolic and physiological purposes. L-arginine metabolism in pig colonocytes has not received much attention. Nonetheless, absorptive colonic epithelial cells isolated from the proximal colon of weaned pigs are characterized by a higher NOS activity when compared with the activity measured in enterocytes isolated from the jejunum. In contrast, arginase activity was found to be similar between pig colonocytes and enterocytes (M'Rabet-Touil et al. 1993). In the large intestine lumen, arginine released from proteins and peptides which are transferred from the ileon to the caecum may presumably serve as a precursor for the synthesis of agmatine, putrescine, and nitric oxide by the microbiota (Blachier et al. 2007). However, to the best of our knowledge, there is no published data on the production of these bacterial metabolites in the pig colonic lumen.

5.4 Metabolic Fate of Branched-Chain Amino Acids (Leucine, Isoleucine, and Valine) in Pig Enterocytes

In milk-fed piglets, 32 % of leucine in the diet are extracted by the portal-drained viscera in the first pass, with 21 % of the extracted leucine being utilized for the intestinal mucosal protein synthesis (Stoll et al. 1998). Overall, it has been estimated that 44 % of total branched-chain amino acids (BCAAs) are extracted by first-pass splanchnic metabolism in neonatal piglets (Elango et al. 2002). The catabolism of BCAAs in enterocytes isolated from developing piglets has been studied. In these cells, BCAAs are extensively transaminated and between 15 and 50 % of decarboxylated branched-chain alpha-ketoacids are oxidized depending on the age of piglets (Chen et al. 2002). Enterocytes isolated from postweaning pigs also actively degrade BCAAs (Chen et al. 2002). Other essential amino acids (i.e., histidine, lysine, methionine, phenylalanine, threonine, and tryptophan) are apparently less catabolized in developing piglet enterocytes (Chen et al. 2002, 2009) and in weaned pigs (Chen et al. 2002, 2007). Metabolism of essential amino acids by colonic epithelial cells remains to be determined.

5.5 Metabolic Fate of Sulfur-Containing Amino Acids in Pig Enterocytes

In piglets, the net portal balance of methionine represents 48 % of intake suggesting that a relatively large part of the dietary methionine is consumed by the intestine (Stoll et al. 1999). Accordingly, parenteral methionine requirement is approximately 69 % of the enteral requirement in neonatal piglets (Shoveller et al. 2003). In addition, the parenteral methionine requirement is approximately 70 % of the enteral requirement when measured in the presence of an excess of dietary cysteine suggesting that cysteine is equally effective in sparing dietary methionine whether fed enterally or parenterally (Shoveller et al. 2003). A recent study has shown that, in neonatal piglets, sulfur-containing amino acid deficiency results in small intestine atrophy with lower goblet cells and lower glutathione content (Bauchart-Thevret et al. 2009). These effects were associated with upregulation of the intestinal methionine cycle activity. The piglet gastrointestinal tract was reported to consume approximately 20 % of the dietary methionine (Riedijk et al. 2007). Furthermore, the piglet gastrointestinal tract appears to be a site for net homocysteine production and to be a significant site for whole-body transmethylation and transsulfuration, these two metabolic pathways being responsible for a majority of methionine utilization by the gastrointestinal tract (Riedijk et al. 2007). However, as previously said, methionine is less catabolized in pig enterocytes raising the view that methionine may be substantially consumed in other cells of the portal-drained viscera and/or of the intestinal mucosa and/or may be degraded by the small intestine microbiota (Blachier et al. 2007).

5.6 Metabolic Fate of Threonine in Pig Enterocytes

As noted above, the metabolic capacity of pig enterocyte for threonine catabolism is virtually absent (Chen et al. 2009). The pig intestinal mucins are glycoproteins very rich in threonine, serine, and proline (Fogg et al. 1996). In a model of minipigs with ileitis, it has been demonstrated that intestinal inflammation increases gastrointestinal threonine uptake and mucin synthesis (Rémond et al. 2009). In neonatal piglets, the threonine requirement during total parenteral nutrition is approximately 45 % of the mean enteral requirement (Bertolo et al. 1998). In growing pigs, it has been shown that the catabolism through the threonine dehydrogenase pathway does not account for the relatively high first-pass extraction rate of dietary threonine by the portal-drained viscera (Le Floc'h and Seve 2005). Dietary threonine extraction by the small intestine is likely to reduce threonine availability for other tissues when young piglets are fed a diet marginally deficient in threonine (Hamard et al. 2009). Lastly, a moderate threonine deficiency was responsible for an alteration of the intestinal functionality in terms of paracellular permeability (Hamard et al; 2010). In piglets, although the absolute amount of systemic and dietary threonine utilized by the portal-drained viscera was reduced in protein-restricted diet, the percentage of dietary threonine intake used by the portal-drained viscera did not differ between groups, with the value being above 85 % (Schaart et al. 2005). As expected, dietary rather than systemic threonine is preferentially utilized for protein synthesis in the piglet small intestine mucosa. Thus, the portal-drained viscera requirement for threonine is high and the high rate of utilization by the piglet intestinal mucosa is largely due to the incorporation of this amino acid in the proteins of the mucosa (Schaart et al. 2005).

5.7 Metabolic Fate of Lysine and Phenylalanine in Pig Enterocytes

In enterocytes isolated from the small intestine, lysine is less catabolized (Chen et al. 2009). In milk protein-fed piglets, when expressed as a percentage of the enteral tracer input, it was calculated that there is a substantial first-pass metabolism of lysine (35 %) (Stoll et al. 1998). However, only 18 % of what is used in the first-pass metabolism is recovered in intestinal mucosal protein. These experiments suggest that lysine catabolism in the intestinal mucosa is quantitatively greater than amino acid incorporation into mucosal proteins. However, from the low capacity of absorptive intestinal cells for lysine catabolism, it is likely that other cell types are able to use lysine. The in vivo catabolism of lysine by the intestinal microbiota may indeed be substantial. However, evidences have been produced in favor of de novo synthesis of lysine by the pig microbiota, a part of it being available for the host (Torrallardona et al. 1996; Backes et al. 2002). The net result of lysine utilization and production by the intestinal microbiota remains to be determined. Interestingly, it has been demonstrated that dietary lysine used by the portal-drained viscera is driven by its luminal bioavailability; and this utilization is markedly stimulated immediately after the meal ingestion (Bos et al. 2003). Additionally, this study has shown that not only the quantity of portal-drained viscera utilization is increased, but there is also a preferential first-pass use of dietary lysine rather than systemic arterial lysine during bolus when compared with continuous feeding. In a situation of protein restriction, neonatal piglets use a large amount of the dietary supply of lysine (van Gaudoever et al. 2000). In milk-fed piglets, when expressed as a percentage of the enteral tracer, it has been determined that there is a marked first-pass metabolism of phenylalanine (35 %) with 18 % of what is used in the total first-pass metabolism of this amino acid being recovered in mucosal proteins. Once again, since phenylalanine is virtually not catabolized by piglet enterocytes, it is likely that the intestinal microbiota represents a significant contributor for the catabolism of this amino acid.

References

Alexander AN, Carey HV (2002) Insulin-like growth factor-I stimulates Na+-dependent glutamine absorption in piglet enterocytes. Dig Dis Sci 47(5):1129–1134

Backes G, Hennig U, Petzke KJ, Elsner A, Junghans P, Nurnberg G, Metges CC (2002) Contribution of intestinal microbial lysine to lysine homeostasis is reduced in minipigs fed a wheat gluten-based diet. Am J Clin Nutr 76(6):1317–1325

Bauchart-Thevret C, Stoll B, Chacko S, Burrin DG (2009) Sulfur amino acid deficiency upregulates intestinal methionine cycle activity and suppresses epithelial growth in neonatal pigs. Am J Physiol Endocrinol Metab 296(6):E1239–E1250

Bertolo RF, Chen CZ, Law G, Pencharz PB, Ball RO (1998) Threonine requirement of neonatal piglets receiving total parenteral nutrition is considerably lower than that of piglets receiving an identical diet intragastrically. J Nutr 128(10):1752–1759

Bindelle J, Pieper R, Montoya CA, Van Kessel AG, Leterme P (2011) Nonstarch polysaccharide-degrading enzymes alter the microbial community and the fermentation patterns of barley cultivars and wheat products in an in vitro model of the porcine gastrointestinal tract. FEMS Microbiol Ecol 76(3):553–563

Blachier F, Darcy-Vrillon B, Sener A, Duee PH, Malaisse WJ (1991a) Arginine metabolism in rat enterocytes. Biochim Biophys Acta 1092(3):304–310

Blachier F, M'Rabet-Touil H, Darcy-Vrillon B, Posho L, Duee PH (1991b) Stimulation by D-glucose of the direct conversion of arginine to citrulline in enterocytes isolated from pig jejunum. Biochem Biophys Res Commun 177(3):1171–1177

Blachier F, M'Rabet-Touil H, Posho L, Morel MT, Bernard F, Darcy-Vrillon B, Duee PH (1992) Polyamine metabolism in enterocytes isolated from newborn pigs. Biochim Biophys Acta 1175 (1):21–26

Blachier F, M'Rabet-Touil H, Posho L, Darcy-Vrillon B, Duee PH (1993) Intestinal arginine metabolism during development. Evidence for de novo synthesis of L-arginine in newborn pig enterocytes Eur J Biochem 216(1):109–117

Blachier F, Guihot-Joubrel G, Vaugelade P, Le Boucher J, Bernard F, Duee P, Cynober L (1999) Portal hyperglutamatemia after dietary supplementation with monosodium glutamate in pigs. Digestion 60(4):349–357

Blachier F, Mariotti F, Huneau JF, Tome D (2007) Effects of amino acid-derived luminal metabolites on the colonic epithelium and physiopathological consequences. Amino Acids 33(4):547–562

Blachier F, Boutry C, Bos C, Tome D (2009) Metabolism and functions of L-glutamate in the epithelial cells of the small and large intestines. Am J Clin Nutr 90(3):814S–821S

Bos C, Stoll B, Fouillet H, Gaudichon C, Guan X, Grusak MA, Reeds PJ, Tome D, Burrin DG (2003) Intestinal lysine metabolism is driven by the enteral availability of dietary lysine in piglets fed a bolus meal. Am J Physiol Endocrinol Metab 285(6):E1246–E1257

Bos C, Stoll B, Fouillet H, Gaudichon C, Guan X, Grusak MA, Reeds PJ, Burrin DG, Tome D (2005) Postprandial intestinal and whole body nitrogen kinetics and distribution in piglets fed a single meal. Am J Physiol Endocrinol Metab 288(2):E436–E446

Brunton JA, Bertolo RF, Pencharz PB, Ball RO (1999) Proline ameliorates arginine deficiency during enteral but not parenteral feeding in neonatal piglets. Am J Physiol 277(2 Pt 1): E223–E231

Buttgereit F, Brand MD (1995) A hierarchy of ATP-consuming processes in mammalian cells. Biochem J 312(Pt 1):163–167

Calignano A, Whittle BJ, Di Rosa M, Moncada S (1992) Involvement of endogenous nitric oxide in the regulation of rat intestinal motility in vivo. Eur J Pharmacol 229(2–3):273–276

Chakravarthi S, Jessop CE, Bulleid NJ (2006) The role of glutathione in disulphide bond formation and endoplasmic-reticulum-generated oxidative stress. EMBO Rep 7(3):271–275

Chen YH, Sung PH, Sung K (2002) Synthesis of proline-derived dipeptides and their catalytic enantioselective direct aldol reactions: catalyst, solvent, additive and temperature effects. Amino Acids 38(3):839–845

Chen LX, Yin YL, Jobgen WS, Jobgen SC, Knabe DA, Hu WX, En GY (2007) In vitro oxidation of essential amino acids by jejunal muscosal cells of growing pigs. Livest Sci 109(1–3):19–23

Chen L, Li P, Wang J, Li X, Gao H, Yin Y, Hou Y, Wu G (2009) Catabolism of nutritionally essential amino acids in developing porcine enterocytes. Amino Acids 37(1):143–152

Darcy-Vrillon B, Morel MT, Cherbuy C, Bernard F, Posho L, Blachier F, Meslin JC, Duee PH (1993) Metabolic characteristics of pig colonocytes after adaptation to a high fiber diet. J Nutr 123(2):234–243

Darcy-Vrillon B, Posho L, Morel MT, Bernard F, Blachier F, Meslin JC, Duee PH (1994) Glucose, galactose, and glutamine metabolism in pig isolated enterocytes during development. Pediatr Res 36(2):175–181

Darcy-Vrillon B, Cherbuy C, Morel MT, Durand M, Duee PH (1996) Short chain fatty acid and glucose metabolism in isolated pig colonocytes: modulation by NH4+. Mol Cell Biochem 156 (2):145–151

Darragh AJ, Hodgkinson SM (2000) Quantifying the digestibility of dietary protein. J Nutr 130 (7):1850S–1856S

Darragh AJ, Cranwell PD, Moughan PJ (1994) Absorption of lysine and methionine from the proximal colon of the piglet. Br J Nutr 71(5):739–752

Davis TA, Nguyen HV, Garcia-Bravo R, Fiorotto ML, Jackson EM, Lewis DS, Lee DR, Reeds PJ (1994) Amino acid composition of human milk is not unique. J Nutr 124(7):1126–1132

Dillon EL, Knabe DA, Wu G (1999) Lactate inhibits citrulline and arginine synthesis from proline in pig enterocytes. Am J Physiol 276(5 Pt 1):G1079–G1086

Duée PH, Dary-Vrillon B, Blachier F, Morel MT (1995) Fuel selection in intestinal cells. Proc Nutr Soc 54(1):83–94

Eklou-Lawson M, Bernard F, Neveux N, Chaumontet C, Bos C, Davila-Gay AM, Tome D, Cynober L, Blachier F (2009) Colonic luminal ammonia and portal blood L-glutamine and L-arginine concentrations: a possible link between colon mucosa and liver ureagenesis. Amino Acids 37(4):751–760

Elango R, Pencharz PB, Ball RO (2002) The branched-chain amino acid requirement of parenterally fed neonatal piglets is less than the enteral requirement. J Nutr 132(10):3123–3129

Fan MZ, Adeola O, McBurney MI, Cheeseman CI (1998) Kinetic analysis of L-glutamine transport into porcine jejunal enterocyte brush-border membrane vesicles. Comp Biochem Physiol A Mol Integr Physiol 121(4):411–422

Fan MZ, Matthews JC, Etienne NM, Stoll B, Lackeyram D, Burrin DG (2004) Expression of apical membrane L-glutamate transporters in neonatal porcine epithelial cells along the small intestinal crypt-villus axis. Am J Physiol Gastrointest Liver Physiol 287(2):G385–G398

Flynn NE, Wu G (1996) An important role for endogenous synthesis of arginine in maintaining arginine homeostasis in neonatal pigs. Am J Physiol 271(5 Pt 2):R1149–R1155

Flynn NE, Wu G (1997) Glucocorticoids play an important role in mediating the enhanced metabolism of arginine and glutamine in enterocytes of postweaning pigs. J Nutr 127(5):732–737

Fogg FJ, Hutton DA, Jumel K, Pearson JP, Harding SE, Allen A (1996) Characterization of pig colonic mucins. Biochem J 316(Pt 3):937–942

Ginty DD, Osborne DL, Seidel ER (1989) Putrescine stimulates DNA synthesis in intestinal epithelial cells. Am J Physiol 257(1 Pt 1):G145–G150

Hata F, Ishii T, Kanada A, Yamano N, Kataoba T, Tabeuchi T, Yagasaki O (1990) Essential role of nitric oxide in descending inhibition in rat proximal colon. Biochem Biophys Res Commun 172(3):1400–1406

Htoo JK, Araiza BA, Saver WC, Rademacher M, Zhang Y, Cervantes M, Zijlstra RT (2007) Effect of dietary protein content on ileal amino acid digestibility, growth performance, and formation of microbial metabolites in ileal and cecal digesta of early-weaned pigs. J Anim Sci 85 (12):3303–3312

Janeczko MJ, Stoll B, Chang X, Guan X, Burrin DG (2007) Extensive gut metabolism limits the intestinal absorption of excessive supplemental dietary glutamate loads in infant pigs. J Nutr 137(11):2384–2390

Kelly D, King TP, Brown DS, McFadyen M (1991) Polyamide profiles of porcine milk and of intestinal tissue of pigs during suckling. Reprod Nutr Dev 31(1):73–80

Kemp M, Go YM, Jones DP (2008) Non equilibrium thermodynamics of thiol/disulfide redox systems: a perspective on redox system biology. Free Radic Biol Med 44(6):921–937

Klein RM, McKenzie JC (1983a) The role of cell renewal in the ontogeny of the intestine. I. Cell proliferation patterns in adult, fetal, and neonatal intestine. J Pediatr Gastroenterol Nutr 2 (1):10–43

Klein RM, McKenzie JC (1983b) The role of cell renewal in the ontogeny of the intestine. II. Regulation of cell proliferation in adult, fetal, and neonatal intestine. J Pediatr Gastroenterol Nutr 2(2):204–228

Konturek SJ, Brzozowski T, Majka J, Czarnobilski K (1992) Role of nitric oxide and prostaglandins in sucralfate-induced gastroprotection. Eur J Pharmacol 211(2):277–279

Kubes P (1992) Nitric oxide modulates epithelial permeability in the feline small intestine. Am J Physiol 262(6 Pt 1):G1138–G1142
Kubes P (1993) Ischemia-reperfusion in feline small intestine: a role for nitric oxide. Am J Physiol 264(1 Pt 1):G143–G149
Le Floc'h N, Seve B (2005) Catabolism through the threonine dehydrogenase pathway does not account for the high first-pass extraction rate of dietary threonine by the portal drained viscera in pigs. Br J Nutr 93(4):447–456
MacKendrick W, Caplan M, Hsueh W (1993) Endogenous nitric oxide protects against platelet-activating factor-induced bowel injury in the rat. Pediatr Res 34(2):222–228
Madej M, Lundh T, Lindberg JE (2002) Activity of enzymes involved in energy production in the small intestine during suckling-weaning transition of pigs. Biol Neonate 82(1):53–60
Miller MJ, Zhang XJ, Sadowska-Krowicka H, Chotinaruemol S, McIntyre JA, Clark DA, Bustamante SA (1993) Nitric oxide release in response to gut injury. Scand J Gastroenterol 28(2):149–154
Motyl T, Ploszaj T, Wojtasik A, Kukulska W, Podgurniak M (1995) Polyamines in cow's and sow's milk. Comp Biochem Physiol B Biochem Mol Biol 111(3):427–433
M'Rabet-Touil H, Blachier F, Morel MT, Darcy-Vrillon B, Duee PH (1993) Characterization and ontogenesis of nitric oxide synthase activity in pig enterocytes. FEBS Lett 331(3):243–247
M'Rabet-Touil H, Blachier F, Hellio N, Robert V, Cherbuy C, Darcy-Vrillon B, Duee PH (1995) Transglutaminase activity in enterocytes isolated from pig jejunum. Mol Cell Biochem 146 (1):49–54
M'Rabet-Touil H, Lerminiaux H, Dvee PH, Blachier F (1996) Evidence for increased anionic arginase activity in pig enterocytes during development. Biochem Mol Biol Int 38(1):197–204
Murphy JM, Murch SJ, Ball RO (1996) Proline is synthesized from glutamate during intragastric infusion but not during intravenous infusion in neonatal piglets. J Nutr 126(4):878–886
Quintero E, Guth PH (1992) Nitric oxide-mediated gastric hyperemia decreases ethanol-induced gastric mucosal injury in uremic rats. Dig Dis Sci 37(9):1324–1328
Reeds PJ, Burrin DG, Jahoor F, Wykes L, Henry J, Frazer EM (1996) Enteral glutamate is almost completely metabolized in first pass by the gastrointestinal tract of infant pigs. Am J Physiol 270(3 Pt 1):E413–E418
Reeds PJ, Burrin DG, Stoll B, Jahoor F, Wykes L, Henry J, Frazer ME (1997) Enteral glutamate is the preferential source for mucosal glutathione synthesis in fed piglets. Am J Physiol 273(2 Pt 1): E408–E415
Rémond D, Buffière C, Gobin JP, Mirand PP, Obled C, Papet I, Dardevet D, Williamson G, Brevillé D, Faure M (2009) Intestinal inflammation increases gastrointestinal threonine and mucin synthesis in enterally fed minipigs. J Nutr 139(4):720–726
Riedijk MA, Stoll B, Chacko S, Schierbeek H, Sunehag AL, van Goudoever JB, Burrin DG (2007) Methionine transmethylation and transsulfuration in the piglet gastrointestinal tract. Proc Natl Acad Sci USA 104(9):3408–3413
Schaart MW, Schierbeek H, van der Schoor SR, Stoll B, Burrin DG, Reeds PJ, van Goudoever JB (2005) Threonine utilization is high in the intestine of piglets. J Nutr 135(4):765–770
Shoveller AK, Brunton JA, Pencharz PB, Ball RO (2003) The methionine requirement is lower in neonatal piglets fed parenterally than in those fed enterally. J Nutr 133(5):1390–1397
Smith MW, James PS (1976) Amino acid transport by the helicoidal colon of the new-born pig. Biochim Biophys Acta 419(2):391–394
Stark ME, Szurszewski JH (1992) Role of nitric oxide in gastrointestinal and hepatic function and disease. Gastroenterology 103(6):1928–1949
Stein TE, Duffy SJ, Wickstrom S (1990) Differences in production values between high- and low-productivity swine breeding herds. J Anim Sci 68(12):3972–3979
Stoll B, Burrin DG, Henry J, Yu H, Jahoor F, Reeds PJ (1998) Dietary amino acids are the preferential source of hepatic protein synthesis in piglets. J Nutr 128(9):1517–1524
Stoll B, Burrin DG, Henry J, Yu H, Jahoor F, Reeds PJ (1999) Substrate oxidation by the portal drained viscera of fed piglets. Am J Physiol 277(1 Pt 1):E168–E175

Torrallardona D, Harris CI, Coates ME, Mf F (1996) Microbial amino acid synthesis and utilization in rats: incorporation of 15N from 15 NH_4Cl into lysine in the tissue of germ-free and conventional rats. Br J Nutr 76(5):689–700

Van Gaudoever JB, Stoll B, Henry JF, Burrin DG, Reeds PJ (2000) Adaptative regulation of intestinal lysine metabolism. Proc Natl Acad Sci USA 97(21):11620–11625

Vaugelade P, Posho L, Darcy-Vrillon B, Bernard F, Morel MT, Duee PH (1994) Intestinal oxygen uptake and glucose metabolism during nutrient absorption in the pig. Proc Soc Exp Biol Med 207(3):309–316

Wang JY, McCormack SA, Viar MJ, Johnson LR (1991) Stimulation of proximal small intestinal mucosal growth by luminal polyamines. Am J Physiol 261(3 Pt 1):G504–G511

Widdowson EM, Colombo VE, Artavanis CA (1976) Changes in the organs of pigs in response to feeding for the first time 24 h after birth. Boil Neonate 28:272–281

Wiese F, Simon O, Weyrauch KD (2003) Morphology of the small intestine of weaned piglets and a novel method for morphometric evaluation. Anat Histol Embryol 32(2):102–109

Windmueller HG, Spaeth AE (1976) Metabolism of absorbed aspartate, asparagine, and arginine by rat small intestine in vivo. Arch Biochem Biophys 175(2):670–676

Wu G (1995) Urea synthesis in enterocytes of developing pigs. Biochem J 312(Pt 3):717–723

Wu G (1997) Synthesis of citrulline and arginine from proline in enterocytes of postnatal pigs. Am J Physiol 272(6 Pt 1):G1382–G1390

Wu G, Knabe DA (1994) Free and protein-bound amino acids in sow's colostrum and milk. J Nutr 124(3):415–424

Wu G, Knabe DA (1995) Arginine synthesis in enterocytes of neonatal pigs. Am J Physiol 269(3 Pt 2): R621–R629

Wu G, Borbolla AG, Knabe DA (1994) The uptake of glutamine and release of arginine, citrulline and proline by the small intestine of developing pigs. J Nutr 124(12):2437–2444

Wu G, Flynn NE, Yan W, Barstow DG Jr (1995a) Glutamine metabolism in chick enterocytes: absence of pyrroline-5-carboxylase synthase and citrulline synthesis. Biochem J 306(Pt 3):717–721

Wu G, Knabe DA, Yan W, Flynn NE (1995b) Glutamine and glucose metabolism in enterocytes of the neonatal pig. Am J Physiol 268(2 Pt 2):R334–R342

Chapter 6
Development of Micro-ecological System in Small and Large Intestine of Piglets

Shengping Wang, Gang Liu, Lichuang Cai, Jiannan Ding, and Yulong Yin

Contents

The piglet gastrointestinal (GI) tract can be considered as a big "warehouse" since it is populated with as many as 100 trillion of microorganism cells (1×10^{14}) belonging to 500–1,000 species. Bacteria make up most of the gut microbiota, which include 300–1,000 different species (Sears 2005). However, about 99 % bacteria belong to 30 or 40 species. Fungi and protozoa also make up a part of the piglet gut microbiota, but little is known about their activities. The gross weight of bacteria in piglet GI tract represents as much as 1.5 kg.

S. Wang • G. Liu • Y. Yin (✉)
Chinese Academy of Sciences, Institute of Subtropical Agriculture, Research Center for Healthy Breeding Livestock & Poultry, Hunan Engineering & Research Center of Animal & Poultry Science, Key Laboratory of Agro-ecological Processes in Subtropical Region, Scientific Observing and Experimental Station of Animal Nutrition and Feed Science in South-Central, Ministry of Agriculture, Changsha, Hunan 410125, People's Republic of China
e-mail: yinyulong@isa.ac.cn

L. Cai • J. Ding
Biological Resources Institute, Jiangxi Academy of Sciences, Nanchang 330029, P.R. China

F. Blachier et al. (eds.), *Nutritional and Physiological Functions of Amino Acids in Pigs*,
DOI 10.1007/978-3-7091-1328-8_6, © Springer-Verlag Wien 2013

Most of the gut bacteria benefit the piglet, for example Bifidobacterium and Lactobacillus. However, some other gut bacteria such as *Clostridium perfringens*, *Pseudomonas*, *Wales* bacteria, and *Staphylococcus aureus* are potentially pathogenic. The normal gut microbiota maintains a stable gut micro-ecological environment (Hou et al. 2011; Wang et al. 2011a, b; Kong et al. 2009). Microbiota activity can impact piglet metabolism and immunity. However, the normal gut microbiota is not static, and the microbiota composition can change according to diet, age, environment, and physiology. The gut microbiota can move from one stable state into another stable state in order to adapt to the changes of internal and external environment (Björkstén et al. 2001).

Recent studies suggest that the gut microbiota plays an important role in piglet health and nutrition by maintaining a stable gut environment, by providing a barrier for colonization of pathogens, and by exerting important metabolic functions (fermentation of non-digestible fibers and amino acids originating from undigested proteins; salvage of energy as short-chain fatty acids; and production of folic acid, vitamin K, vitamin H, B vitamins, essential amino acids, and antibiotic peptides). The microbiota is also able to stimulate the development of the immune system and to participate in the maintenance of piglet energy homeostasis. Disturbances of the gut microbiota can result in GI tract malfunction, resulting in phenomena such as diarrhea, enteritis, and indigestion. So, the relationship between the gut microbiota and piglet is not merely commensal, but rather a symbiotic relationship (Sears 2005).

Because the gut microbiota is involved in many aspects of piglet health and nutrition, it is important to understand how the composition of this microbiota is established. The gut microbiota forms an essential part of a complex ecosystem that plays an important role in piglet health, nutrition, and metabolism. Research related to the study of the gut microbiota has involved an increasing number of laboratories all around the world in recent years. The major objective of this present chapter is to provide insights into new development in piglet gut microbiota research, including the composition, development, and factors influencing the gut microbiota.

6.1 The Composition and Diversity of the Gut Microbiota in Piglet

The piglet gut hosts a dynamically evolving microbial ecosystem that consists of various bacterial populations. These bacteria are either permanent intestinal residents such as *Lactobacillus* and *Streptococcus* or transient inhabitants, for example allochthonous members, introduced from the environment. It is believed that intensive selection and coevolution result in the genetic diversity of the intestine. Horizontal gene transfer (HGT) is assumed to function as the principal evolutionary force in shaping the host microbiome (Palmer et al. 2007).

The piglet gut microbiota is characterized by a relatively small number of bacterial and archaeal species, compared to other environments such as soil or stromatolite. The 16S rRNA gene sequence analysis revealed that the piglet gut microbiota is made of only nine major bacterial divisions, including Bacteroidetes, Firmicutes, Fusobacteria, Proteobacteria, Verrucomicrobia, Cyanobacteria, Spirochaetes, VadinBE97, and Actinobacteria (Rajilić-Stojanović et al. 2007; Palmer et al. 2007). However, there are different bacterial cell density and bacterial species in different sites along the digestive tract (Figs. 6.1 and 6.2). Generally, the number and diversity of bacteria flora are increasing from stomach to foregut and then to hindgut. The spatial distribution of the piglet gut microbiota is also different, not at random, but rather organized. There are noticeable difference for the gut microbiota distribution between the mucus layers and the luminal content (Fig. 6.2). For example, *Lactobacillus* is the major bacteria adhering to the mucus layer or to the epithelial cells, while *Streptococcus* mainly distributes within the gut lumen. The diversity of the piglet gut microbiota in the gut wall is lower than that in the gut lumen. In addition, the piglet gut microbiota maintains a dynamic equilibrium with both temporary variation and continuous stability. For example, some *Lactobacillus* species such as *Lactobacillus reuteri*, *Lactobacillus amylovorus*, *Lactobacillus acidophilus*, and *Lactobacillus mucosae* and some Streptococcus species such as *Streptococcus hyointestinalis*, *Streptococcus alactolyticus*, and *Streptococcus gallolyticus* can live in the piglet gut for the whole life, but the number of these bacteria may change with age, diet, or environment (Wang et al. 2009). However, Clostridium has a high temporal variation. Indeed, some *Clostridium* species rapidly emerge, and then rapidly disappear. The possible reason is that *Clostridium* may be susceptible to the change of internal and external environment.

6.2 The Development of the Gut Microbiota in Piglet

The piglet is axenic in the sow uterus, and is suddenly plunged at birth into a complex bacterial environment. During its passage through the maternal vagina, the piglet is in contact with bacteria for the first time, and then meets the enormous bacterial population from the sows' feces and rearing environment. Thus, the newborn piglet is very likely to meet potentially pathogens, and accordingly, the survival of the piglet depends on the very early establishment of defense systems. Therefore a fast and stable establishment of the gut microbiota will play a very important role in keeping the health and growth of the piglets.

6.2.1 The Nature of the First Established Strains

The establishment of the microbiota is very rapid in the piglet digestive cavities. Only few bacteria are enumerated at the third hour of life. However, after 10–12 h, the population density reaches 10^8–10^9 bacterial cells per gram in feces. After 24 h, the piglet already possesses a dominant microbiota established at a level which will not change until weaning (Ducluzeau 1983). There are differences in the order of implantation of the main bacterial groups according to the animal species. In the

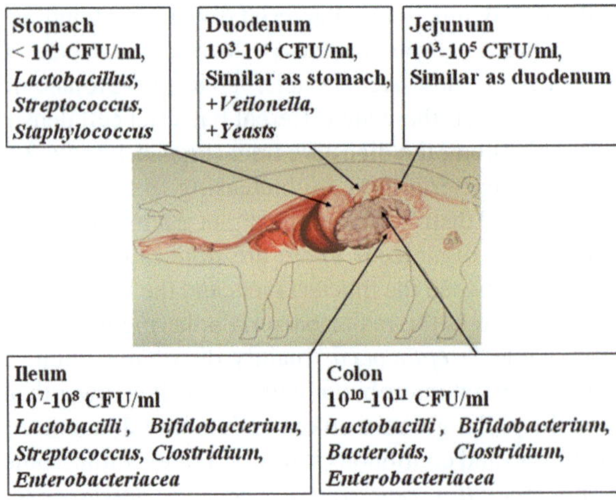

Fig. 6.1 Bacterial cells density and major bacterial species for digesta in the different site along the digestive tract of pig

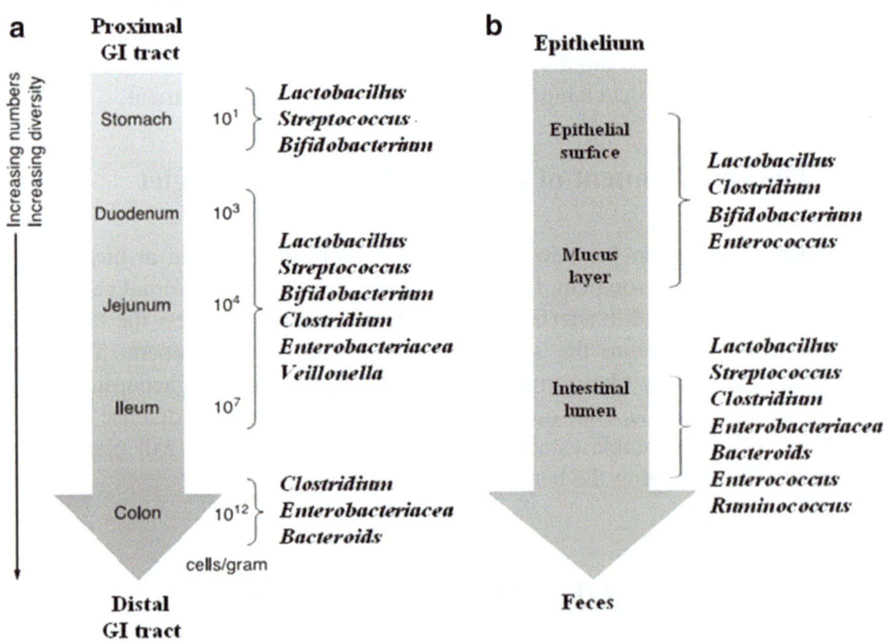

Fig. 6.2 (**a**) Bacterial cells density and major bacterial species for digesta in the different site along the digestive tract of pig; (**b**) Bacterial cells density and major bacterial species for epithelium in the different site along the digestive tract of pig

young mouse, facultatively anaerobic bacteria appear very early; and strictly anaerobic bacteria, which will eventually become dominant, will appear later at the third week after birth (Ducluzeau 1983). The opposite is observed in the young hare and rabbit (Ducluzeau et al. 1975). The piglet represents an intermediate case, with a simultaneous establishment from the first day of both facultatively anaerobic bacteria (mainly *Lactobacillus*, *Streptococcus*, and *Escherichia*) and strictly anaerobic bacteria belonging to *Clostridium*, *Bifidobacterium*, *Fusobacterium*, *Peptostreptococcus*, and sometimes Bacteroides genera (Wilbur et al. 1960; Pesti 1962).

6.2.2 Origin of the Piglet Gut Microbiota

During the process of birth, the newborn piglet meets successively several different microbial ecosystems: the sow vagina, the sow feces, and the rearing environment. It may be considered a priori that each of these ecosystems is liable to contribute to the creation of the piglet gut microbiota. In fact, the newborn piglet is provided with powerful selective systems allowing only some bacteria species to develop among all the bacteria. The first bacteria which become established in the digestive tract actually come from the sow or from the environment, but they are not the most abundant in the ecosystems met by the piglet.

6.2.3 Mechanisms Controlling the Establishment of the Gut Microbiota in Piglet

One of the main factors liable to account for the selection exerted by the newborn piglet among all the bacteria met is its dietary regime. Within 7 days after birth the piglet receives the mother's milk as feed, which is favorable to the growth of some bacteria and unfavorable to others in the piglet gut. For example, it is observed in the piglet born to gnotobiotic sow harboring only *Escherichia coli* and *Shigella flexneri* that these bacteria are implanted in a high number after 12 h (Hayashi et al. 2002). Conversely, when the sow harbors strictly anaerobic bacteria such as bacteroides and Clostridium, these bacteria are detected in the piglet between 10 and 15 days of age, i.e., when the piglet begins to consume solid food in addition to milk. To take another example, many of *Lactobacillus sp.* and *Lactobacillus salivarius* are implanted in the piglet ileum when the piglet receives liquid milk. However, after weaning these bacteria gradually decrease, and *Streptococcus alactolyticus* as well as *Streptococcus hyointestinalis* increase if the piglets receive solid food (Wang et al. 2009).

The immune status of the sow, which affects the composition of the colostrum given to the piglet, might also influence the order of the establishment of the bacteria in the piglet intestine. Colostrum, undoubtedly, plays a role in the fight against microbial infection of digestive origin, but it has never been demonstrated that it has an effect on the microflora in the gut lumen. Immunization of the sow against different bacterial species did not prevent the further establishment of these bacteria in the piglet (Ducluzeau 1983; Hooper and Macpherson 2010). Ducluzeau studied the establishment of the fecal flora between birth and 48 h of age in a litter of conventional piglets divided into two groups: in the first one the animals were left with their dam and suckled colostrums and milk, while in the other, animals were taken off at birth, were housed in a cage inside the dam's pen, and received sterile cow's milk. Despite these experimental differences, no difference was observed in the systematic order of implantation of the first bacteria between both groups of piglets (Hooper et al. 2002).

Bacterial interactions certainly play an important role in the order of establishment of the piglet gut microbiota. The early development of some bacteria in the digestive tract prevents proliferation of other bacteria. A strain of *E. coli* is capable to eliminate from the digestive tract a strain of *Lactobacillus casei* derived from a commercial preparation and previously established in the axenic newborn (Dethlefsen et al. 2006). In human newborn it was shown that the inoculation of a strain of *E. coli* without plasmid within 2 h following birth led to the elimination or to a marked decrease of the *E. coli* population carrying plasmids of resistance to antibiotics, while these strains were frequently dominant in the control.

In the piglet, many studies indicate that single supplement with Lactobacillus in diet decreases the number of *E. coli* in the gut lumen, and concomitantly decreases the diarrhea rate among animals (Konstantinov et al. 2008).

The order of establishment of the bacteria in the piglet intestine is probably important for its survival. Many studies have shown that the supplement with antibiotics in the diet enables to increase the rate of survival in piglets and to decrease diarrhea episodes due to toxicogenic bacteria such as *Clostridium* and *E. coli* (Rohde et al. 2009). However, the use of antibiotics may result in the formation of resistant bacteria. In addition, the disturbance by the supplement of antibiotics of the initial establishment of the gut microbiota may lead to some negative effect in the piglets like an impairment of the piglet immune response, thus producing cross infection. The practical application of the bacterial interaction concept was found to be more useful in the above case than the use of antibiotics (Rohde et al. 2009). If the young hare are inoculated within the hour following birth with a complex flora derived from a healthy young hare, the animal is totally protected from neonatal diarrhea without further pathological manifestation at weaning. This protective flora, preserved in gnotobiotic mice, has never been simplified up to now to obtain a mixture of pure strains liable to be cultured in vitro (Ducluzeau et al. 1981a, 1981b). A protective effect was also demonstrated in the axenic newborn piglet: the early inoculation of a selected strain of *E. coli* from pig origin was found to protect the animals from the further development of toxinogenic strains of *E. coli* provided that the latter were not capable of adhering to the gut mucosa (Duval-Iflah et al. 1983).

6.2.4 Lactobacilli Community Development

There are several hundred bacterial species in the piglet GI tract. All the members of the piglet gut microbiota are needed for the gut to develop its specific intestinal functions (Hooper and Gordon 2001). However, Lactobacilli are the most prevalent

a

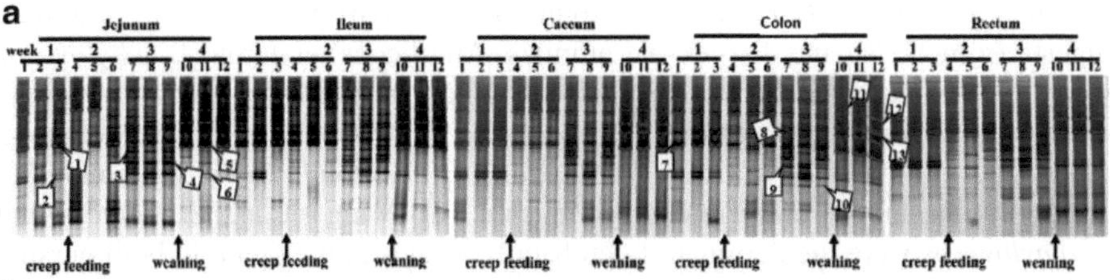

b

N	Closest Relative	Identity %	Clone ID	Accession	N	Closest Relative	Identity %	Clone ID	Accession
1	*Lactobacillus delbrueckii subsp. lactis*	98	YRJ9	AY662527	7	*Lactobacillus delbrueckii subsp. lactis*	96	YRC4	AY662519
2	*Lactobacillus sp. oral clone* CX036	95	YRJ28	AY662528	8	*Lactobacillus delbrueckii subsp. bulgaricus*	97	YRC89	AY662523
3	*Lactobacillus crispatus* strain TSK V38-1	99	YRJ39	AY662530	9	*Lactobacillus delbrueckii*	92	YRC69	AY662521
4	*Lactobacillus delbrueckii*	94	YRJ58	AY662531	10	*Lactobacillus amylovorus*	98	YRC86	AY662522
5	*Lactobacillus reuteri*(DSM20016T)	97	YRJ100	AY662532	11	*Lactobacillus acidophilus johnsonii*	95	YRC100	AY662525
6	*Lactobacillus reuteri*(DSM20016T)	98	YRJ37	AY662529	12	*Lactobacillus sp.* G35	88	YRC95	AY662524
					13	*Lactobacillus reuteri*(DSM20016T)	97	YRC37	AY662520

Fig. 6.3 (**a**) Succession of the Lactobacillus community of digesta during the first month in the piglet intestinal tract. Changes of feeding strategy are indicated by *arrows*. The bands identified from the 16S rDNA clone libraries are numbered and are indicated by *arrowheads*. (**b**) Closest relative as determined by comparative sequence analysis, identity with this relative, clone ID, and accession number for each band identified in (**a**)

bacteria, and are often considered beneficial because they are important for maintaining a healthy and stable microbiota in the GI tract. It has been postulated that Lactobacilli has several health-promoting effects, including immuno-stimulation, alleviation of food intolerance and allergy, and prevention of diarrhea and intestinal infection (Servin 2004; Bengmark 2000; Salminen et al. 1998). Consequently, a normal Lactobacilli community development in time and space in the gut is significantly important for the health and growth of the piglet.

The experiments conducted by Yao et al. analyzed *Lactobacillus* diversity and development of neonatal piglet gut microbiota during the first 4 weeks of life by a combination of denaturing gradient gel electrophoresis (DGGE) and 16S ribosomal DNA sequence (Fig. 6.3a). The bands in DGGE profile represent the majority of the dominant *Lactobacillus* populations in the community, and their appearance and disappearance reflect approximate changes in the *Lactobacillus* community composition. The intensity of a band provides a rough estimate of the proportion of the corresponding population in a sample. The *Lactobacillus* communities in each compartment changed from simple to complex and returned to simple during this 4-week period. The most complex banding pattern appeared at week 3. There were several new bands (or new lactobacilli) only observed at this time, and disappeared again at week 4 (Yao et al. 2011).

Yao et al. also compared the mobility of the cloned fragments with the mobility of fragments amplified from the original samples by DGGE, and the DNA sequences of clones that corresponded to dominant bands were determined (Fig. 6.3b). Six dominant DGGE bands generated from jejunal digesta were matched with sequences that show 94–98 % similarity to those derived from *L. reuteri*, *L. delbrueckii*, and *L. crispatus*. Seven dominant DGGE bands generated from colon digesta were matched with sequences that show 88–99 % similarity to those derived from *L. reuteri*, *L. delbrueckii*, *L. amylovorus*, and *L. acidophilus*.

Amplicons related to *L. reuteri* were found in all DGGE fingerprints from jejunal digesta from 1-, 3-, and 4-week-old animals. Amplicons related to *L. amylovorus* were present in all DGGE fingerprints from colon digesta from 1-, 3-, and 4-week-old animals. Amplicons related to *L. delbrueckii* were found before weaning, *L. crispatus* in the stage of creep feeding, and *L. acidophilus* after weaning (Yao et al. 2011). In the experiments conducted by Wang et al., *L. reuteri* and *L. amylovorus* were the dominant Lactobacilli species in the ileum in both nursing and weaned piglet (Wang et al. 2009).

In conclusion, *Lactobacillus* communities follow successional changes associated with piglet age and diet shifting. Creep feeding stabilizes the *Lactobacillus* community of weaning piglets. Within the *Lactobacillus* community, some members like *L. reuteri* and *L. amylovorus*/*L. sobrius* might be permanent colonizers, while *L. delbrueckii*, *L. acidophilus*, *L. salivarius*, *L. mucosae*, and *L. crispatus* might be transient members of the *Lactobacillus* communities in the piglet's GI tract.

6.3 Factors Influencing the Composition of Piglet GI Tract Microbiota

As said above, the piglet is born without microorganisms. Starting at birth, the gut microbiota must develop from a simple unstable community into a complex community, which comprises microorganisms (bacteria, archaea, fungi, viruses, and protozoa) that represent superior competitors for a complex intestinal ecosystem (Palmer et al. 2007). It is generally considered that the genotype of host and the diet are the major factors defining a microbial intestinal niche. However, many other factors also influence mammal intestinal microbiota such as interactions between individual colonizers, age, weaning, as well as transient microorganisms (pathogens and probiotics) (Deng et al. 2007; Yin and Tan 2010).

6.3.1 The Genotype of the Piglet Influences Gut Microbiota

The genotype of the piglet is considered as a factor influencing variation of gut microbiota between individuals. Differential action of genes that control the immune system is involved in such an effect (Ley et al. 2006). For instance, the piglet immune system modulates the intestinal microbiota composition by restricting microbial penetration through the host mucosal barrier, and by secreting different antimicrobial products such as peptides, as well as antimicrobial enzymes (Cobb et al. 2004; Macdonald and Monteleone 2005). The genotype of the piglet determines the availability of specific attachment sites and host-derived resources, which influences the intestinal microbiota. So, monozygotic twins exhibit fewer differences in the intestinal microbiota than their unrelated marital partners. However, there are small differences between the intestinal microbiota of the monozygotic twins, suggesting that maternal transmission (Ley et al. 2005) is another determining factor. The initial colonizing microbiota influences the eventual microbial composition of the intestine.

6.3.2 Diet Influences Piglet Gut Microbiota

It was demonstrated that diet can have an important influence on the composition of intestinal microbiota. Some studies have shown a clear correlation between diet and the presence of specific bacterial groups. For example, a diet rich in inulin and related fibers promote an increase in *Bifidobacter*ia (Roberfroid 1998). Supplementation with astragalus polysaccharide and related polysaccharide in diet increases the number of *Lactobacillus* and *Bifidobacteria.* When nursing piglet is provided breast milk, there are many *Lactobacillus salivarius* and *Lactobacillus mucosae* in intestine. However, when the diet is evolving from liquid breast milk to solid foods, the two species of *Lactobacillus* are largely replaced by *Lactobacillus amylovorus* (Wang et al. 2009). In addition, mucin and other host-derived products, as well as the competition and cooperation between bacteria, are thought to be important influencing factors of intestinal microbiota (Dethlefsen et al. 2006).

6.3.3 Weaning Influences Piglet Gut Microbiota

Recent cultivation-independent studies provided a more comprehensive understanding of the development of the microbiota in pigs from birth to adulthood. Previously, such studies were hampered by the fact that a large proportion of the intestinal microbiota is inaccessible to cultivation approaches (Zoetendal et al. 2004 and Zoetendal et al. 2006). Cultivation-based studies have shown that lactic acid bacteria, enterobacteria, and streptococci were the most important first colonizers of the piglet gut (Stewart et al. 1997). Analysis of 16S rRNA gene clone libraries indicated that ileal samples of 2-day-old piglets harbored a consortium of *E. coli-*, *Shigella flexneri-*, *L. sobrius-*, *L. reuteri-*, and *L. acidophilus*-related sequences (Konstantinov et al. 2006). While populations of *L. sobrius*, *L. reuteri*, and *L. acidophilus* remained stable and abundant before weaning, their numbers dropped significantly after weaning. The experiment conducted by Wang et al. also led to similar results which are shown in Table 6.1 (Wang et al. 2009). Remarkable shifts in the microbiota composition and metabolic activities are found immediately after weaning (PW) for both ileal and colonic microbiota. Similar sequential changes in fecal microbiota were also found in other studies (Konstantinov et al. 2003 and Inoue et al. 2005). Hence, the early PW period is characterized by compositional and functional instability of the predominant piglet gut microbiota. In addition, several potentially beneficial lactobacilli are significantly suppressed during the weaning transition. Possibly, lactobacilli, having complex nutritional requirements, are affected by the restricted feed intake after weaning. This microbiota instability followed by the rapid decline of previously predominant lactobacilli may have greater contributions to PW intestinal disturbances than previously appreciated. Hence, the microbiota can be seen as an important potential target in dietary interventions targeting piglet GI tract

Table 6.1 Major microbiota in ileal digesta of piglets

Groups	Species	Rate of clones/% Nursing piglets aged 21 days	Nursing piglets aged 28 days	Weaned piglets aged 28 days
Lactobacillus	*L. amylovorus*	41.5 (71/171)	25.8 (50/194)	31.6 (54/171)
	L. mucosae	4.7 (8/171)	3.1 (6/194)	0.6 (1/171)
	L. reuteri	12.3 (21/171)	18.6 (36/194)	16.4 (28/171)
	L. salivarius	9.4 (16/171)	6.2 (12/194)	1.2 (2/171)
	Lactobacillus sp.	12.9 (22/171)	10.8 (50/194)	
	Uncultured Lactobacillus sp.	6.4 (11/171)	4.6 (9/194)	0.6 (1/171)
	L. acidophilus	0.6 (1/171)	1.5 (3/194)	1.8 (3/171)
	L. coleohominis		0.5 (1/194)	
Streptococcus	*S. alactolyticus*		5.2 (10/194)	7.0 (12/171)
	S. gallolyticus	0.1 (2/171)	0.1 (2/194)	
	S. hyointestinalis	0.1 (2/171)	10.8 (21/194)	13.5 (23/171)
	Streptococcus sp.		0.5 (1/194)	0.6 (1/171)
	S. suis	0.1 (2/171)	3.6 (7/194)	5.3 (9/171)

disturbances, specifically aiming at supporting the predominant microbiota stability while maintaining indigenous lactobacilli communities.

Finally, age, antibiotics, and probiotics have crucial influences on the mammal intestinal microbiota. It is noted that the effect differs between antibiotics or probiotics. The composition of intestinal microbiota in adults has generally been considered to be stable as long as the host is not subjected to stressful circumstances. Some studies reported, however, that the elderly subjects have fewer bifidobacteria and more fungi and enterobacteria in their intestinal luminal content when compared with younger subjects (Hopkins et al. 2001). Lastly, *Clostridium difficile* is isolated more frequently in elderly mammals than in young subjects (Hopkins et al. 2001).

6.4 Conclusion and Perspectives

Piglet gut microbiota displays remarkable metabolic and immune versatility. Microbiota is involved in the synthesis of a variety of molecules with enormous importance, and also regulates key metabolic pathways and processes that are vital to the growth, development, health, and homeostasis of piglet. These findings exemplified the power of basic research on gut microbiota, nutrition, and immunity for the discovery of new knowledge and for solving significant practical problems in piglet production. However, in order to enlarge the available knowledge on this topic in the context of piglet production, it appears necessary to clarify the composition and development of piglet gut microbiota.

Studies of piglet gut microbiota have been largely based on traditional approaches (e.g., culture-dependent method), polymerase chain reaction-denaturing gradient gel electrophoresis (PCR-DGGE), 16S rRNA gene sequencing, as well as terminal-restriction fragment length polymorphism (T-RLFP). These techniques have played historically significant roles in the development of the field. However, recent advances of meta-genome sequencing, transcriptomics, metabolomics, and proteomics have provided new insights into the complex interactions between the piglet and gut microbiota. Using these approaches, the composition and development of the piglet will be gradually discovered. Furthermore, the relationship between the gut microbiota and the piglet both at the intestinal mucosa level and at the peripheral level needs further investigations. It is becoming increasingly evident that the gut microbiota may shape the host metabolic and immune network activity. It would be important to capitalize on these revolutionary methods in future endeavors in order to rapidly and extensively expand our knowledge on the composition and metabolic capacities of the intestinal microbiota as well as the consequences of the modifications of such parameters for development, nutrition, immunity, and health in piglet and other species.

References

Bengmark S (2000) Bacteria for optimal health. Nutrition 16:611–615

Björkstén B, Sepp E, Julge K, Voor T, Mikelsaar M (2001) Allergy development and the intestinal microflora during the first year of life. J Allergy Clin Immunol 108:516–520

Cobb BA, Wang Q, Tzianabos AO, Kasper DL (2004) Polysaccharide processing and presentation by the MHCII pathway. Cell 117:677–687

Deng ZY, Zhang JW, Li J, Fan YW, Cao SW, Huang RL, Yin YL, Zhong HY, Li TJ (2007) Effect of polysaccharides of cassiae seeds on the intestinal microflora of piglets. Asia Pac J Clin Nutr 16(Suppl 1):143–147

Dethlefsen L, Eckburg PB, Bik EM, Relman DA (2006) Assembly of the human intestinal microbiota. Trends Ecol Evol 21:517–523

Ducluzeau R (1983) Implantation and development of the gut flora in the newborn animal. Ann Rech Vet 14:354–359

Ducluzeau R, Dubos F, Hudault S, Nicolas JL, Dabard J, Raibaud P (1981a) Microbial barrier against various enteropathogenic strains in the digestive tract of gnotoxenic animals. Application to the treatment of Clostridium difficile diarrhoea in the young hare. 11th International Symposium on gnotobiology, Tokyo, p. 59

Ducluzeau R, Raibaud P, Dubos F, Clara A, Lhuillery C (1981b) Remnant effect of some dietary regimens on the establishment of two Clostridium strains in the digestive tract of gnotobiotic mice. Am J Clin Nutr 34:520–526

Ducluzeau R, Dubos F, Martinet L, Raibaud P (1975) Digestive tract microflora in healthy and diarrheic young hares born in captivity. Effect of intake of different antibiotics. Ann Biol Anim Biochim Biophys 15:529–535

Duval-Iflah Y, Chappuis JP, Ducluzeau R, Raibaud P (1983) Intraspecific interactions between Escherichia coli strains in human newborns and in gnotobiotic mice and piglets. Prog Food Nutr Sci 7(3–4):107–116

Hayashi H, Sakamoto M, Benno Y (2002) Phylogenetic analysis of the human gut microbiota using 16S rDNA clone libraries and strictly anaerobic culture-based methods. Microbiol Immunol 46:535–548

Hooper LV, Midtvedt T, Gordon JI (2002) How host–microbial interactions shape the nutrient environment of the mammalian intestine. Annu Rev Nutr 22:283–307

Hooper LV, Macpherson AJ (2010) Immune adaptations that maintain homeostasis with the intestinal microbiota. Nat Rev Immunol 10:159–169

Hooper LV, Gordon JI (2001) Commensal host-bacterial relationships in the gut. Science 292:1115–1118

Hopkins MJ, Sharp R, Macfarlane GT (2001) Age and disease related changes in intestinal bacterial populations assessed by cell culture, 16S rRNA abundance, and community cellular fatty acid profiles. Gut 48(2):198–205

Hou ZP, Wang WJ, Liu ZQ et al (2011) Effect of lactoferricin B and cecropin P1 against enterotoxigenic *Escherichia coli* in vitro. J Food Agric Environ 9(2):271–274

Inoue R, Inoue T, Tsukahara NN, Ushida K (2005) Development of the intestinal microbiota in the piglet. J Gen Appl Microbiol 51:257–265

Kong XF, Zhang YZ, Wu X, Yin YL, Tan ZL, Feng Y, Yan FY, Bo MJ, Huang RL, Li TJ (2009) Fermentation characterization of Chinese yam polysaccharide and its effects on the gut microbiota of rats. Int J Microbiol 13 pages

Konstantinov SR, Konstantinov AA, Awati BA, Williams BG, Miller P, Jones CR, Stokes ADL, Akkermans HS, de Vos WM (2006) Post-natal development of the porcine microbiota composition and activities. Environ Microbiol 8:1191–1199

Konstantinov SR, Zhu WY, Williams BA, Tamminga S, Vos WM, Akkermans AD (2003) Effect of fermentable carbohydrates on faecal bacterial communities as revealed by DGGE analysis of 16S rDNA. FEMS Microbiol Ecol 43:225–235

Konstantinov SR, Smidt H, Akkermans ADL, Casini L, Trevisi P, Mazzoni M, De Filippi S, Bosi P, De Vos WM (2008) Feeding of *Lactobacillus sobrius* reduces *Escherichia coli* F4 levels in the gut and promotes growth of infected piglets. FEMS Microbiol Ecol 66:599–607

Ley RE, Turnbaugh PJ, Klein S, Gordon JI (2006) Microbial ecology: human gut microbes associated with obesity. Nature 444:1022–1023

Ley RE, Bäckhed F, Turnbaugh P et al (2005) Obesity alters gut microbial ecology. Proc Natl Acad Sci U S A 102:11070–11075

Macdonald TT, Monteleone G (2005) Immunity, inflammation, and allergy in the gut. Science 307:1920–1925

Palmer C, Bik EM, Digiulio DB, Relman DA, Brown PO (2007) Development of the human infant intestinal microbiota. PLoS Biol 5:1556–1773

Pesti L (1962) Qualitative and quantitative examination of the intestinal bacterium flora of healthy pigs. Zentralbl Bakteriol I Abt Orig 189:282–293

Rajilić-Stojanović M, Smidt H, Vos WM (2007) Diversity of the human gastrointestinal tract microbiota revisited. Environ Microbiol 9:2125–2136

Roberfroid MB (1998) Prebiotics and synbiotics: concepts and nutritional properties. Br J Nutr 80:197–202

Rohde CL, Bartolini V, Jones N (2009) The use of probiotics in the prevention and treatment of antibiotic-associated diarrhea with special interest in Clostridium difficile-associated diarrhea. Nutr Clin Pract 24(1):33–40

Salminen S, Deighton MA, Benno Y, Gorbach SL (1998) Lactic acid bacteria in health and disease. In: Salminen A, von Wright A (eds) Lactic acid bacteria: microbiology and functional aspects, 2nd edn. Marcel Dekker, New York, pp 211–253

Sears CL (2005) A dynamic partnership: Celebratingour gut flora. Anaerobe 11:247–251

Servin AL (2004) Antagonistic activities of lactobacilli and bifidobacteria against microbial pathogens. FEMS Microbiol Rev 28:405–440

Stewart CS (1997) Microorganisms in hindgut fermentors. In: Mackie RI, White BA, Isaacson RE (eds) Gastrointestinal microbiology II. Gastrointestinal microbes and host interactions, vol 2. Chapman & Hall, New York, NY, pp 142–186

Wang SP, Bo MJ, Kong XF et al (2009) 16S rRNA gene-based analysis of ileal bacterial community and phylogeny in nursing and weaned piglets. Animal Husbandry and Feed Science 1:12–17

Wang S-P, Yang L, Tang X-S, Cai L-C, Liu G, Kong X-F, Blachier F, Yin Y-L (2011a) Dietary supplementation with high-dose *Bacillus subtilis* or *Lactobacillus reuteri* modulates cellular and humoral immunities and improves performance in weaned piglets. J Food Agric Environ 9:181–187

Wang S, Blachier F, Zhao F, Yin Y (2011b) Intestinal microbiota: development, metabolism and functions. J Food Agric Environ 9(121–129)

Wilbur RD, Catron DV, Quinn LY, Speer VC, Hays VW (1960) Intestinal flora of pigs as influenced by diet and age. J Nutr 71:168–175

Yao W, Zhu WY, Smidt H, Verstegen MWA (2011) Cultivation-independent analysis of the development of the Lactobacillus spp. Community in the intestinal tract of newborn piglets. Agricultural Sciences in China 10(3):438–447

Yin YL, Tan BE (2010) Manipulation of dietary nitrogen, amino acids and phosphorus to reduce environmental impact of swine production and enhance animal health. J Food Agric Environ 8:447–462

Zoetendal EG, Collier CT, Koike S, Mackie RI, Gaskins HR (2004) Molecular ecological analysis of the gastrointestinal microbiota: a review. J Nutr 134:465–472

Zoetendal EG, Zoetendal EEV, de Vos WM (2006) A microbial world within us. Mol Microbiol 59:1639–1650

Part III
Physiological Functions of Amino Acids in Pigs

Chapter 7
Structure, Metabolism and Functions of Amino Acids: An Overview

Wenkai Ren, Yinghui Li, Yulong Yin, and François Blachier

Contents

W. Ren • Y. Li • Y. Yin (✉)
Institute of Subtropical Agriculture, The Chinese Academy of Sciences, 410125 Hunan, People's Republic of China
e-mail: yinyulong@isa.ac.cn

F. Blachier
UMR 914 INRA/AgroParisTech, Nutrition Physiology and Alimentary Behavior, 16 rue Claude Bernard, 75005 Paris, France

F. Blachier et al. (eds.), *Nutritional and Physiological Functions of Amino Acids in Pigs*, DOI 10.1007/978-3-7091-1328-8_7, © Springer-Verlag Wien 2013

7.1 Introduction

Amino acids are classically considered as the building blocks for the synthesis of proteins. However, amino acids are also used as precursors for the biosynthesis of numerous compounds with various physiological functions (Wu et al. 2007; Kim et al. 2007; Li et al. 2007). For instance, compounds like glutathione, creatine, carnitine, taurine, and heme of hemoglobin are synthesized in the body from amino acids in different metabolic pathways. Some amino acids like glutamine, aspartate, and lysine are nitrogen and carbon precursors for the synthesis of ribo- and deoxyribonucleotides which are precursors for DNA and RNA synthesis. Furthermore, some amino acids like arginine and glutamine are precursors for amino acids which are not present in proteins like ornithine and citrulline and which play important roles in detoxification metabolism (ammonia metabolism in the urea cycle, e.g.) and in interorgan metabolism. In addition, some amino acids like glutamine, glutamate, and aspartate are precursors of the tricarboxylic acid cycle intermediates and represent oxidative substrates in numerous cell types which use them as major energy substrates (Yao et al. 2012). Amino acids can also be used indirectly as energy source, notably in the postprandial phase, as precursors for neoglucogenesis and ketogenesis.

In addition, and more recently, amino acids by themselves or after conversion in secondary metabolites have been associated with "signal functions" associated with metabolic and physiologic modifications. Just to take an example, L-arginine, after its enzymatic conversion by one isoform of nitric oxide synthase into nitric oxide (NO) and L-citrulline in endothelial cells, allows the action of NO which acts as a regulator of the arterial pressure. It is also feasible to consider that some amino acids, which are acting as neurotransmitter (like glutamate) or are precursor of neurotransmitter (like tryptophan which is a precursor of serotonine) or of biogenic amines (like histidine which is a precursor for histamine production), are amino acids with signaling functions.

Furthermore, an amino acid-like L-arginine is able to exert both metabolic regulatory (by activating the first step of ammonia detoxification in the urea cycle) and secretagogue functions (by stimulating for instance insulin and growth hormone secretion). Last but not least, the role of leucine as a stimulator of the protein synthesis through its action on the intracellular signaling pathways represents a very interesting example of the signaling functions of the amino acids (Yin et al. 2010; Li et al. 2011).

Proteins of all biological organisms, from the virus and bacteria to humans, are made up from the same set of 20 amino acids. Nineteen of these are α-amino acids with a primary amino group ($-NH_3^+$) and a carboxyl group ($-COO^-$) attached to a central carbon atom, aspartate ammonia-lyase, which is called the α-carbon atom (C_α) because it is adjacent to the carboxyl group (Fig. 7.1). The remaining groups attached to the C_α atom are a hydrogen atom (H) and a variable side chain or R' group. The one exception to this general structure is proline, a cyclic amino acid, which has a secondary amino group and is thus an α-imino acid. The only difference between any two amino acids is in their different side chains. Each side chain has

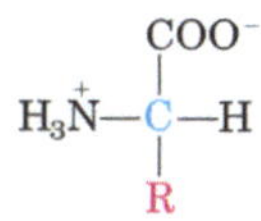

Fig. 7.1 General structure of an amino acid. Chemical structure of an amino acid showing the four different groups around the central α-carbon atom (with the exception of proline). The *R* group or side chain (*red*) attached to the α-carbon (*blue*) is different in each amino acid

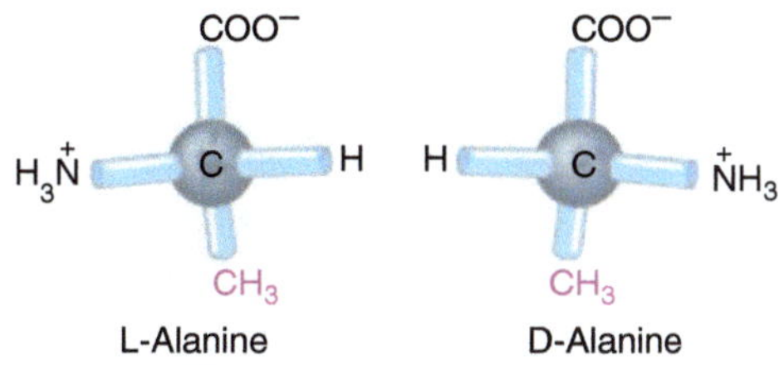

Fig. 7.2 Stereoisomerism in amino acids. The two stereoisomers of alanine, L- and D-alanine, are nonsuperimposable mirror images of each other (enantiomers)

distinct properties, including charge, hydrophobicity, and polarity. It is the arrangement of amino acids, with their distinct side chains, that is mainly responsible for protein structure and related functions.

All the 20 amino acids, except glycine (Gly or G), have four different groups arranged in tetrahedra around the central C_α atom which is thus known as an asymmetric center or chiral center and has the property of chirality. We note in Fig. 7.1 that the α-carbon is asymmetric, bonded to four different substituent groups: a carboxyl group, an amino group, an *R* group, and a hydrogen atom. Because of the tetrahedral arrangement of the bonding orbitals around the α-carbon atom of amino acids, the four different substituent groups can occupy two different arrangements in space, which are nonsuperimposable mirror images of each other. These two forms are called enantiomers or stereoisomers. Enantiomers are physically and chemically indistinguishable by most techniques, but can be distinguished on the basis of their different optical rotation of plane-polarized light. Molecules are classified as dextrorotatory (D; Greek *dextro* = right) or levorotatory (L; Greek *levo* = left) depending on whether they rotate the plane of plane-polarized light in a clockwise or an anticlockwise manner (Fig. 7.2). Surprisingly, only the L-amino acids are found in proteins, while D-isomers have been found only in small peptides of bacterial cell walls and in some peptide antibiotics.

Amino acids joined together by peptide bonds form the primary structure of a protein. The amino group of one molecule reacts with the carboxyl group of the other in a condensation reaction resulting in the elimination of water and the formation of a dipeptide (Fig. 7.3). A short sequence of amino acids is called a peptide, with the term polypeptide applied to longer chains of amino acids, usually of known sequence and length. When joined in a series of peptide bonds, amino acids are called "residues" to distinguish between the free form and the form found in protein. The peptide bond has a partial double-bond character.

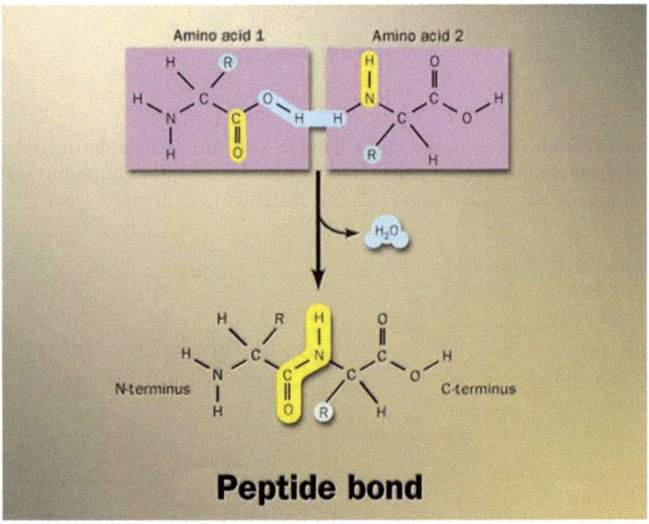

Fig. 7.3 Formation of a peptide bond. The peptide bond is a chemical, covalent bond formed between the α-amino group of one amino acid and the α-carboxyl group of another. Once two amino acids are joined together via a peptide bond to form a dipeptide, there is still a free amino group at one end and a free carboxyl group at the other, each of which can in turn be linked to further amino acids

7.2 Classification of Amino Acids

There are reasons to believe that amino acids are the oldest nutrients that have existed on earth. They have been used as the source of life and are sometimes called as the building blocks of life. There are as many as hundred thousand kinds of proteins that constitute the body, and these proteins are made with only 20 amino acids arranged in various combinations. The combination of the amino acids in proteins depends on the genetic code. In addition to the amino acids found in proteins, some amino acids not involved in the proteosynthesis are found in living organisms.

The names of the amino acids are often abbreviated, either to three letters or to a single letter. Thus, for example, alanine is abbreviated to Ala or A (Table. 7.1).

The hydrophobicity index informs on the relative hydrophobicity of amino acids. Higher positive value indicates stronger hydrophobicity. Hydrophilic amino acids have negative values. In a protein, hydrophobic amino acids are more likely to be located in the protein interior, whereas hydrophilic amino acids are more likely to face the aqueous environment.

7.2.1 Common Proteinogenic Amino Acids

Some of the 20 standard amino acids are called essential amino acids, because they cannot be synthesized by the body from other compounds through available metabolic pathways in the body, but instead must be provided by dietary sources.

Table. 7.1 Names, symbols, chemical structures, and hydrophobicity indices of the 20 amino acids found in proteins

Name	Symbol 3 Letter	Symbol 1 Letter	*R* group	Hydrophobicity
Aspartate	Asp	D		−3.5
Glutamate	Glu	E	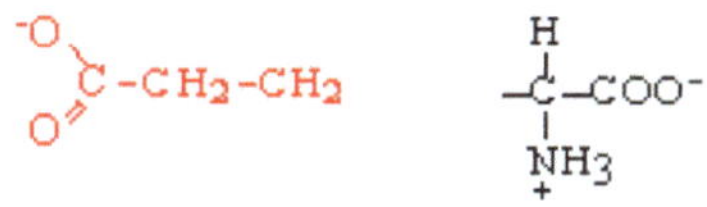	−3.5
Lysine	Lys	K	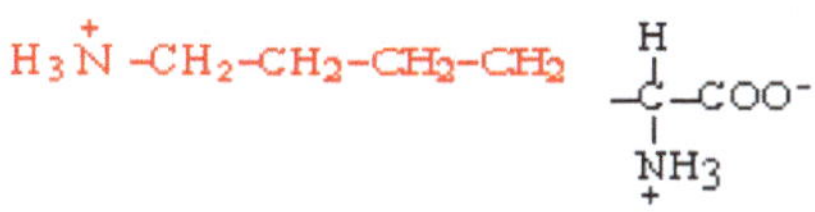	−3.9
Arginine	Arg	R	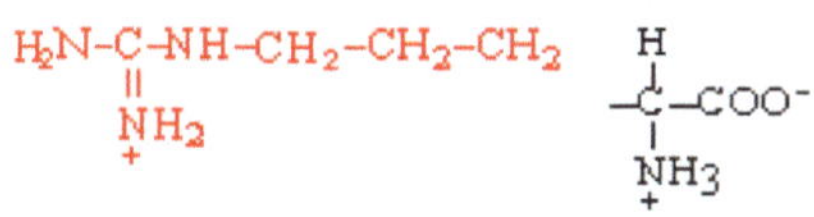	−4.5
Histidine	His	H	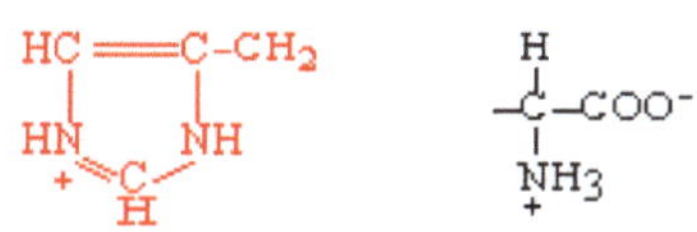	−3.2
Tyrosine	Tyr	Y	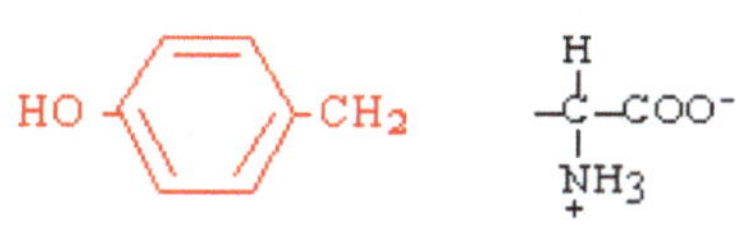	−1.3
Tryptophan	Trp	W	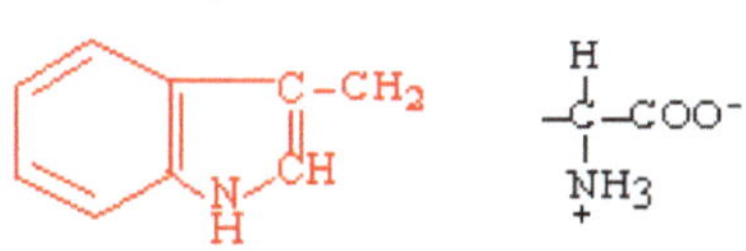	−0.9
Phenylalanine	Phe	F	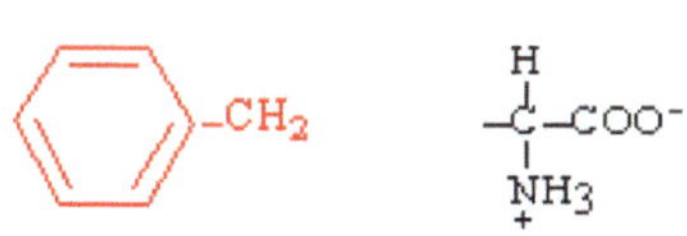	2.8
Cysteine	Cys	C		2.5

(continued)

Table. 7.1 (continued)

Name	Symbol 3 Letter	Symbol 1 Letter	*R* group	Hydrophobicity
Methionine	Met	M	$CH_3{-}S{-}CH_2{-}CH_2$ — H–C(–COO⁻)–NH_3^+	1.9
Serine	Ser	S	$HO{-}CH_2$ — H–C(–COO⁻)–NH_3^+	−0.8
Threonine	Thr	T	$CH_3{-}CH(OH)$ — H–C(–COO⁻)–NH_3^+	−0.7
Asparagine	Asn	N	$NH_2(O{=})C{-}CH_2$ — H–C(–COO⁻)–NH_3^+	−3.5
Glutamine	Gln	Q	$NH_2(O{=})C{-}CH_2{-}CH_2$ — H–C(–COO⁻)–NH_3^+	−3.5
Glycine	Gly	G	H — H–C(–COO⁻)–NH_3^+	−0.4
Alanine	Ala	A	CH_3 — H–C(–COO⁻)–NH_3^+	1.8
Valine	Val	V	$(CH_3)_2CH$ — H–C(–COO⁻)–NH_3^+	4.2
Leucine	Leu	L	$(CH_3)_2CH{-}CH_2$ — H–C(–COO⁻)–NH_3^+	3.8

(continued)

Table. 7.1 (continued)

Name	Symbol		*R* group	Hydrophobicity
	3 Letter	1 Letter		
Isoleucine	Ile	I		4.5
Proline	Pro	P		−1.6

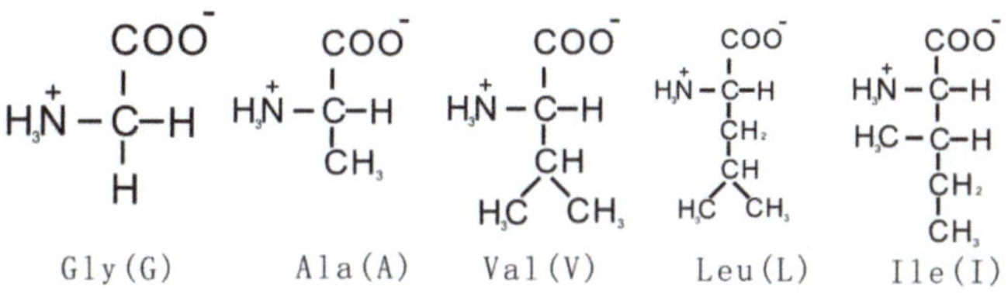

Fig. 7.4 Neutral aliphatic amino acids

In humans, the nine essential amino acids are lysine, leucine, isoleucine, methionine, phenylalanine, threonine, tryptophan, valine, and (in children) histidine. It is thus necessary to provide these essential amino acids from proteins in food in well-balanced, appropriate amounts and, in some cases, to provide some individual amino acids which are limiting for protein biosynthesis. Some amino acids are considered as conditionally indispensable since in some physiological situation (for instance in neonates) and in some pathological situations, the needs for some amino acids are increased to such an extent that endogenous synthesis capacity is not able to provide enough amino acids related to these needs.

Based on the chemical structure of the *R* groups, the 20 amino acids of proteins can be divided into aliphatic amino acids, aromatic amino acids, and heterocyclic amino acids, and among them, aliphatic amino acids are the most.

7.2.1.1 Aliphatic Amino acids

1. Neutral amino acids (Fig. 7.4):
   (a) Glycine (Gly or G): The smallest amino acid with the simplest structure, has an hydrogen atom in the side-chain position, and thus does not exist as a pair of stereoisomers since there are two identical groups (hydrogen atoms) attached to the C_α atom. In addition, Gly has minimal steric hindrance and in protein structure Gly offers the most flexibility.

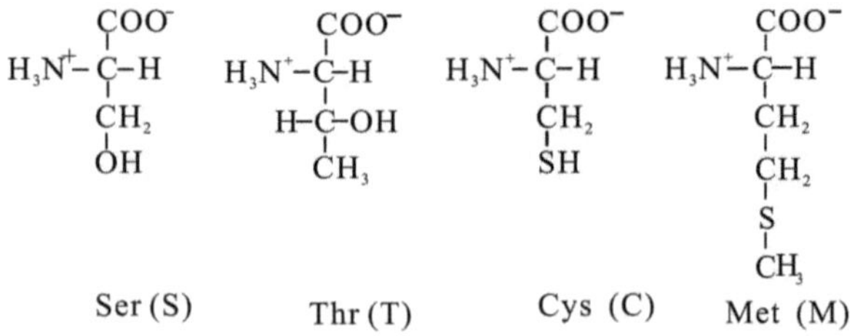

Fig. 7.5 Hydroxyl-containing or sulfur-containing amino acids

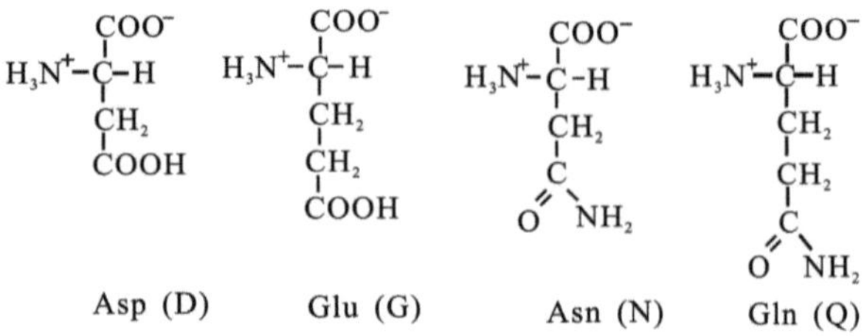

Fig. 7.6 Acidic amino acids and their amide derivatives

   (b) Alanine (Ala or A).
   (c) Valine (Val or V).
   (d) Leucine (Leu or L).
   (e) Isoleucine (Ile or I).

2. Hydroxyl-containing or sulfur-containing amino acids (Fig. 7.5):

   (a) Serine (Ser or S).
   (b) Threonine (Thr or T).
   (c) Cysteine (Cys or C): Capable of reacting with another cysteine to form a disulfide bond.
   (d) Methionine (Met or M).

3. Acidic amino acids and their amide derivatives (Fig. 7.6):

   (a) Aspartate (Asp or D).
   (b) Glutamate (Glu or E).
   Asp and Glu have carboxyl in their *R* groups. They have net negative charge at pH 7.0, thus usually named as aspartate and glutamate (conjugate base names, instead of aspartic acid and glutamic acid, unionized form).
   (c) Asparagine (Asn or N).
   (d) Glutamine (Gln or Q).

4. Basic amino acids (Fig. 7.7):

   (a) Lysine (Lys or K).
   (b) Arginine (Arg or R).

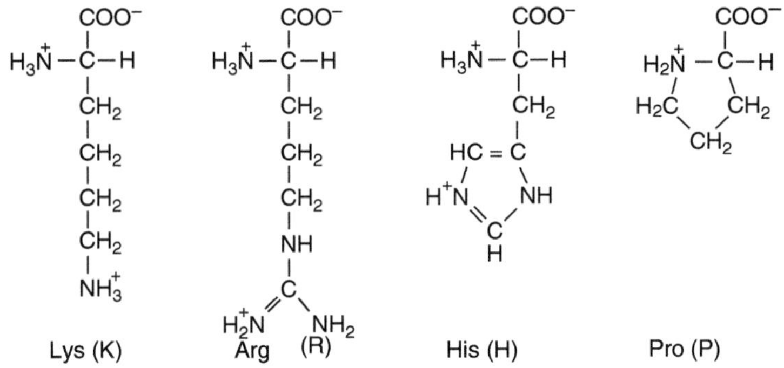

Fig. 7.7 Basic amino acids and heterocyclic amino acids

Fig. 7.8 Aromatic amino acids

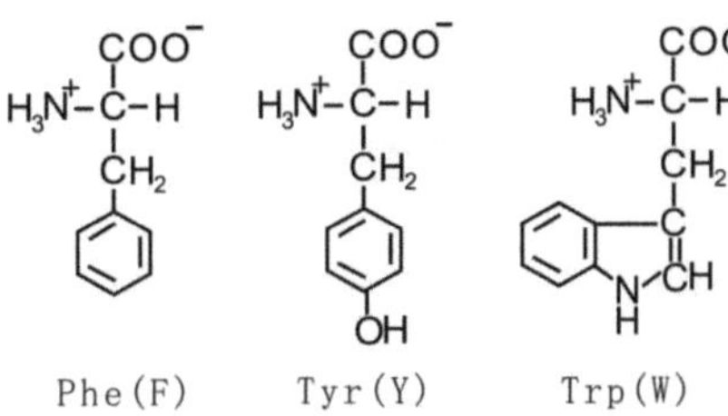

7.2.1.2 Aromatic Amino Acids (Fig. 7.8)

1. Phenylalanine (Phe or F): Which contains a benzene ring.
2. Tyrosine (Tyr or Y): Which contains a benzene ring.
3. Tryptophan (Trp or W): Which contains an indole ring.

They are jointly responsible for the light absorption of proteins at 280 nm, a property used as a measure of the concentration of proteins.

7.2.1.3 Heterocyclic Amino Acids (Fig. 7.7)

1. Histidine (His or H).
2. Proline (Pro or P): Has an imino group, instead of an amino group, forming a ring structure, being rigid in conformation. It is often found in the bends of folded protein chains and often present on the surface of proteins. It offers the least flexibility.

Based on the polarity of *R* groups, the 20 amino acids of proteins can be classified as follows: (1) nonpolar, (2) polar but uncharged, (3) positively charged, and (4) negatively charged (Table. 7.2).

Table 7.2 The 20 standard amino acids of proteins

Nonpolar amino acids	*Polar-uncharged amino acids*
Gly	Ser
Ala	Cys
Pro	Thr
Val	Asn
Leu	Gln
Ile	Tyr
Met	
Phe	
Trp	
Positively charged amino acids	*Negatively charged amino acids*
Lys	Asp
Arg	Glu
His	
Nonpolar amino acids	*Polar-uncharged amino acids*
Gly	Ser
Ala	Cys
Pro	Thr
Val	Asn
Leu	Gln
Ile	Tyr
Met	
Phe	
Trp	
Positively charged amino acids	*Charged amino acids*
Lys	Asp
Arg	Glu
His	

1. Nonpolar amino acids
   The *R* groups in this class of amino acids are nonpolar and hydrophobic. The bulky side chains of alanine, valine, leucine, and isoleucine, with their distinctive shapes, are important in promoting hydrophobic interactions within protein structures. Glycine has the simplest amino acid structure. Where it is present in a protein, the minimal steric hindrance of the glycine side chain allows much more structural flexibility than the other amino acids. Proline represents the opposite structural extreme. The secondary amino (imino) group is held in a rigid conformation that reduces the structural flexibility of the protein at that point. Methionine, one of the two sulfur-containing amino acids, has a nonpolar thioether group in its side chain. Phenylalanine and tryptophan, with their aromatic side chains, are relatively nonpolar (hydrophobic).

2. Polar-uncharged amino acids
   The *R* groups of these amino acids are more water soluble, or hydrophilic, than those of the nonpolar amino acids, because they contain functional groups that form hydrogen bonds with water. This class of amino acids includes serine, threonine, cysteine, asparagine, glutamine, and tyrosine. The polarity of serine, threonine, and tyrosine is contributed by their hydroxyl groups; that of cysteine and methionine by their sulfur atom; and that of asparagine and glutamine by their amide groups.
3. Positively charged amino acids
   The amino acids in which the *R* groups have significant positive charge at pH 7.0 are lysine, which has a second amino group at the e position on its aliphatic chain; arginine, which has a positively charged guanidino group; and histidine, containing an imidazole group. Histidine is the only standard amino acid having a side chain with a pK_a near neutrality. This group of amino acids is sometimes refereed as cationic amino acids.
4. Negatively charged amino acids
   The two amino acids having *R* groups with a net negative charge at pH 7.0 are aspartate and glutamate, each with a second carboxyl group. These amino acids are the parent compounds of asparagine and glutamine, respectively. This group of amino acids is sometimes referred as anionic amino acids.

7.2.2 Uncommon Proteinogenic Amino Acids

In addition to the 20 usual amino acids that are common in all proteins, other amino acids have been found as components of only certain types of proteins (Fig. 7.9). Each of these is derived from one of the 20 usual amino acids, in a modification reaction that occurs after the standard amino acid has been inserted into a protein. For example, 4-hydroxyproline, a derivative of proline, is found in the fibrous protein collagen of connective tissues and 4-hydroxylysine is found in plant cell-wall proteins. *N*-methyllysine is found in myosin, a contractile protein of muscle. Another important unusual amino acid is γ-carboxyglutamate, found in the blood-clotting protein prothrombin as well as in certain other proteins that bind Ca^{2+} for their biological function. More complicated is the nonstandard amino acid desmosine, a derivative of lysine residues, found in the fibrous protein elastin. Selenocysteine contains selenium rather than the oxygen of serine, and is found in glutathione peroxidase and a few other proteins.

7.2.3 Nonprotein Amino Acids

Some 300 additional amino acids have been found in cells and have a variety of functions but are not substituents of proteins (Fig. 7.10). Ornithine and citrulline which are also present in proteins deserve special attention because they are key intermediates in the biosynthesis of arginine and in the urea cycle.

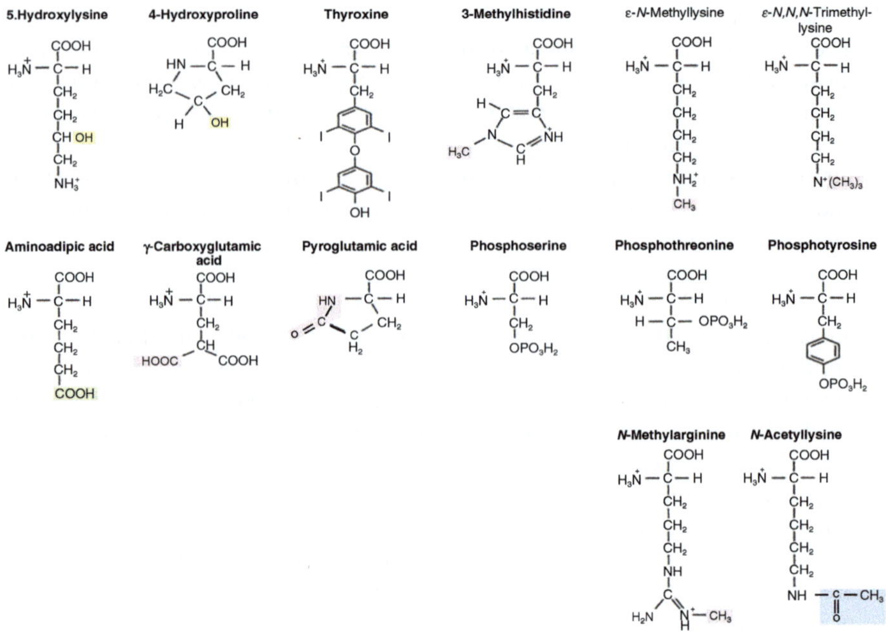

Fig. 7.9 Some of the uncommon proteinogenic amino acids

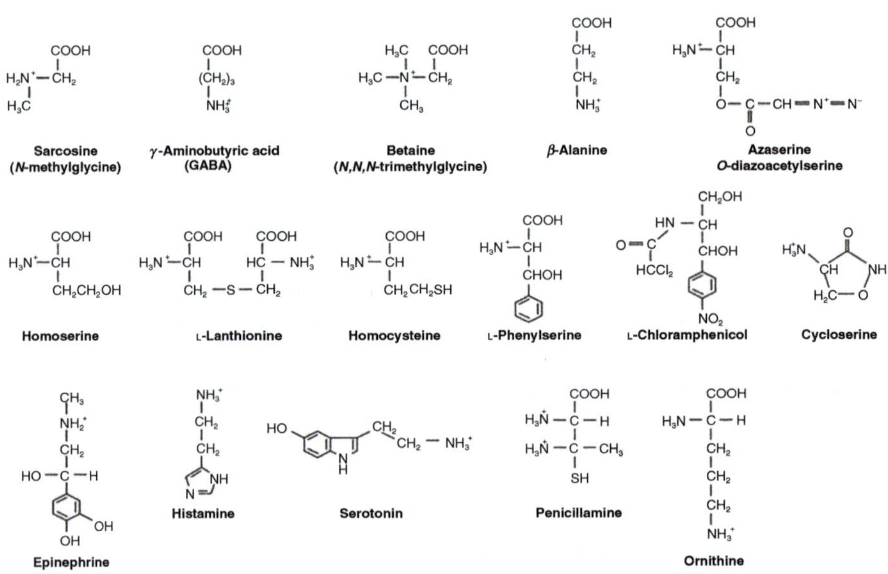

Fig. 7.10 Some of the amino acids and nitrogenous metabolites not found in proteins

7.3 Function of Amino Acids

The aim of this paragraph is to briefly recapitulate the metabolism of each amino acid together with their main biological functions.

7.3.1 Alanine

Alanine transaminase catalyzes the reversible conversion of L-alanine and alpha-ketoglutarate into pyruvate and glutamate. Alanine is a substrate for the neoglucogenesis.

7.3.2 Arginine

L-arginine is a precursor for biosynthesis of other amino acids present in proteins (glutamate and proline) or not present in proteins (ornithine, citrulline). Arginine is a precursor of urea, polyamines, creatine, and nitric oxide. Polyamines are involved in cell mitosis. Creatine is synthesized from arginine, glycine, and methionine but can also be provided by the diet. Creatine and creatine phosphate are mainly found in skeletal muscles. Phosphocreatine plays a central role in muscle contraction. Nitric oxide has been implicated in numerous functions, but has been initially shown to be produced in vascular endothelial cells and to be implicated in vasorelaxation. Arginine is used in urea cycle for ammonium detoxification.

7.3.3 Asparagine

L-asparagine can be converted into aspartate and ammonium.

7.3.4 Aspartic Acid

Aspartate transaminase is responsible for the reversible conversion of aspartate and alpha-ketoglutarate into oxaloacetate and glutamate. Aspartate is involved in the urea cycle for the conversion of citrulline into argininosuccinate. Aspartate is a precursor for pyrimidine synthesis.

7.3.5 Cysteine

Cysteine can be produced from the metabolic conversion of methionine and serine. Cysteine is a precursor for taurine biosynthesis and is one of the amino acids which are precursors for glutathione and hydrogen sulfide (H_2S). Cysteine is a precursor for pyruvate synthesis and for the synthesis of the nucleotidic coenzyme CoA. Taurine is a beta-amino sulfonic acid which is not incorporated into proteins. The circulating concentrations of taurine are much dependent on the dietary intake (Laidlaw et al. 1998). Taurine functions are numerous. This amino acid is involved in the synthesis of conjugated biliary acids. Taurine has been implicated as a regulator of the cellular volume, and as such, as an agent with cytoprotective effect. A protective role of taurine against deleterious effects of the "oxidative stress" has been proposed and may be due to the reaction between taurine and hypochlorous acid (Schuller-Levis and Park, 2004). Glutathione is a tripeptide which is synthesized in cells from glutamate, cysteine, and glycine. Circulating glutathione is usually micromolar (Jones et al. 2000). This compound from alimentary origin can be absorbed and the jejunum represents the main site of absorption (Hagen et al. 1990). Glutathione can exist in cells both in reduced (GSH) and oxidized form (GSSG) with reduced glutathione being much prominent. The GSH/GSSG couple represents the main component fixing the intracellular redox state. In mammals, the amount of cysteine used for glutathione synthesis is very high since it represents as much as 30–50% of the total body utilization (Fukagawa et al. 1996). Glutathione is involved in several physiological functions including the elimination of electrophilic compounds, control of intracellular reactive oxygen species, etc.

7.3.6 Glutamic Acid

Glutamic acid is metabolized in transamination pathways. This amino acid can be converted in the presence of pyruvate into alpha-ketoglutarate and alanine, and in the presence of oxaloacetate into aspartate and alpha-ketoglutarate. Glutamic acid can also undergo conversion to alpha-ketoglutarate and ammonium. As such, glutamate is an oxidative substrate in several cell types. Glutamic acid can be synthesized from other amino acids like arginine, proline, and histidine. Glutamine is a precursor of glutamate and ammonium. Inversely, glutamate is a precursor for glutamine synthesis. Lastly, glutamate is one of the amino acid precursors for glutathione synthesis (Wang et al. 2008).

7.3.7 *Glutamine*

Glutamine is metabolized into glutamate, ammonium, alanine, aspartate, citrulline and proline. Glutamine is a major oxidative substrate in numerous cell types. Glutamine is a nitrogen donor for the biosynthesis of nucleotides.

7.3.8 *Glycine*

Glycine is a precursor for the biosynthesis of serine and for the synthesis of glutathione, creatine, purines, and heme in hemoglobin.

7.3.9 *Histidine*

Histidine is abundant in hemoglobin. Histidine is a precursor of glutamate and histamine. This latter compound is involved in allergy and in inflammatory reactions.

7.3.10 *Isoleucine, Leucine, and Valine*

The amino acids with branched chain undergo the transamination in the presence of alpha-ketoglutarate allowing the synthesis of glutamate and of the corresponding alpha-ketoacids (alpha-ketoisocaproate from leucine, alpha-keto-beta-methyl-valerate from isoleucine, and alpha-ketoisovalerate from valine). Each of these ketoacids undergoes then an oxidative decarboxylation by a ketoacid dehydrogenase (BCKAD). The products of decarboxylation undergo then several metabolic steps leading to the synthesis of acetyl-CoA, acetoacetic acid, propionyl-CoA, and succinyl CoA. It is well known that branched-chain amino acids can be in competition with other amino acids (notably tryptophan and tyrosine) for transport through cellular membranes (Anderson and Johnston 1983). Although branched-chain amino acids are not direct precursors for neurotransmitter synthesis, they can affect the transport of other amino acids across the blood–brain barrier, and then the concentrations of some neurotransmitters in the central nervous system (Fernstrom, 1973).

7.3.11 *Lysine*

Lysine is a precursor for acetoacetyl-CoA, carnitine, and cadaverine. Carnitine is a metabolite with important functions in the intermediary metabolism. Carnitine is involved in the transport of fatty acids with long chain from the cytosol to the mitochondrial matrix where the beta-oxidation occurs. In the body, carnitine is originating mainly from dietary source, with endogenous synthesis playing a minor role (Vaz and Wanders, 2002). Carnitine is mainly synthesized in the liver from lysine and methionine. Cadaverine is a polyamine produced by the intestinal microbiota.

7.3.12 *Methionine*

Methionine is a precursor of succinyl-CoA, homocysteine, cysteine, creatine, and carnitine. In addition, methionine is a precursor for S-adenosyl-methionine synthesis which is involved in polyamine, creatine, and phosphatidylcholine (one of the membrane phospholipids) metabolism.

7.3.13 *Phenylalanine*

This amino acid is a precursor for the synthesis of tyrosine and acetoacetyl-CoA.

7.3.14 *Proline*

Proline is a precursor for glutamate synthesis in several cell types. Proline can undergo hydroxylation leading to the production of hydroxyproline. Proline and hydroxyproline are abundant in collagens.

7.3.15 *Serine*

Serine is one of the precursors for the biosynthesis of cysteine and for the synthesis of the sphingolipid sphingosine. This amino acid can also be metabolized into pyruvate.

7.3.16 *Threonine*

This amino acid is a precursor for the synthesis of glycine and acetyl-CoA. This amino acid is very abundant in mucins (Wang et al. 2007).

7.3.17 *Tryptophan*

Tryptophan is a precursor of numerous metabolites including the neurotransmitter serotonine (5-hydroxytryptamine), tryptamine, the hormone melatonin, and niacin (vitamin B3). In addition, tryptophan is a precursor for the synthesis of acetyl-CoA (Yao et al. 2011).

7.3.18 *Tyrosine*

Tyrosine is a precursor for active substances including the neurotransmitters epinephrine and norepinephrine, the pigment melanin, and the hormone thyroxine. Tyrosine is also a precursor for the production of acetoacetyl-CoA.

It is worth to note that although metabolites derived from amino acids play major role in the body physiology, the metabolic pathways involved in their synthesis represent a very minor utilization of the amino acid precursors when compared with the utilization of the amino acids for protein synthesis.

References

Anderson GH, Johnston JL (1983) Nutrient control of brain neurotransmitter synthesis and function. Can J Physiol Pharmacol 61:271–281

Fernstrom JD (1973) Correlations between brain tryptophan and plasma neutral amino acid levels following food consumption in rats. Life Sci 13:517–524

Fukagawa NK, Ajami AM, Young VR (1996) Plasma methionine and cysteine kinetics in response to an intravenous glutathione infusion in adult humans. Am J Physiol 270:E209–E214

Hagen TM, Wierzbicka AH, Sillau AM, Bowmann BB et al (1990) Bioavailability of dietary glutathione: effects on plasma concentration. Am J Physiol 259:G524–G529

Jakobs BS, Wanders RJ (1995) Fatty acid beta-oxidation in peroxisomes and mitochondria: the first, unequivocal evidence for the involvement of carnitine in shuttling propionyl-CoA from peroxisomes to mitochondria. Biochem Biophys Res Commun 213:1035–1041

Jones DP, Carlson JL, Mody VC, Cai J et al (2000) Redox state of glutathione in human plasma. Free Radic Biol Med 28:625–635

Kim SW, Mateo RD, Yin YL, Wu GY (2007) Functional amino acids and fatty acids for enhancing production performance of sows and piglets. Asian-Australas J Anim Sci 20:295–306

Laidlaw SA, Shultz TD, Cecchino JT, Kopple JD (1998) Plasma and urine taurine levels in vegans. Am J Clin Nutr 47:660–663

Li P, Yin YL, Li DF, Kim SW, Wu GY (2007) Amino acids and immune function. Br J Nutr 98:237–252

Li F, Yin Y, Tan B, Kong X, Guoyao W (2011b) Leucine nutrition in animals and humans: mTOR signaling 3 and beyond. Amino Acids 41:1185–1193

Schuller-Levis GB, Park E (2004) Taurine and its chloramines: modulators of immunity. Neurochem Res 29:117–126

Vaz FM, Wanders RJ (2002) Carnitine biosynthesis in mammals. Biochem J 361:417–429

Wang X, Qiao SY, Yin YL, Yu LG, Wang ZY, Wu GY (2007) A deficiency or excess of dietary threonine reduces protein synthesis in jejunum and skeletal muscle of young pigs. J Nutr 137:1442–1446

Wang JJ, Chen LX, Li P, Li XL, Zhou HJ, Wang FL, Li DF, Yin YL, Wu GY (2008) Gene expression is altered in piglet small intestine by weaning and dietary glutamine supplementation. J Nutr 138:1025–1032

Wu GY, Bazer FW, Davis TA, Johnson GA, Kim SW, Knabe DA, Spencer TE, Yin YL (2007) Important roles for arginine-family amino acids in swine nutrition and production. Livest Sci 122:8–22

Yao K, Fang J, Yin YL, Feng ZM, Tang ZR, Wu GY (2011) Tryptophan metabolism in animals: important roles in nutrition and health. Front Biosci S3:286–297

Yao K, Wang L, Ding BY, Fu DB, Liu YL, Zhu HL, Liu J, Li YT, Kang P, Yin YL, Wu GY, Hou YQ (2012) Alpha-ketoglutarate inhibits glutamine degradation and enhances protein synthesis in intestinal porcine epithelial cells. Amino Acids 42(6):2491–2500

Yin YL, Yao K, Liu ZJ, Gong M, Ruan Z, Deng D, Tan BE, Liu ZQ, Wu GY (2010) Supplementing L-leucine to a low-protein diet increases tissue protein synthesis in weanling pigs. Amino Acids 39:1477–1486

Chapter 8
Synthesis and Degradation of Proteins in Pigs

Fugui Yin, Yulong Yin, and Yongqing Hou

Contents

8.1 Introduction

Growth of animals is the complex result of competition between anabolic and catabolic process, which implies constant changes and remodeling through synthesis of new proteins and breakdown of existing proteins (Jobgen et al. 2006; Tan et al. 2009). Together, these processes are called protein turnover and produce muscle growth or hypertrophy when synthesis is greater than breakdown and muscle wasting when synthesis is less than breakdown (Norton and Layman 2006). Protein turnover requires large amounts of ATP. However, this costly metabolic cycle fulfills key obligatory

---

F. Yin • Y. Yin (✉)
Chinese Academy of Sciences, and Ministry of Agriculture, Observation and Experiment Station of Animal Nutrition and Feed Science in South-Central China; Hunan Provincial Engineering Research Center for Healthy Livestock and Poultry Production; Key Laboratory of Agri-Ecological Processes in Subtropical Region, Institute of Subtropical Agriculture, Changsha, Hunan 410125, People's Republic of China
e-mail: yinyulong@isa.ac.cn

Y. Hou
Hubei Key Laboratory of Animal Nutrition and Feed Science, Wuhan Polytechnic University, Wuhan 430023, China

F. Blachier et al. (eds.), *Nutritional and Physiological Functions of Amino Acids in Pigs*,
DOI 10.1007/978-3-7091-1328-8_8, © Springer-Verlag Wien 2013

functions, including protein homeostasis, cell turnover, removal of aged and damaged proteins, synthesis of new proteins like heat-shock and immunological proteins, gluconeogenesis from amino acids, wound healing, tissue repair, adaptation to nutritional and pathological alterations, and immune responses (Wu 2009). In pigs, although the protein syntheses increase in all tissues, the greatest response occurs in skeletal muscle in response to feeding stimulation. The elevated postprandial protein synthesis in skeletal muscle of pigs therefore increases the protein deposition during the post-absorptive period allowing growth and development. A sharp increase of circulating glucose, insulin, and amino acids, especially some nutritional indispensable amino acids, is observed after meal (Yin et al. 2010, 2011; Kong et al. 2012), in accordance with higher protein deposition in pigs (Drew et al. 2012). The mechanism responsible for the stimulation of protein synthesis by feeding was therefore focused on the roles of postprandial circulating glucose, insulin, and amino acids (Kong et al. 2012). The protein degradation is a process where proteins are broken into smaller peptides as well as free amino acids, the latter being either reused for new protein synthesis or further degraded into several metabolites, several of them being able to generate ATP for energy use. In addition, the protein degradation plays important roles in animal physiological process, especially in the cellular signal transduction system as well as in the maintenance of the integrity of the proper folded state of protein.

8.2 Protein Syntheses in Pigs

Protein biosynthesis is the process in which cells build or manufacture proteins. This process which refers to a multistep pathway begins with amino acid synthesis and transcription of nuclear DNA into messenger RNA, which are then used as inputs for translation. In pigs, the neonatal period is characterized by rapid growth and elevated rates of protein synthesis, a phenomenon which decreases sharply with age. The declined efficiency of protein synthesis and deposition with age in the whole animal depends on two factors: (1) the energy cost of protein synthesis and (2) the extent by which the efficiency of protein synthesis in muscle is done (Bates and Millward 1981). The elevated capacity for protein synthesis in the young pig is driven by the high ribosomal content together with the increased efficiency of the translation process, allowing accelerated protein synthesis rates (Davis et al. 2008). In accordance with the elevated protein synthesis rate, the deposition of protein therefore is increased. Protein degradation is modest during early life. Hormonal and some nutritional factors, such as glucose and amino acids, are thought to be the most potent signal molecules that regulate the postnatal muscle protein synthesis. However, the feeding can be considered as a basic and active stimulator of muscle protein synthesis in neonatal pigs even if the response to such a stimulation tends to decrease with age. The feeding-induced stimulation of muscle protein synthesis is modulated by an enhanced sensitivity to the postprandial increase in insulin and amino acids. The developmental decline in response to insulin and amino acids parallels a marked decrease in the feeding-induced activation of the translation initiation factors that regulate the binding of mRNA to the 40S ribosomal complex.

The abundance and activation of many known positive regulators of the nutrient- and insulin-signaling pathways that are involved in translation initiation are high, whereas those of many negative regulators are low in skeletal muscle of younger pigs (Davis et al. 2008).

8.2.1 Signaling Pathways in Response to Protein Synthesis

The mammal target of rapamycin (mTOR) is a major protein kinase that modulates translation initiation components. The phosphorylation of mTOR on Ser^{2448} activates the kinase and is stimulated by both insulin and amino acids (Fig. 8.1) (O'Connor et al. 2003a, b; Davis 2008). The mTOR system consists of (1) rapamycin-sensitive complex 1 (mTOR1, mTOR raptor (regulatory associated protein of TOR), and G protein b-subunit-like protein); and (2) rapamycin-insensitive complex 2 (mTOR2) (mTOR rictor (rapamycin-insensitive companion of TOR), mitogen-activated-protein kinase-associated protein 1, and G protein b-subunit-like protein) (Wu 2009). These two complexes are structurally and functionally distinct in cells. The mTOR1 regulates mRNA translation by phosphorylation of the 70-kDa S6 kinase-1 (S6K1) and eIF4E-binding protein 1 (4E-BP1). S6K1 activation results in phosphorylation of the ribosomal subunit, ribosomal protein S6, which may lead to an increase in the translation of mRNAs that encode proteins found in the protein synthetic machinery (Kimball and Jefferson 2004). Phosphorylation of the repressor protein, 4E-BP1, permits the dissociation of eIF4E, allowing it to bind to eIF4G. This active complex of eIF4E.eIF4G mediates the binding of mRNA to the 40S ribosomal complex in the initiation of mRNA translation (Davis et al. 2000; Kimball et al. 2000; Suryawan et al. 2001). In addition to mRNA binding to the ribosome, translation initiation requires initiator methionyl-tRNA (met-tRNAi) binding to the start codon, a step mediated by eIF2 (Wilson et al. 2009). In its GTP form, eIF2 binds the ribosome and locates the mRNA start site, causing eIF2 hydrolysis and disassociation; eIF2B then recharges eIF2 with a GTP (Wek et al. 2006). The phosphorylation of the α-subunit of eIF2 transforms it into a competitive inhibitor of eIF2B, preventing activation of eIF2 and, consequently, reducing met-tRNAi binding to the ribosome (Bush et al. 2003; O'Connor et al. 2004). The mTOR2 phosphorylates protein kinase B/Akt and may function to regulate cell proliferation, differentiation, migration, and cytoskeletal reorganization (Sarbassov et al. 2005).

8.2.2 Feeding Induces Protein Syntheses in Pig

The feeding causes complex physiological and biochemical responses in animals. The circulating hormones (e.g., insulin) and nutritional metabolites (e.g., glucose and amino acids) are relatively elevated even with no food intake in pigs, and therefore ameliorate the whole-body nutritional metabolic status. Although with

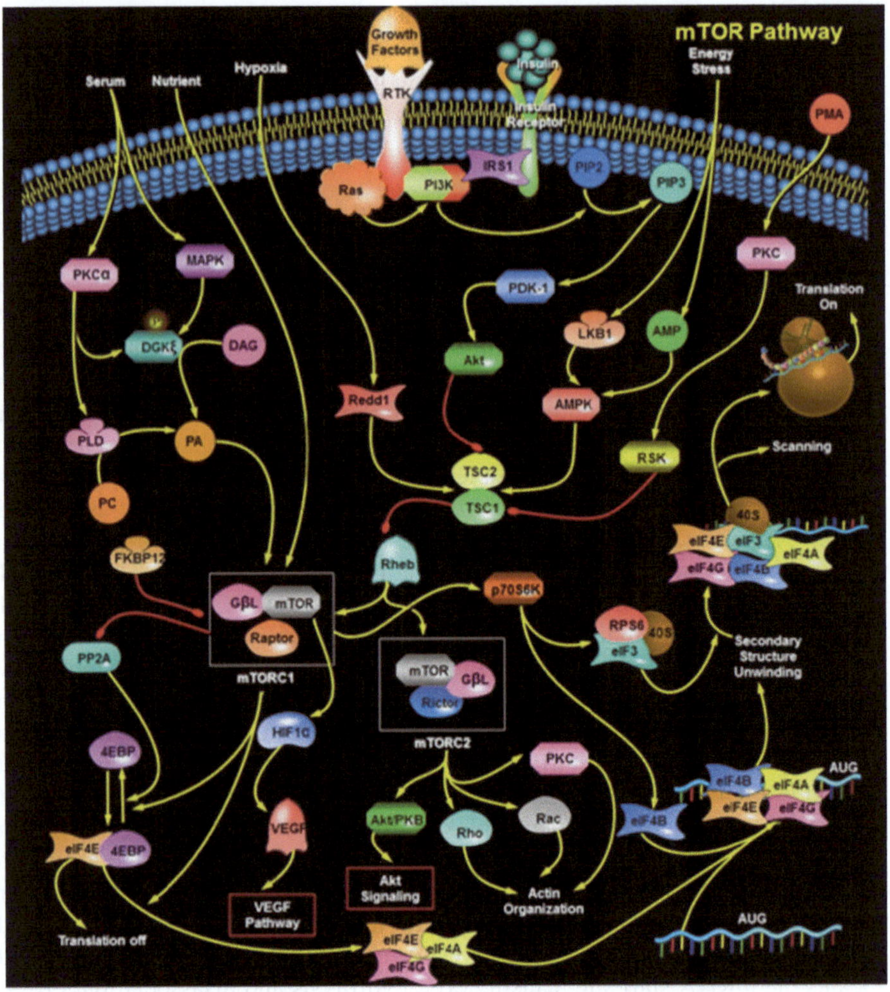

Fig. 8.1 Insulin- and nutrition (main are amino acid and glucose)-signaling pathway leading to the stimulation of translation initiation. Activation of the insulin-signaling pathway is initiated by the binding of insulin to its receptor. This activates the insulin receptor and insulin receptor substrate-1 (IRS-1), followed by the activation of phosphoinositide-3 kinase (PI 3-K). Activated PI 3-K then stimulates the activation of phosphoinositide-dependent kinase 1 (PDK-1) and protein kinase B (PKB). Phosphorylation of PKB inactivates tuberous sclerosis complex 1/2 (TSC1/2), thereby inducing the activation of mammalian target of rapamycin (mTOR). Amino acids as well as insulin can activate mTOR, which exists in a complex with raptor and G protein β-subunit-like protein (GβL). Activated mTOR phosphorylates ribosomal protein S6 kinase 1 (S6K1) and eukaryotic initiation factor (eIF) 4E-binding protein-1 (4EBP1). Phosphorylation of S6K1 enhances the activation of ribosomal subunit S6 (rpS6), which increases the translation of specific mRNA. Phosphorylated 4EBP1 releases eIF4E from an inactive eIF4E–4EBP1 complex, allowing the formation of the active eIF4G–eIF4E complex, which mediates the binding of mRNA to the 43S ribosomal complex. Insulin signaling can be attenuated by protein tyrosine phosphatase-1B (PTP-1B), which dephosphorylates the insulin receptor and IRS-1; phosphatase and tensin

some differences in response to different dietary treatments (Deng et al. 2009; Deng et al. 2010), the postprandial circulating concentrations of glucose, amino acids, and insulin raised and are maintained at relatively high levels for 4 h after the meal and then returned to the baseline whatever pigs consumed their daily allowance under the feeding procedures of "six-time intake per day" (Yin et al. 2010; 2011; Deng et al. 2009; Deng et al. 2010), "three-time intake per day" (Liu et al. 2007), or "two-time intake per day" (Regmi et al. 2010) (Tables 8.1 and 8.2, Figs. 8.2 and 8.3). Therefore the whole-body fractional protein syntheses rate may be higher within the postprandial 4 h than during the time from postprandial 4th hour till next meal. In young pigs, the feeding stimulates protein synthesis in most tissues, and the efficiency by which dietary amino acids are used for protein deposition is high (Burrin et al. 1992). Notably, the most profound postprandial increase in protein synthesis occurs in skeletal muscle, which is of high metabolic significance, because skeletal muscle comprises a large proportion of the body mass in pigs. Feeding promotes the protein synthesis largely through glucose-, amino acid-, or insulin-induced formation of the active eukaryotic translation initiation factor (eIF) 4E.eIF4G complex or mTOR pathways (Kimball et al. 2000; Avruch et al. 2006; Jeyapalan et al. 2007; Suryawan et al. 2007). The phosphorylation of mTOR peaked within 30 min after starting the feeding, followed by the phosphorylation of the downstream targets of mTORC1, 4E-BP1, eIF4G, and S6 (Fig. 8.4), and then regulation of the binding of the mRNA with the 43S pre-initiation complex (Wilson et al 2009). These physiological and biochemical processes paralleled the stimulation of skeletal muscle protein synthesis and led to the enhancement of protein synthesis in pigs. In addition, the rate of translation initiation is up-regulated earlier than elongation after feeding. Although the elongation is also increased after meal consumption, this increase is delayed relative to the change in initiation. Interestingly, increased rates of elongation did not further increase muscle protein synthesis in response to feeding, which therefore indicates that initiation is the rate-limiting step in translation (Wilson et al. 2009). Therefore, the increased rates of initiation, but not elongation, are responsible for the increase in muscle protein synthesis in response to feeding in pigs.

8.2.3 Glucose, Amino Acids, and Insulin Stimulate Protein Syntheses in Pig

By using an insulin–amino acid–glucose clamp technique, the protein synthesis in skeletal muscle of young pigs has been found to be stimulated by glucose, insulin,

Fig. 8.1 (continued) homologue deleted on chromosome 10 (PTEN), which inactivates PI 3-K; and protein phosphatase 2A (PP2A), which acts on PKB and S6K1. An increase in adenosine monophosphate (AMP) levels enhances AMP kinase activation, resulting in the activation of TSC1/2 complex and the decreased activation of mTOR (Davis et al. 2008)

Table 8.1 Serum amino acid concentrations after first meal in response to different dietary treatment in pigs (mmol/L)

Item	Sampling time 8:30 HDRS	8:30 MDRS	8:30 LDRS	9:50 HDRS	9:50 MDRS	9:50 LDRS	10:50 HDRS	10:50 MDRS	10:50 LDRS	11:50 HDRS	11:50 MDRS	11:50 LDRS	Pooled SEM	Time effect, *P* value
Nutritionally indispensable amino acids														
Arginine	0.089	0.087	0.083	0.179^{a}	0.128^{b}	0.094^{b}	0.163^{a}	0.107^{b}	0.084^{c}	0.070	0.068	0.064	0.004	<0.001
Cysteine	0.055^{a}	$0.045^{a,b}$	0.038^{b}	0.086^{a}	0.064^{b}	0.068b	0.059	0.045	0.046	0.037	0.034	0.034	0.001	<0.001
Histidine	0.038	0.036	0.032	0.062^{a}	0.041^{b}	0.045^{b}	0.038	0.032	0.044	0.032	0.028	0.040	0.001	0.018
Isoleucine	0.050	0.047	0.048	0.090^{a}	0.055^{b}	0.060^{b}	0.061^{a}	$0.050^{a,b}$	0.048^{b}	0.049^{a}	0.034^{b}	$0.038^{a,b}$	0.001	0.001
Leucine	0.205^{a}	0.215^{a}	0.173^{b}	0.310^{a}	0.260^{b}	0.245^{b}	0.271^{a}	$0.254^{a,b}$	0.204^{b}	0.148	0.154	0.142	0.004	<0.001
Lysine	0.287^{a}	$0.261^{a,b}$	0.233^{b}	0.583^{a}	0.574^{a}	0.297^{b}	0.320^{a}	0.233^{b}	0.227^{b}	0.242^{a}	0.220^{a}	0.158^{b}	0.009	0.004
Methionine	0.037	0.031	0.031	0.055^{a}	0.038^{b}	0.033^{b}	0.061^{a}	0.034^{b}	0.036^{b}	0.040^{a}	0.029^{b}	0.028^{b}	0.002	0.004
Phenylalanine	0.085	0.089	0.083	0.137^{a}	0.116^{b}	0.102^{b}	0.120^{a}	0.100^{b}	0.103^{b}	0.082	0.071	0.074	0.002	0.001
Threonine	0.452^{a}	0.380^{b}	0.375^{b}	0.525^{a}	0.422^{c}	0.460^{b}	0.384^{a}	$0.342^{a,b}$	0.308^{b}	0.285^{a}	0.294^{a}	0.247^{b}	0.010	<0.001
Tryptophan	0.235^{a}	0.197^{b}	0.183^{b}	0.288^{a}	0.237^{b}	0.223^{b}	0.246^{a}	0.213^{b}	0.200^{b}	0.200^{a}	0.166^{b}	0.143^{b}	0.007	0.009
Tyrosine	0.104^{a}	$0.093^{a,b}$	0.084^{b}	0.164^{a}	0.118^{b}	0.117^{b}	0.152^{a}	0.114^{b}	0.093^{b}	0.101^{a}	$0.076^{a,b}$	0.068^{b}	0.002	<0.001
Valine	0.112	0.094	0.100	0.166^{a}	0.098^{c}	0.129^{b}	0.138^{a}	0.101^{b}	0.100^{b}	0.121^{a}	0.074^{b}	0.087^{b}	0.002	0.011
Nutritionally dispensable amino acids														
Alanine	0.387^{a}	0.334^{b}	$0.350^{a,b}$	0.450^{a}	0.389^{b}	0.384^{b}	0.476^{a}	$0.465^{a,b}$	0.414^{b}	0.406^{a}	0.390^{a}	0.366^{b}	0.011	0.001
Aspartate	0.025	0.022	0.028	0.030	0.032	0.030	0.049	0.042	0.042	0.034^{a}	0.026^{b}	0.022^{b}	0.001	0.001
Glutamate	0.241	0.227	0.226	0.248^{b}	$0.278^{a,b}$	0.298^{a}	0.324^{a}	$0.318^{a,b}$	0.279^{b}	0.310^{a}	0.205^{c}	0.260^{b}	0.007	0.010
Glycine	0.463^{a}	$0.432^{a,b}$	0.410^{b}	0.541^{a}	$0.520^{a,b}$	0.487^{b}	0.549^{a}	$0.545^{a,b}$	0.522^{b}	0.471^{a}	0.473^{a}	0.427^{b}	0.015	0.014
Proline	0.337^{a}	0.359^{a}	0.256^{b}	0.623^{a}	0.577^{b}	0.555^{b}	0.541^{a}	0.492^{b}	0.435^{c}	0.205	0.232	0.233	0.012	<0.001
Serine	0.105^{a}	$0.095^{a,b}$	0.084^{b}	0.117	0.124	0.122	$0.089^{a,b}$	0.095^{a}	0.080^{b}	0.088	0.084	0.071	0.001	<0.001

[a–c]Within the same sampling time, values in a row sharing different superscript letters differ ($P < 0.05$); $n = 6$

HDRS-fed group high digestion rate starch-fed group, *LDRS* low digestion rate starch-fed group, *MDRS* moderate digestion rate starch, *SEM* standard error of the mean

Source: Yin et al. (2010) (with minor modification, the pigs were fed at 8:00 am)

Table 8.2 Serum amino acid concentrations after second feeding in response to different dietary treatment in pigs (mmol/L, continued from Table 8.1)

	Sampling time													
	12:30			13:30			14:30			15:30				Time effect,
Item	HDRS	MDRS	LDRS	HDRS	MDRS	LDRS	HDRS	MDRS	LDRS	HDRS	MDRS	LDRS	Pooled SEM	*P* value
Nutritionally indispensable amino acids														
Arginine	0.112^{a}	0.079^{b}	0.100^{a,b}	0.181^{a}	0.132^{b}	0.128^{b}	0.141^{a}	0.115^{a,b}	0.080^{b}	0.074	0.065	0.069	0.004	<0.001
Cysteine	0.036^{a}	0.029^{b}	0.036^{a}	0.056^{a}	0.043^{b}	0.041^{b}	0.037	0.034	0.034	0.034	0.033	0.030	0.001	<0.001
Histidine	0.051	0.043	0.046	0.063^{a}	0.041^{b}	0.049^{b}	0.041	0.042	0.050	0.037	0.031	0.040	0.001	0.018
Isoleucine	0.057^{a}	0.037^{b}	0.056^{a}	0.090^{a}	0.055^{b}	0.060^{b}	0.066	0.056	0.047	0.049^{a}	0.036^{b}	0.041^{a,b}	0.001	0.001
Leucine	0.192	0.184	0.177	0.271^{a}	0.278^{a}	0.204^{b}	0.213^{b}	0.254^{a}	0.208^{b}	0.159	0.179	0.167	0.004	<0.001
Lysine	0.302^{a}	0.272^{b}	0.234^{b}	0.583^{a}	0.574^{a}	0.297^{b}	0.491^{a}	0.497^{a}	0.222^{b}	0.276	0.263	0.269	0.009	0.004
Methionine	0.044^{a}	0.032^{b}	0.034^{b}	0.055^{a}	0.038^{b}	0.033^{b}	0.046^{a}	0.037^{a,b}	0.032^{b}	0.039^{a}	0.030^{b}	0.030^{b}	0.002	0.004
Phenylalanine	0.096	0.087	0.096	0.133^{a}	0.100^{b}	0.103^{b}	0.105^{a}	0.100^{a,b}	0.088^{b}	0.087	0.077	0.082	0.002	0.001
Threonine	0.392^{a}	0.316^{b}	0.345^{a,b}	0.452^{a}	0.380^{b}	0.375^{b}	0.374^{a}	0.411^{a}	0.126^{b}	0.276^{a,b}	0.326^{a}	0.257^{b}	0.010	<0.001
Tryptophan	0.243^{a}	0.192^{b}	0.171^{b}	0.280^{a}	0.226^{b}	0.223^{b}	0.252^{a}	0.217^{b}	0.208^{b}	0.190^{a}	0.156^{b}	0.147^{b}	0.007	0.009
Tyrosine	0.128^{a}	0.090^{b}	0.094^{b}	0.164^{a}	0.114^{b}	0.093^{b}	0.146^{a}	0.116^{a,b}	0.092^{b}	0.112^{a}	0.083^{a,b}	0.074^{b}	0.002	<0.001
Valine	0.127^{a}	0.082^{b}	0.117^{a}	0.166^{a}	0.098^{c}	0.129^{b}	0.138^{a}	0.100^{b}	0.101^{b}	0.120^{a}	0.074^{b}	0.089^{a,b}	0.002	0.011
Nutritionally dispensable amino acids														
Alanine	0.473^{a}	0.434^{b}	0.452^{a,b}	0.480^{b}	0.474^{b}	0.569^{a}	0.582^{a}	0.594^{a}	0.486^{b}	0.530^{a}	0.341^{b}	0.374^{b}	0.011	0.001
Aspartate	0.033	0.029	0.036	0.035	0.033	0.044	0.033	0.042	0.041	0.036	0.027	0.029	0.001	0.001
Glutamate	0.273^{a}	0.226^{b}	0.282^{a}	0.291	0.266	0.299	0.305	0.319	0.345	0.319^{a}	0.227^{b}	0.252^{a,b}	0.007	0.010
Glycine	0.491^{a,b}	0.525^{a}	0.479^{b}	0.551^{b}	0.624^{a}	0.590^{a,b}	0.587^{b}	0.684^{a}	0.483^{c}	0.499^{a}	0.473^{a,b}	0.444^{b}	0.015	0.014
Proline	0.561^{a}	0.527^{b}	0.537^{b}	0.727^{a}	0.781^{a}	0.512^{b}	0.654^{b}	0.714^{a}	0.330^{c}	0.252^{a}	0.269^{a}	0.218^{b}	0.012	<0.001
Serine	0.100	0.101	0.089	0.126^{a}	0.115^{a,b}	0.093^{b}	0.108^{a,b}	0.115^{a}	0.084^{b}	0.089	0.088	0.077	0.001	<0.001

$^{a–c}$Within the same sampling time, values in a row sharing different superscript letters differ ($P < 0.05$); $n = 6$

HDRS high digestion rate starch-fed group, *LDRS-fed group* low digestion rate starch-fed group, *MDRS* moderate digestion rate starch-fed group, *SEM* standard error of the mean

Source: Yin et al. (2010) (with minor modification, the pigs were fed at 12:00 am)

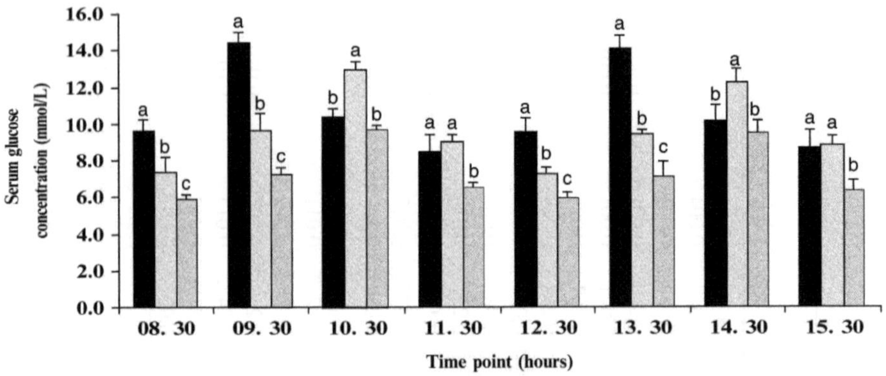

Fig. 8.2 Variation in postprandial serum systemic circulating glucose in two feeding cycles. All pigs were fed at 8:00 and 12:00 pm, respectively. (■), HDRS-fed group; ( ), MDRS-fed group; ( ), LDRS-fed group. [a–c]mean values (*n*, 8 per group) within the same sampling time with unlike letters were significantly different ($P < 0.05$). *HDRS* high digestion rate starch, *LDRS* low digestion rate starch, *MDRS* moderate digestion rate starch. *Source*: Yin et al. (2011)

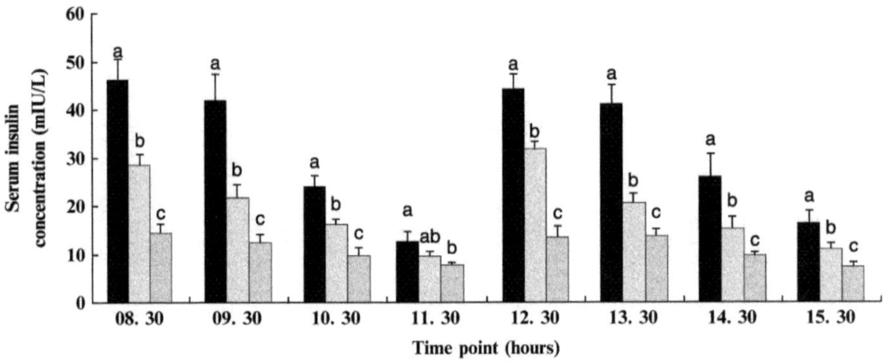

Fig. 8.3 Variation in postprandial serum systemic circulating insulin concentration in two feeding cycles. All pigs were fed at 8:00 am and 12:00 pm, respectively. (■), HDRS group; ( ), MDRS group; ( ), LDRS group. [a–c]mean values (*n*, 8 per group) within the same sampling time with unlike letters were significantly different ($P < 0.05$). *HDRS* high digestion rate starch, *LDRS* low digestion rate starch, *MDRS* moderate digestion rate starch. *Source*: Yin et al. (2011)

and amino acids used independently or together (O'Connor et al. 2003a, b; Jeyapalan et al. 2007). Raising glucose alone increased protein synthesis more in fast-twitch glycolytic muscles than in other tissues like the liver, intestine, pancreas, spleen, kidney, and lung (Jeyapalan et al. 2007). Raising either insulin or amino acids alone increases protein synthesis in most tissues, especially in skeletal muscle (Davis et al. 2001; O'Connor et al. 2003a, b). Interestingly, elevating both insulin and amino acids or raising glucose, insulin, and amino acids together to postprandial levels increases protein synthesis in several visceral tissues, such as pancreas, kidney, spleen, and liver, as well as in skeletal muscles (Davis et al. 2000). Since the circulating

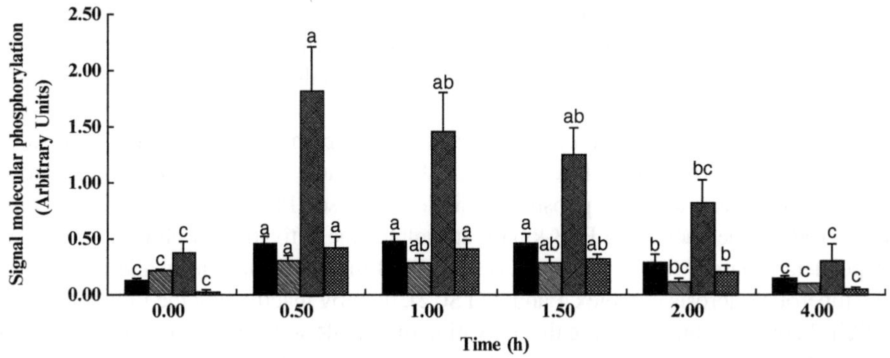

Fig. 8.4 Time course of the changes in phosphorylation of important molecular signal in protein syntheses in skeletal muscle of neonatal pigs after a meal. All pigs were fed at 8:00 am. (■), mTOR; (▨), 4EBP1; (▩), eIF4G; (▦), S6. [a–c]The multiple comparison was conducted within the same parameter with different sampling time, unlike letters indicated a significantly difference between sampling time ($P < 0.05$). All results are corrected for total protein. *mTOR* the mammal target of rapamycin, *4EBP1* 4E-binding protein-1, *eIF4G* eukaryotic initiation factor (eIF) 4G, *S6* S6 kinase-1. *Source*: Wilson et al. (2009)

concentration of glucose, amino acids, and insulin did not peak simultaneously, the maximal stimulation of protein synthesis after meal may occur during the time when circulating concentrations of these metabolites and hormones raised and are maintained at a moderate level, normally 30–120 min in the postprandial phase, rather than that at their own individual peak point.

Glucose stimulation of protein synthesis in fast-twitch glycolytic is associated with increased phosphorylation of protein kinase B (PKB) and enhanced formation of the active eIF4E.eIF4G complex. This enables mRNA to bind to the 43S pre-initiation complex, thereby enhancing translation initiation (Wang and Proud 2006). However, eIF4E.4E-BP1 and eIF4E.eIF4G complexes gradually return to the baseline by 4 h in the postprandial phase, thus slowly than the decline in protein synthesis. This suggests that assembly of eIF4E.eIF4G is not the rate-limiting step in the feeding-induced stimulation of skeletal muscle protein synthesis (Wilson et al. 2009).

The signaling pathway by which amino acids stimulate protein synthesis is less well understood. Intracellular amino acids may directly or indirectly affect signaling components that stimulate protein synthesis. The amino acid-induced signaling component converts Rheb-guanosine diphosphate (GDP) to Rheb-guanosine triphosphate (GTP) and the association between Rheb and mTOR to promote mTOR activation (Saucedo et al. 2003; Long et al. 2005; Kimball 2007). Some amino acids (e.g., arginine) have been shown to be of particular importance in mediating the increase in translation initiation and protein synthesis (Yao et al. 2008). Indeed, either infusion of leucine or arginine or increased dietary leucine or arginine intake acutely increases protein synthesis in skeletal muscle of neonatal pigs by an mTORC1-mediated process (Suryawan et al. 2008; Yao et al. 2008; Yao et al. 2011; Li et al. 2011). The circulating leucine and arginine levels increased by 30 min after the initiation of eating and are maintained at a relative higher level until 240 min postprandial, mirroring the increase in rates of protein synthesis (Wilson et al. 2009; Yin et al. 2010). However, little is known about the mechanisms by which amino acids directly or indirectly modulate the activation of signaling components upstream of mTOR.

Insulin initiates the protein synthesis-signaling pathway by binding to insulin receptor (InR), which therefore activates the InR and insulin receptor substrate-1 (IRS-1). This activates the phosphoinositide 3-kinase (PI 3-kinase) (Di Guglielmo et al. 1998). Then activated PI 3-kinase stimulates the activation of phosphoinositide-dependent kinase 1 (PDK-1) and PKB. The PKB then phosphorylates and inactivates the tuberous sclerosis complex 1 and 2 (TSC2), thereby inactivating the function of the TSC1/2 complex, and inducing the activation of mTOR signal pathway (Di Guglielmo et al. 1998). Insulin signaling can be attenuated by the action of a number of phosphatases including protein tyrosine phosphatase-1B (PTP-1B) which dephosphorylates the InR and IRS-1, phosphatase and tensin homologue which inactivates PI 3-kinase, and PP2A which attenuates the action of PKB and S6K1 (Goberdhan and Wilson 2003). In pigs, feeding rapidly increases the activation of the InR, IRS-1, PI 3-kinase, PDK-1, and PKB in skeletal muscle. Activation of PKB is rapid and peaked at submaximal insulin concentrations, returning to baseline by 90 min, prior to the reduction in mTORC1 activation (Tremblay et al. 2005). The discrepancy between PKB and mTOR activation reflects the ability of both amino acids and insulin (Wang et al. 2006) to stimulate mTORC1 rather than PKB (Suryawan et al. 2007). In addition, physiological changes in the levels of amino acids have no effect on the activation of these early steps in the insulin-signaling pathway.

8.3 Protein Degradation in Pigs

Protein degradation is a part of a major mechanism by which cells regulate the concentration of particular proteins and eliminate mis-folded proteins. Some peptides derived from protein degradation could be reused for new protein syntheses while others could be degraded into free amino acids. Generally, a large percentage of free amino acids released from protein by degradation can be further hydrolyzed into small molecules, such as nitric oxidize (NO), ammonia (NH_3), carbon dioxides (CO_2), and water (H_2O).

The protein degradation plays important roles in animal physiological process, especially in the cellular signal transduction as well as in maintaining the integrity of the proper folded state of protein.

8.3.1 Main Signaling Pathway in Protein Degradation

There are two pathways for intracellular protein degradation, the proteasome and lysosome pathways. Intracellular proteins are mainly degraded by the ubiquitin–proteasome pathway (Lecker et al. 2006). The organelles contain several acid-optimal proteases, including cathepsins B, D, and H and many other acid hydrolases. Some cytosolic proteins are degraded in lysosomes after being engulfed in autophagic vacuoles that fuse with lysosomes (Baehrecke 2005). In most cells, this process is accelerated by the lack of insulin or essential amino acids and in liver by glucagon (Gronostajski et al. 1984). Another cytosolic proteolytic system in mammalian cells is the Ca^{2+}-activated (ATP-independent) proteolytic process, which involves the cysteine proteases termed calpains. These proteases seem to be activated when cells are injured and cytosolic Ca^{+} rises; therefore they may play an important role in tissue injury, necrosis, and autolysis (Goll et al. 2003). In addition, the caspases are also an important family of cytosolic proteases that cleave proteins after aspartic acid residues. These enzymes, which are cysteine proteases, are critical in destruction of cell constituents during apoptosis (Salvesen and Dixit 1997).

The ubiquitin–proteasome pathway consists of concerted actions of enzymes that link chains of the polypeptide cofactor, ubiquitin, onto proteins to mark them for degradation (Glickman and Ciechanover 2002). This tagging process leads to their recognition by the 26S proteasome, a very large multicatalytic protease complex that degrades ubiquitinated proteins to small peptides (Baumeister et al. 1998). Three enzymatic components are required to link chains of ubiquitin onto proteins that are destined for degradation: E1 (ubiquitin-activating enzyme), E2s (ubiquitin-carrier or conjugating proteins) which prepare ubiquitin for conjugation, and E3 (ubiquitin-protein ligase). The E3 is a key enzyme in the process because it recognizes the protein substrate and catalyzes the transfer of activated ubiquitin to it. Till now, more than thousands of proteins have shown to be degraded by this pathway.

8.3.2 Degradation of Amino Acids in Pigs

All amino acids contain at least one nitrogen atom, which forms their α-amino group. Some amino acids also contain additional nitrogen atoms in their side chains. Theoretically, nitrogen is useless in energy metabolism and needs to be eliminated. Recent studies in young pigs indicated that both nutritional essential amino acids and non-nutritional essential amino acids derived from dietary protein sources are degraded extensively by the small intestine in first pass, with less than approximately 20 % of the extracted amino acids being utilized for intestinal mucosal protein synthesis (Stoll and Burrin 2006). Among dietary amino acids, glutamate, glutamine, aspartate, and proline are degraded faster than others in the small intestine (Yao et al. 2012). There are two key processes in the process of amino acid degradation: (1) transamination removes the α-amino group from one amino acid and transfers it to α-ketoglutarate, which leads to the accumulation of glutamate Yao et al. 2012, and (2) release of nitrogen from glutamate and its conversion to urea, which is accomplished by the urea cycle in the liver. Removal of nitrogen is typically an early step in amino acid degradation and leaves behind the carbon skeleton. Particularly, glutamine is the main energy source for the intestinal epithelium, and the total rate of utilization of glutamine by the gut may be greater than that for others (Wu 2010). Intestinal degradation of amino acids has profound impacts on nutrition and health in pigs. Firstly, degradation of glutamine, glutamate, as well as aspartate provides the necessary ATP to maintain gut integrity and function (Watford 2008). Secondly, the extensive catabolism of amino acids by the small intestine significantly decreases the potentially neurotoxic effects of the excitatory neurotransmitter glutamate. Thirdly, the endogenous synthesis of nonessential amino acids (for example, citrulline, arginine, proline, and alanine) mainly relies on the transformations of amino acids in the intestine and thus modulating the availability of dietary amino acids to extra-intestinal tissues (Wu 1998; Bergen and Wu 2009).

References

Avruch J, Hara K, Lin Y, Liu M, Long X, Ortiz-Vega S, Yonezawa K (2006) Insulin and amino-acid regulation of mTOR signaling and kinase activity through the Rheb GTPase. Oncogene 25(48):6361–6372

Baehrecke EH (2005) Autophagy: dual roles in life and death? Nat Rev Mol Cell Biol 6(6): 505–510

Bates PC, Millward DJ (1981) Characteristics of skeletal muscle growth and protein turnover in a fast-growing rat strain. Br J Nutr 46(1):7–13

Baumeister W, Walz J, Proteolysis C (1998) The proteasome: paradigm review of a self-compartmentalizing protease. Cell 92:367–389

Bergen WG, Wu G (2009) Intestinal nitrogen recycling and utilization in health and disease. J Nutr 139(5):821–825

Burrin DG, Shulman RJ, Reeds PJ, Davis TA, Gravitt KR (1992) Porcine colostrum and milk stimulate visceral organ and skeletal muscle protein synthesis in neonatal piglets. J Nutr 122(6):1205–1213

Bush JA, Kimball SR, O'Connor PMJ, Suryawan A, Orellana RA, Nguyen HV, Jefferson LS, Davis TA (2003) Translational control of protein synthesis in muscle and liver of growth hormone-treated pigs. Endocrinology 144(4):1273–1283

Davis TA (2008) Insulin and amino acids are critical regulators of neonatal muscle growth. Nutrition today 43(4):143–149

Davis TA, Nguyen HV, Suryawan A, Bush JA, Jefferson LS, Kimball SR (2000) Developmental changes in the feeding-induced stimulation of translation initiation in muscle of neonatal pigs. Am J Physiol Endocrinol Metab 279(6):E1226–E1234

Davis TA, Fiorotto ML, Beckett PR, Burrin DG, Reeds PJ, Wray-Cahen D, Nguyen HV (2001) Differential effects of insulin on peripheral and visceral tissue protein synthesis in neonatal pigs. Am J Physiol Endocrinol Metab 280(5):E770–E779

Davis TA, Suryawan A, Orellana RA, Nguyen HV, Fiorotto ML (2008) Postnatal ontogeny of skeletal muscle protein synthesis in pigs. J Anim Sci 86(14 suppl):E13–E18

Deng D, Yao K, Chu WY, Li TJ, Huang RL, Yin YL, Liu HQ, Zhang JS, Wu GY (2009) Impaired translation initiation activation and reduced protein synthesis in weaned piglets fed a low-protein diet. J Nutr Biochem 20:544–552

Deng J, Wu X, Bin S, Li TJ, Huang R, Liu Z, Liu Y, Ruan Z, Deng Z, Hou Y, Yin YL (2010) Dietary amylose and amylopectin ratio and resistant starch content affects plasma glucose, lactic acid, hormone levels 1 and protein synthesis in splanchnic tissues. Z Tierphysiol Tierernahr Futtermittelkd 94:220–226

Di Guglielmo GM, Baass PC, Authier F, Posner BI, Bergeron JJM (1998) Insulin receptor internalization and signalling. Mol Cell Biochem 182(1):59–63

Drew MD, Schafer TC, Zijlstra RT (2012) Glycemic index of starch affects nitrogen retention in grower pigs. J Anim Sci 90(4):1233–1241

Glickman MH, Ciechanover A (2002) The ubiquitin-proteasome proteolytic pathway: destruction for the sake of construction. Physiol Rev 82(2):373–428

Goberdhan DCI, Wilson C (2003) PTEN: tumour suppressor, multifunctional growth regulator and more. Hum Mol Genet 12(suppl 2):R239–R248

Goll DE, Thompson VF, Li H, Wei W, Cong J (2003) The calpain system. Physiol Rev 83(3): 731–801

Gronostajski RM, Goldberg AL, Pardee AB (1984) The role of increased proteolysis in the atrophy and arrest of proliferation in serum-deprived fibroblasts. J Cell Physiol 121(1):189–198

Jeyapalan AS, Orellana RA, Suryawan A, O'Connor PMJ, Nguyen HV, Escobar J, Frank JW, Davis TA (2007) Glucose stimulates protein synthesis in skeletal muscle of neonatal pigs through an AMPK- and mTOR-independent process. American Journal of Physiology - Endocrinology And Metabolism 293(2):E595–E603

Jobgen WS, Fried SK, Fu WJ, Meininger CJ, Wu G (2006) Regulatory role for the arginine–nitric oxide pathway in metabolism of energy substrates. J Nutr Biochem 17(9):571–588

Kimball S (2007) The role of nutrition in stimulating muscle protein accretion at the molecular level. Biochem Soc Trans 35:1298–1301

Kimball SR, Jefferson LS (2004) Molecular mechanisms through which amino acids mediate signaling through the mammalian target of rapamycin. Curr Opin Clin Nutr Metab Care 7(1):39–44

Kimball SR, Jefferson LS, Nguyen HV, Suryawan A, Bush JA, Davis TA (2000) Feeding stimulates protein synthesis in muscle and liver of neonatal pigs through an mTOR-dependent process. Am J Physiol Endocrinol Metab 279(5):E1080–E1087

Kong XF, Tan BE, Yin YL, Li XL, Jaeger LA, Bazer FW, Wu GY (2012) Arginine stimulates the mTOR signaling pathway and protein synthesis in porcine trophectoderm cells. J Nutr Biochem 23(9):1178–1183

Lecker SH, Goldberg AL, Mitch WE (2006) Protein degradation by the ubiquitin–proteasome pathway in normal and disease states. J Am Soc Nephrol 17:1807–1819

Li F, Yin Y, Tan B, Kong X, Wu G (2011) Leucine nutrition in animals and humans: mTOR signaling and beyond. Amino Acids 41:1185–1193

Liu JG, Zhang P, Bin SY (2007) Effects of different dietary starch constituents on level of blood glucose and insulin of weaned piglets (article in Chinese). Food Sci 28:315–319 (in Chinese)

Long X, Lin Y, Ortiz-Vega S, Yonezawa K, Avruch J (2005) Rheb binds and regulates the mTOR kinase. Curr Biol 15(8):702–713

Norton LE, Layman DK (2006) Leucine regulates translation initiation of protein synthesis in skeletal muscle after exercise. J Nutr 136(2):533S–537S

O'Connor PMJ, Bush JA, Suryawan A, Nguyen HV, Davis TA (2003a) Insulin and amino acids independently stimulate skeletal muscle protein synthesis in neonatal pigs. Am J Physiol Endocrinol Metab 284(1):E110–E119

O'Connor PM, Kimball SR, Suryawan A, Bush JA, Nguyen HV, Jefferson LS, Davis TA (2003b) Regulation of translation initiation by insulin and amino acids in skeletal muscle of neonatal pigs. Am J Physiol Endocrinol Metab 285(1):E40–E53

O'Connor PM, Kimball SR, Suryawan A, Bush JA, Nguyen HV, Jefferson LS, Davis TA (2004) Regulation of neonatal liver protein synthesis by insulin and amino acids in pigs. Am J Physiol Endocrinol Metab 286(6):E994–E1003

Regmi PR, Matte JJ, van Kempen T, Zijlstra RT (2010) Starch chemistry affects kinetics of glucose absorption and insulin response in swine. Livest Sci 134(1–3):44–46

Salvesen GS, Dixit VM (1997) Caspases: intracellular signaling by proteolysis. Cell 91(4): 443–446

Sarbassov DD, Guertin DA, Ali SM, Sabatini DM (2005) Phosphorylation and regulation of Akt/PKB by the rictor-mTOR complex. Science 307(5712):1098–1101

Saucedo LJ, Gao X, Chiarelli DA, Li L, Pan D, Edgar BA (2003) Rheb promotes cell growth as a component of the insulin/TOR signalling network. Nat Cell Biol 5(6):566–571

Stoll B, Burrin DG (2006) Measuring splanchnic amino acid metabolism in vivo using stable isotopic tracers. J Anim Sci 84(13 suppl):E60–E72

Suryawan A, Nguyen HV, Bush JA, Davis TA (2001) Developmental changes in the feeding-induced activation of the insulin-signaling pathway in neonatal pigs. Am J Physiol Endocrinol Metab 281(5):E908–E915

Suryawan A, Orellana RA, Nguyen HV, Jeyapalan AS, Fleming JR, Davis TA (2007) Activation by insulin and amino acids of signaling components leading to translation initiation in skeletal muscle of neonatal pigs is developmentally regulated. Am J Physiol Endocrinol Metab 293(6): E1597–E1605

Suryawan A, Jeyapalan AS, Orellana RA, Wilson FA, Nguyen HV, Davis TA (2008) Leucine stimulates protein synthesis in skeletal muscle of neonatal pigs by enhancing mTORC1 activation. Am J Physiol Endocrinol Metab 295(4):E868–E875

Tan B, Yin YL, Liu ZQ, Li XG, Xu HJ, Kong XF, Huang RL, Tang WJ, Shinzato I, Smith SB, Wu GY (2009) Dietary L-arginine supplementation increases muscle gain and reduces body fat mass in growing-finishing pigs. Amino Acids 37:169–175

Tremblay F, Gagnon AM, Veilleux A, Sorisky A, Marette A (2005) Activation of the mammalian target of rapamycin pathway acutely inhibits insulin signaling to Akt and glucose transport in 3 T3-L1 and human adipocytes. Endocrinology 146(3):1328–1337

Wang X, Proud CG (2006) The mTOR pathway in the control of protein synthesis. Physiology (Bethesda) 21:362–369

Wang L, Rhodes CJ, Lawrence JC (2006) Activation of mammalian target of rapamycin (mTOR) by insulin is associated with stimulation of 4EBP1 binding to dimeric mTOR complex 1. J Biol Chem 281(34):24293–24303

Watford M (2008) Glutamine metabolism and function in relation to proline synthesis and the safety of glutamine and proline supplementation. J Nutr 138(10):2003S–2007S

Wek RC, Jiang HY, Anthony TG (2006) Coping with stress: eIF2 kinases and translational control. Biochem Soc Trans 34(1):7–11

Wilson FA, Suryawan A, Orellana RA, Kimball SR, Gazzaneo MC, Nguyen HV, Fiorotto ML, Davis TA (2009) Feeding Rapidly Stimulates Protein Synthesis in Skeletal Muscle of Neonatal Pigs by Enhancing Translation Initiation. J Nutr 139(10):1873–1880

Wu G (1998) Intestinal mucosal amino acid catabolism. J Nutr 128(8):1249–1252

Wu G (2009) Amino acids: metabolism, functions, and nutrition. Amino Acids 37(1):1–17

Wu G (2010) Recent advances in swine amino acid nutrition. J Anim Sci Biotechnol 1:49–61

Yao K, Yin Y, Chu W, Liu Z, Deng D, Li T, Huang R, Zhang J, Tan B, Wang W, Wu G (2008) Dietary Arginine Supplementation Increases mTOR Signaling Activity in Skeletal Muscle of Neonatal Pigs. J Nutr 138(5):867–872

Yao K, Guan S, Li TJ, Yin YL (2011) Dietary L-arginine supplementation enhances intestinal development and expression of vascular endothelial growth factor in weanling piglets. Br J Nutr 105:703–709

Yao K, Wang L, Ding BY, Fu DB, Liu YL, Zhu HL, Liu J, Li YT, Kang P, Yin YL, Wu GY, Hou YQ (2012) Alpha-ketoglutarate inhibits glutamine degradation and enhances protein synthesis in intestinal porcine epithelial cells. Amino Acids 42(6):2491–2500

Yin F, Zhang Z, Huang J, Yin Y (2010) Digestion rate of dietary starch affects systemic circulation of amino acids in weaned pigs. Br J Nutr 103:1404–1412

Yin F, Yin Y, Zhang Z, Xie M, Huang J, Huang R, Li T (2011) Digestion rate of dietary starch affects the systemic circulation of lipid profiles and lipid metabolism-related gene expression in weaned pigs. Br J Nutr 106:369–377

Chapter 9
Factors That Affect Amino Acid Metabolism in Pigs

Qinghua He, Yulong Yin, Yongqing Hou, Guixin Qin, Hui Sun, Jianhong Liu, Bo Liu, and Zongkun Zheng

Contents

Q. He • J. Liu • B. Liu • Z. Zheng
Department of Food Science and Engineering, College of Chemistry and Chemical Engineering, Shenzhen University, Shenzhen, Guangdong 518060, People's Republic of China

Y. Yin (✉)
Chinese Academy of Sciences, Institute of Subtropical Agriculture, Research Center for Healthy Breeding Livestock & Poultry, Hunan Engineering & Research Center of Animal & Poultry Science, Key Laboratory of Agro-ecological Processes in Subtropical Region, Scientific Observing and Experimental Station of Animal Nutrition and Feed Science in South-Central, Ministry of Agriculture, Changsha, Hunan 410125, People's Republic of China
e-mail: yinyulong@isa.ac.cn

Y. Hou
Laboratory of Animal Nutrition and Feed Science, Wuhan Polytechnic University, Wuhan, China

G. Qin • H. Sun
Jilin Agricultural University, No. 2888 Xincheng Street, Changchun, Jilin Province 130118, People's Republic of China

F. Blachier et al. (eds.), *Nutritional and Physiological Functions of Amino Acids in Pigs*,
DOI 10.1007/978-3-7091-1328-8_9, © Springer-Verlag Wien 2013

9.1 Introduction

Dietary protein is digested and decomposed into free amino acids and peptides in the gastrointestinal lumen, because protein cannot be absorbed as such and needs to be converted into amino acids and peptides (Krehbiel and Matthews 2003). These amino acids and peptides will flow into intestinal venous blood from intestinal lumen, and will be partly used for protein synthesis and deposition through hepatic metabolism. It is a very important issue in pig nutrition research to understand the digestion and metabolism of dietary proteins and amino acids as well as the regulation involved in such process (Yin and Tan 2010). This issue is also important for the accurate calculation and supply in diet formulation, which will be helpful to decrease the nitrogen excretion in pig production (Zhang et al. 2012). In addition, the researches about this issue might be helpful to resolve the other relative problems, including antibiotics and unconventional feedstuff in the feed industry. However, this is a complicated issue, which needs to resolve two problems. One is to know what factors affect the amino acid metabolisms, which need the deep and intensive understanding about the amino acid metabolic processes and its regulations in pigs. Another one is to know how to regulate the amino acid metabolism by nutritional methods according to the above impacting factors. The ultimate goal is to exert economic and environmental values of dietary amino acids in pig production. Since the impacting factors and regulating means of amino acid metabolism are not only various and complicated but also associated with each other, we will elaborate on the progress of these two aspects in this chapter.

9.2 Factors That Affect Amino Acid Metabolisms in Pigs

The extensive catabolism of dietary amino acids by luminal microbes or by the intestine may result in an alteration in nutritional efficiency of amino acids. However, knowledge about digestion, absorption, and metabolism of amino acids and epithelial cell transport of amino acids in the gastrointestinal tract remains incomplete.

9.2.1 *Effects of Gut Microbiota on Intestinal Amino Acid Metabolism and Nitrogen Emissions in Pigs*

Nutritional value of food nutrients and other functional elements are related to composition and content, but also to the digestion, absorption, and metabolism of these elements in the gastrointestinal tract. Intestinal tract is well known to be the key place for digestion, absorption, and metabolism of dietary amino acids. Gut microbiota is a complex assortment of microorganisms inhabiting the mammalian gastrointestinal tract (Wang et al. 2011). Because the gut microbiota participates in the metabolic activity of amino acids together with the intestinal mucosa, it represents an important environmental factor that affects the nitrogen harvested from the diet and also nitrogen deposition in host (Kong et al. 2009).

Studies have shown that the gut microbiota plays an important role in the regulation of energy and lipid metabolism in host (Velagapudi et al. 2010) and increases efficiency of energy harvest from diet (Turnbaugh et al. 2006). Recent studies also indicated that human intestinal microbiota from children was sharpened by a modern western diet and a rural diet (De Filippo et al. 2010). It has been hypothesized that the gut microbiota has coevolved with the polysaccharide-rich diet of individuals from rural African village, allowing hosts to maximize energy intake from fibers. Most certainly, the gut microbiota has a tremendous influence on catabolism and de novo synthesis of amino acids and thus can modulate digestion, absorption, and metabolism of amino acids in diet (Libao-Mercado et al. 2009; Fang et al. 2010).

Since gut microbiota can hydrolyze urea and deaminate amino acids in the intestine, these processes usually result in metabolism and loss of amino acids (especially essential amino acids (EAA)), which otherwise should have been used for the synthesis and deposition of body protein. Researches in animal nutrition have indicated that EAA including branched-chain amino acids (leucine, isoleucine, and valine), histidine, lysine, methionine, phenylalanine, threonine, and tryptophan can be degraded by intestinal mucosa in pigs (Dai et al. 2005). However, there is still a lack of direct evidences to validate the effects of porcine small intestinal microbiota on the catabolism of EAA. The key point is to differentiate intestinal cells and gut microbes for their respective capacity to oxidize EAA. Based on this consideration, experiments were carried out to investigate this issue (Yin et al. 2004). The oxidation of amino acids in piglet intestinal mucosal cell was investigated in vitro under the conditions of inhibition or without inhibition. The results obtained indicate that only branched-chain amino acids were exclusively degraded in mucosal cell. In contrast, other EAA, including histidine, lysine, methionine, phenylalanine, threonine, and tryptophan, were not oxidized exclusively in intestinal cells. The oxidation of these amino acids may thus be carried out partly by gut microbes (Chen et al. 2007, 2009).

The balance of intestinal microbiota also has important impacts on the amino acid metabolism in the intestines. The activity of gut microbes affects the intestinal structure and functions, thus affecting the digestion and absorption of amino acids. In addition, it has to be taken into account that when the intestinal mucosa is facing the harmful effect of some bacteria, intestinal mucosa may need more energy resources.

Carbon- and nitrogen-containing substances can be fermented by intestinal microbes and used for body protein synthesis. Some studies have showed that microbial nitrogen is the main component of endogenous nitrogen in ileal digesta and feces of pig. Indeed, the endogenous nitrogen is about 47–74 % of total ileal nitrogen emission in growing pigs. Microbial nitrogen is about 30–50 % of the amount of ileal endogenous nitrogen excretion. The endogenous nitrogen is about 75–90 % of fecal nitrogen emissions. Microbial nitrogen is about 70–90 % of the amount of fecal endogenous nitrogen excretion (Yin and Tan 2010). Thus, the increase in the amount of endogenous nitrogen is related to the metabolic activities of microbes in large intestines (Yin et al. 2000b). Thus, the main reason of the nonoptimal conversion of dietary protein to body protein and nitrogen pollution in the pig industry is the microbial nitrogen.

9.2.2 Effects of Intestinal Mucosal Metabolism on Utilization of Amino Acids in Pigs

The researches in recent 20 years have shown that the amino acids are involved in the metabolism of intestinal mucosa. For example, some amino acids, including threonine and valine, are likely representing energy sources for the intestinal mucosa (Dudley et al. 1998; Stoll et al. 1998). Threonine is important for mucin synthesis and maintenance of the gut barrier integrity and it has been reported that as much as 60 % of dietary threonine is retained by the intestine (Wang et al. 2009, 2010; Yin et al. 2010a, 2010b). Recently, the metabolic signatures in the jejunal mucosal tissues obtained from piglets on age of 21 days have been characterized in situ directly by HR-MAS NMR spectroscopy (He et al. 2011a). Marked changes were found in jejunal metabolites, including decreased levels of valine, alanine, glutamine, glutamate, trimethylamine-N-oxide, lactate, creatine, glucose, galactose, phenylalanine, tyrosine, and taurine in intrauterine growth restriction (IUGR) piglets compared with normal-weight piglets ($P < 0.05$). These findings suggest that IUGR piglets can be affected in energy supply, utilization of amino acids, and protein metabolism.

Amino acids also participate in the synthesis of secreted protein in intestinal mucosa and can be transferred into other amino acids by deamination and transamination (Hoerr et al. 1993).

Because intestinal tissues are able to utilize amino acids for their own needs, amino acid metabolism in intestinal mucosa will have obvious impacts on the net portal absorption, thus affecting the utilization of dietary amino acids in the body.

9.2.2.1 Effects of Gut Mucosal Metabolism on the Net Absorption and Composition Patterns of Amino Acids in Portal Vein

The amount of amino acids that disappeared in the digestive tract is not equal to the net amount of absorption in portal vein. The latter is significantly lower than the former (MacRae et al. 1997a, b). Studies found that not all absorbed amino acids entered into the portal vein. Indeed, the net apparent absorption of amino acids in portal vein only account for a part of total intake amino acids. The proportion of threonine is about 40–50 %. The proportion of lysine is about 50–69 %. The proportion of glutamic acid and aspartate is close to zero. Glutamine shows a negative balance. But the proportion of alanine is more than 100 %, which is almost twice higher than that of threonine or lysine (van der Meulen et al. 1997; Deutz et al. 1998).

Yin et al. (2003) found similar results in growing pig study. In this study, pigs were fed two kinds of diets. One contained starch and casein (SC). Another contained low-level starch and sub-flour of wheat (LSW). The plasma samples and intestinal digesta were collected during 8 h after feeding. Studies showed that only 50–60 % of dietary digested and absorbed amino appeared in the portal vein, which can be used for protein synthesis in the peripheral tissues. The remaining amino acids (40–50 %) were used by the portal vein-drained viscera (PDV, gastrointestinal tract, pancreas, spleen, and omental fat).

Thus, the metabolism of intestinal mucosa can affect the net portal absorption and composition of amino acids in the portal vein. The intestinal mucosa and presumably the microbiota can oxidize nonessential amino acids and branched-chain amino acids. The nonessential amino acids can be both degraded and also synthesized in the intestine.

9.2.2.2 Effects of Net Absorption and Composition of Amino Acids in the Portal Vein on Nitrogen Deposition

Protein deposition, which is the main indicator for pig growth, depends on the net portal absorption of the EAA. The regression equations between average daily gain (ADG) and the net portal absorption of amino acids were developed in growing pigs (Huang et al. 2006). In these equations, the net portal absorption of nutrients is independent variables. ADG is the dependent variable. It is showed that there are relatively high correlations ($P = 0.0295$) between the net portal absorption and ADG. The related regression equations are as follows:

$$\text{ADG} = 51.9321\ \text{EAA} + 446.0028\ (R^2 = 0.9419, P = 0.0295)$$

$$\text{ADG} = 305.5017\ \text{Leu} + 491.7915\ (R^2 = 0.9514, P = 0.0246)$$

$$\text{ADG} = 502.1378\ \text{Ile} + 471.1824\ (R^2 = 0.8538, P = 0.0760)$$

$$\text{ADG} = 483.8102\ \text{Val} + 441.0182\ (R^2 = 0.8361, P = 0.0856)$$

$$\text{ADG} = 305.5941\ \text{His} + 568.1355\ (R^2 = 0.8229, P = 0.0929)$$

From these equations, it was deduced that the correlations are relatively high ($P < 0.05$ or 0.1) between ADG and the net portal absorption of total EAA, leucine (Leu), isoleucine (Ile), valine (Val), and histidine (His), while the correlations are low ($P < 0.05$ or 0.1) between ADG and the net portal absorption of glucose and nonessential amino acids. In addition, several independent variables, such as some essential amino acids, including Met, His, and the hexose glucose can be introduced into equations as independent variables when stepwise regressions were performed in growing pigs. In these regression equations, the correlations can be greatly improved ($P < 0.01$). Relationships are as follows:

$$\mathrm{ADG} = 263.7903\ \mathrm{Leu} + 197.6090\ \mathrm{Met} + 482.7015\ (R^2 = 0.9999, P = 0.0097)$$
$$\mathrm{ADG} = 35.9182\ \mathrm{EAA} + 129.3837\ \mathrm{His} + 471.7131\ (R^2 = 0.9998, P = 0.0121)$$
$$\mathrm{ADG} = 1345.0675\ \mathrm{Ile} - 15.5291\ \mathrm{Glucose} + 514.3573\ (R^2 = 0.9999, P = 0.0078)$$

9.2.3 *Effects of Anti-nutritional Factors on Amino Acid Metabolism*

As mentioned above, a major source of nitrogen pollution in the pig industry is related to fermentation of microorganisms in the digestive tract. The anti-nutritional factors in feed are the major factors that cause the intestinal microbial fermentations.

9.2.3.1 Non-starch Polysaccharides

Generally, non-starch polysaccharides (NSPs) in feed are considered as anti-nutritional factors for pigs. These NSPs not only have a great negative impact on digestion and absorption of nutrients in pigs but also are associated with the intestinal microbial fermentations, which is the energy source of microbial fermentation in gut. Additionally, NSP content in feed can affect the digestibility of dietary protein and amino acid. It was reported that the apparent ileal digestibility of crude protein and most of amino acids decreased by 3–5 % in growing pigs when β-glucan in diet increased from 29.0 to 31.8 g/kg (Yin et al. 2000b, c). It was also reported that the apparent ileal digestibility of crude protein and amino acid is reduced by 12 % and 6 %, respectively, when total NSP in diet increased from 83 to 193 g/kg (Yin et al. 2000a).

NSP not only can affect secretion of endogenous nitrogen but also affect deposition and biological value of feed nitrogen. It was reported that cereal fiber content can significantly increase the excretion of endogenous nitrogen and amino acids in pigs (Low and Rainbird 1984; Yin et al. 2000a). Pectin can reduce the amount of nitrogen deposition and nitrogen microbiological value (Xu et al. 2005). It has been suggested that pectin and other gel of polysaccharides (such as methyl cellulose) can affect the secretion of endogenous protein and amino acids, and increase endogenous excretion of nitrogen, which increased the amount of nitrogen coming from the small intestine to the large intestine, and thus led to the decrease of body nitrogen retention and nitrogen biological value.

The main factor responsible for the intestinal secretion of endogenous nitrogen in roughage feed is not the insoluble part of non-starch polysaccharides (INSP) but the soluble part of non-starch polysaccharides (SNSP) (Yin et al. 2000a, b, c, 2004). Although SNSP do not affect the digestive process of feed nitrogen and amino acids (that is the true digestibility), there is a positive linear relationship between SNSP content and intestinal microbial fermentation or endogenous nitrogen (amino acid) excretion in pigs that thereby reduces the amino acid digestibility. However, there is no positive linear relationship between INSP content and intestinal microbial fermentation. INSP does not affect the endogenous excretion of nitrogen and total nitrogen. When soluble NSPs in growing pig diets are more than 0.8 %, the protein digestibility and growth performance are significantly affected. There is a linear relationship between diet SNSP (X) and ileal excretion of nitrogen (y) as follows: $y = 1.7832X - 8.2074$ ($R^2 = 0.99$). Also, there is a linear relationship between the INSP and ileal excretion of nitrogen as follows: $y = 0.026X + 5.0431$ ($R^2 = 0.43$), which has however no significant correlation. Interestingly, the fecal nitrogen excretion (y) has the similar linear relationship with two types of NSPs (X) as follows: $y = 0.2563X + 0.5773$ ($R^2 = 0.89$, SNSP) and $y = 0.3165X + 1.546$ ($R^2 = 0.35$, INSP). This indicates that the two types of NSPs have similar influence on fecal nitrogen excretion (Yin et al. 2004).

Lastly, the dietary NSPs accelerated the basal metabolism of PDV tissues, thus affecting the utilization of amino acids in feedstuffs.

9.2.3.2 Other Anti-nutritional Factors

Phytic acid in plant feedstuff not only form phytate combined to phosphorus but also form insoluble complexes combined with protein and amino acids, thus reducing the utilization of these nutrients. Studies have showed that some other anti-nutritional factors, including insulin-inhibitory factor, tannin, and so on, can increase the endogenous ileal amino acid and nitrogen content (Huang et al. 2001). Tannin can reduce the true digestibility of amino acids by combination with endogenous nitrogen in digestive tract (Jansman et al. 1995).

9.2.4 Effects of Feed Starch on Amino Acid Metabolism

The synchronized supply of energy and amino acids for tissue protein synthesis is conducive to the optimal utilization of amino acids and thus maximum protein synthesis. Because the different sources of dietary starch release glucose at different rates during digestion in the pig body, the supply of glucose and amino acids have different synchronization levels. The different sources of feed starch can affect in the different degrees the protein digestion and utilization, the uptake capacity and proportion of portal vein amino acids, microbial protein synthesis, and levels of nitrogen excretion in feces and urine.

9.2.4.1 Effects of Starch Sources, Composition, and Structure on Feed Digestion and Absorption of Amino Acids

The type and source of starch have significant impact on the digestibility of protein and amino acid absorption, particularly on ileal digestibility (Li et al. 2007; Laplace et al. 2001). It was reported that the ileal digestibility of the protein and some amino acids (cysteine, serine, threonine, phenylalanine, and tyrosine) were significantly reduced when pig ingested the diet including slow digestion of pea starch (Everts et al. 1996). It was also reported that diets including high level of rapidly digestible amylopectin (glutinous rice) and high level of slowly digestible amylose (resistant starch) not only reduce the apparent and true digestibility of dietary protein but also can reduce the apparent and true ileal digestibility of aspartic acid, glutamic acid, serine, histidine, threonine, arginine, tyrosine, methionine, phenylalanine, leucine, isoleucine, and lysine. Furthermore, when the ratio of amylose to amylopectin was equal to 0.23 in the corn-based diet, the ileal amino acid digestibility was significantly better (Li et al. 2008).

It is believed that the different structures of starch can affect differently the digestion of amino acids. Slowly digestible starch can increase the endogenous nitrogen and amino acid secretion and the microbial nitrogen and amino acid content in digesta because slowly digestible starch has relatively slow digestion property. That is the reason why slowly digestible starch reduces amino acid digestibility. Rapidly digestible starch can accelerate glucose absorption, which competitively inhibits the absorption of amino acids, thereby reducing the digestion and absorption of dietary amino acids (Yin et al. 2001a, b, 2011).

9.2.4.2 Effects of the Starch Sources, Composition, and Structures on the Microbial Nitrogen Content and the Flow of Nitrogen in Fecal and Urine

Starch sources, composition, and structures can affect microbial biomass nitrogen and the flow of nitrogen in feces and urine. It was reported that urinary nitrogen excretion has a downward trend with the increased content of resistant starch diet and increased excretion of fecal nitrogen (Li et al. 2008). There is a quantitative relationship between fecal nitrogen excretion (y) and resistant starch content of the diet (X) as follows: $y = 0.785X + 9.1739$ ($R^2 = 0.997$). It was reported that pigs fed resistant starch diet have higher fecal nitrogen content. It is supposed that there are two reasons for such a result: (1) resistant starch may reduce the intestinal absorption of nitrogen and (2) the microbial fermentation in the large intestine increased microbial protein synthesis (Heijnen and Beynen 1997).

9.2.4.3 Effects of Starch Sources, Composition, and Structures on the Net Absorption and Composition Model of the Portal Vein Amino Acids

Studies have shown that starch sources, composition, and structures can affect the net absorption and amino acid composition in the portal vein. The content of essential amino acids (valine, isoleucine, phenylalanine, tryptophan, arginine, serine, cystine, tyrosine, lysine, histidine) in portal vein of pigs fed pea starch diet was higher than that in pigs fed the corn starch diet (van der Meulen et al. 1997).

The diet, including high percentage of rapidly digestible amylopectin (glutinous rice) and slowly digestible amylose (resistant starch), not only reduced the total absorption of amino acids in the portal venous blood after feed intake but also changed the composition pattern of absorbed portal amino acids, as well as reduced the proportion of EAA in the total pool of absorbed amino acids. However, the total absorption of amino acids in the portal venous blood and the composition pattern of amino acids absorbed in the portal blood were significantly improved when the ratio of amylose to amylopectin was 0.23 in the cornstarch diet (Deng et al. 2010). Since the different starch in feed have different digestibility, the different starch composition will not only affect digestion and absorption of dietary amino acids but also affect amino acid metabolism in intestinal mucosa, thus affecting the net portal absorption and composition pattern of amino acids (Huang et al. 2006).

9.2.4.4 Effects of Starch Sources, Composition, and Structure on Amino Acid Metabolism in Visceral Tissues

Starch source, composition, and structure can affect the amino acid metabolism in visceral tissues. It has been reported that pigs fed corn-based diet (the ratio of amylose to amylopectin is 0.23) have the highest protein synthesis rate (FSR) in PDV tissues and liver. The order of FSR in other diets is as follows: brown rice-based diet > starch-based diet (including high percentage of slowly digestible amylose) > glutinous rice-based diet (including high percentage of rapidly digestible amylopectin). The FSR (%/day) of spleen, pancreas, duodenum, jejunum, ileum, and colon in pigs fed glutinous rice-based diet were significantly lower than those in the corn-based group ($P < 0.05$) and brown rice-based group ($P < 0.05$). The FSR of spleen, pancreas, duodenum, jejunum, ileum, and colon in pigs fed glutinous rice-based diet were lower than corn-based group by 84.76 %, 30.34 %, 46.20 %, 32.19 %, 27.16 %, and 36.02 %, respectively. The FSR were lower than brown rice-based group by 81.97 %, 21.72 %, 46.15 %, 30.48 %, 25.14 %, and 32.78 %, respectively. The FSR also were lower than resistant starch group by 60.49 %, 17.79 %, 34.10 %, 23.23 %, 13 %, and 28.9 %, respectively. Additionally, the liver FSR in pigs fed glutinous rice-based diet were lower than corn, brown rice, and resistant starch groups by 14.99 %, 14.38 %, and 9.27 %, respectively (Deng et al. 2010).

9.3 Regulatory Means of Amino Acid Metabolism in Pigs

9.3.1 Diet Formulation

The measurement of true ileal digestible amino acids can eliminate the impacts of measurement conditions. The values of true ileal digestible amino acids can be additive. The data of true ileal digestible amino acids can make diet formulations more accurate, which can reduce the safety margin and dietary crude protein level without reducing animal performance. Some synthetic amino acids can be supplemented in the diets in order to meet the requirements of some limiting amino acids, thereby improving the overall efficiency of amino acids.

The feasibility of reducing nitrogen excretion by amino acid-balanced low-protein diet was investigated in growing pigs, which was formulated based on the true digestible amino acids and synthetic amino acids (Deng et al. 2007a, b, 2009). Experiment designed 5 crude protein levels, namely, 18.2, 16.5, 15.5, 14.5, and 13.6 %. The results showed that fecal nitrogen, urinary nitrogen, and total nitrogen excretion (g/day) decreased with the reduced dietary crude protein levels. Compared with those of crude protein group of 18.2 % (control group), fecal nitrogen were reduced by 7.45, 13.04, 13.82, and 17.39 % in crude protein groups of 16.5 %, 15.5 %, 14.5 %, and 13.6 %, respectively. Urine nitrogen decreased by 19.25, 26.86, 37.43, and 44.64 %. Total nitrogen decreased by 15.13 %, 22.00 %, 29.18 %, and 35.14 %, respectively. In addition, the ratio of urinary nitrogen to fecal nitrogen levels decreased with reduced dietary crude protein ($P < 0.0001$). The nitrogen retention also decreased with decreased dietary crude protein ($P < 0.001$).

However, the percentage of nitrogen deposition to nitrogen intake increased with the reduced dietary crude protein levels ($P < 0.01$). Dietary crude protein levels did not affect nitrogen digestibility ($P > 0.05$). Nitrogen digestibility in each group fluctuated between 83.86 and 82.19 %. Moreover, the apparent digestibility of nitrogen in groups supplemented with synthetic amino acids was lower than that of control group. But the apparent biological value of nitrogen increased with the decrease of dietary crude protein ($P < 0.001$). From the results of growth performance in different crude protein levels, it was indicated that the production performance of growing pigs was not affected by the decrease of dietary crude protein levels from 18.2 to 16.5 % in the whole trial period. On the contrary, the ADG, average daily feed intake, and the feed conversion rate were enhanced to a certain extent, while the production performance decreased with further reduction of protein levels. Carcass measurement indicated that the back fat thickness of pigs increased with the decrease of dietary protein levels ($P < 0.03$). In summary, the production performance, carcass quality, and nitrogen utilization as well as economic and ecological efficiency can be achieved when dietary crude protein levels reduced from 18.2 to 16.5 % (thus at less than 2 %).

9.3.2 Regulation of Functional Carbohydrate in Feed on Digestion and Metabolism of Amino Acids

During several decades, it has been thought that the role of carbohydrates is mainly related to their structure for energy storage material. But in recent years, it has been found that glycosylation in vivo is involved in the synthesis of numerous proteins as well as numerous life processes. In addition, many natural polysaccharides and oligosaccharides can affect various physiological functions and activities of animals and humans. Particularly oligosaccharides can regulate animal growth and metabolism, enhance immune functions, and maintain intestinal health (Huang et al. 2005, 2007; Deng et al. 2007c, d; Kong et al. 2007; Yin et al. 2008). These functions are more and more recognized by scientists. Animal growth, immunity, and gut microflora changes are closely associated with digestion and metabolism of amino acids. Therefore, these functional carbohydrates in feed have an important regulatory role in digestion and metabolism of amino acids.

9.3.2.1 Oligosaccharides Can Improve Digestion and Utilization of Amino Acids and Substitute Antibiotics in Feed

Studies have shown that oligosaccharides can improve the utilization of amino acids in pig diets. For instance, supplementation with chitosan or GMOS in diet decreased plasma urea nitrogen levels, while increased total plasma protein concentration (Tang et al. 2005). Supplementation with 0.1 % GMOS in diet increased the apparent ileal digestibility of most amino acids when compared to the control group (Hou et al. 2005).

Moreover, oligosaccharides can protect the intestinal functions and reduce the gastrointestinal mucosa diastasis, thereby reducing the endogenous ileal amino acid secretion and increasing amino acid digestibility. Supplementation of GMOS and chitosan in the casein-based diets significantly reduced the loss of endogenous amino acids (Tang et al. 2005; Deng et al. 2007c, d).

Long-term supplementation with antibiotics in feed disrupts the diversity in swine intestinal microflora and also the normal immune functions in body. In contrast, oligosaccharides can maintain the normal intestinal microbial community. Compared with antibiotics, oligosaccharides enhanced the cell and humoral immunity in animal body, increased the secretion and expression of growth-related factors, and thus promoted animal growth. It has been reported that oligosaccharides increased the levels of IgA, IgM, IgG, IL-6, IL-1β, and IL-2 in serum and the gene expression of IL-1β in blood mononuclear cells, intestinal mucosa cells, and intestinal lymph node cells compared with antibiotics group. The serum IGF-I and GH levels and muscle IGF-I gene expression in oligosaccharides group were significantly higher than those in the antibiotic group (Hou et al. 2005; Tang et al. 2005).

Oligosaccharide can be viewed as a new nutritional means for the regulation of amino acid metabolism and livestock production and quality. This may help to reduce nitrogen pollution in pig production and reduce the dosage of antibiotics in feed.

9.3.2.2 NSP and Resistant Starch

Though NSP cannot be digested by pigs, NSP can be fermented and used by pig intestinal microbes. Some NSP (such as inulin) can selectively promote the growth of beneficial bacteria. It is considered that inulin is a bifidus factor which has prebiotic effects. Inulin has the advantages to be produced at a lower cost than oligosaccharides. In addition, NSP can play probiotic-like role by digestion into oligosaccharides in the upper digestive tract.

Like oligosaccharides, resistant starch can regulate the intestinal microbial flora in a way that it may promote intestinal health. Microbial fermentation can produce short-chain fatty acids to reduce the intestinal pH, thus suppressing the activities of pH-sensitive pathogens.

9.3.3 Effects of Enzyme (NSP Enzymes and Phytase) on the Amino Acid Metabolism

Some enzymes are able to destroy anti-nutritional factors in feed and to reduce the viscosity of intestinal contents, therefore improving the feed nutrient utilization. Furthermore, several enzymes play an important regulatory role on the intestinal microflora (Garry et al. 2007). These enzymes have shown great potential in regulating feed amino acid metabolism, eliminating feed anti-nutritional factors, and reducing nitrogen excretion. In recent years, as a metabolic regulator of nutrients, enzymes have been widely applied in feed production.

9.3.3.1 Improvement of NSP Enzymes on Digestion and Utilization of Nitrogen (Amino Acids)

As mentioned above, the soluble NSP in carbohydrates is the main factor for increasing the secretion of intestinal endogenous nitrogen (amino acids), and thereby for reducing digestion and utilization of amino acids. Since there is a lack of endogenous enzymes to degrade NSP in the digestive tract of pigs, it is necessary to add exogenous enzymes in order to eliminate or reduce anti-nutritional effects of NSP. Several reports have shown that supplementation with NSP enzymes can improve the digestion and utilization of nitrogen (amino acids) in many diets, especially in diets with high NSP content.

Supplementation of xylanase in wheat- or its by-product (wheat bran, sub-powder)-based diets can improve the digestibility of proteins and reduce the loss of endogenous nitrogen (Yin et al. 2000b). Supplementation with β-glucanase or other enzymes (β-glucanase, xylanase, and protease) in shelling barley (Falcon) significantly increased the apparent ileal digestibility of crude protein and amino acid absorption in growing pigs. The supplementation of xylanase also significantly increased the apparent ileal absorption of cystine, isoleucine, lysine, methionine, and tyrosine (Yin et al. 2000a). Supplementation of β-glucanase and xylanase in naked barley diets can improve digestion and utilization of amino acids as measured by an increased plasma urea nitrogen concentration (Yin et al. 2001a, b, c, d). Moreover, supplementation with arabinoxylanase in wheat bran or rice bran diets can reduce the negative nutritional effect of soluble NSP in the small intestine and thereby increase the digestibility of amino acids in proteins (NRC 1998).

Effect of enzymes depends on the types of diet. Different types of feed have inconsistent type and content of NSP. For example, the main NSP of barley is β-glucan although the main NSP of wheat is arabinoxylans. It has been reported that, in wheat bran diet including the highest content of NSP and xylan, xylanase shows the best improvement of the digestibility of nutrients and reduction of the excretion of endogenous nitrogen compared with other diets (Yin et al. 2000b). In five barley diets supplemented with β-glucanase, xylanase together with complex enzyme β-glucanase have the most obvious improvement on apparent ileal amino acid digestibility of several amino acids. In comparison of five barley diets, complex enzyme mixture had better improvement when dietary β-glucan content was higher (Yin et al. 2001a, b, c, d). The improving effect of enzymes in diet containing high content of soluble NSP (0.8 %) in corn and wheat bran diet was obviously better than that of low content of soluble NSP (0.5 %) in corn and rice bran diet (NRC 1998).

NSP-degrading enzyme can cause hydrolysis or dissolution of NSP in the cell wall, thereby reducing or eliminating gel packets, and thus improving the digestion of amino acids in the small intestine (Li et al. 1996). The increase of amino acid ileal digestibility induced by enzyme was mainly due to the decrease in nitrogen excretion, rather than to the increase of protein true digestibility. Indeed, it has been found that the endogenous nitrogen loss is reduced by 19 % in growing pigs fed wheat bran diets supplemented with xylanase, while the ileal true digestibility of crude protein is similar (approximately 92 %) with or without enzyme (Yin et al. 2000a, b). It has been reported that the total protein in pancreatic exocrine secretion was reduced by 17 % in pigs fed with barley-based diet supplemented with enzyme (Jansman et al. 1995). The above studies suggest that the main role of the supplementation with NSP enzymes in the pig diets is to promote the degradation of dietary water-soluble NSP in the intestine. This would allow to increase the digestion of some nutrients in the small intestine instead of the large intestine, thereby reducing the intestinal fermentation of some substrates, including carbohydrates and amino acids. It would indirectly reduce the excretion of endogenous and microbial amino acids, thus increasing the utilization of amino acids.

9.3.3.2 Improvement of the Utilization of Nitrogen by Phytase

Supplementation of phytase in pig diets can break the union between phytic acid and amino acids, and thus improve utilization of protein and amino acids in feed. It has been reported that ileal apparent digestibility of dry matter, organic matter, and crude protein as well as fecal apparent digestibility of dry material, organic matter, and crude protein are significantly improved in growing pigs supplemented with 500 U/kg phytase. Ileal true digestibility of seven essential and semi-essential amino acids (histidine, arginine, methionine, lysine, isoleucine, leucine, and phenylalanine) and seven nonessential amino acids (aspartic acid, glycine, proline, alanine, threonine, tyrosine, and glutamic acid) were also improved. Nitrogen deposition, nitrogen deposition rate, and net deposition rate of nitrogen were enhanced by 2.8 g/day and 8.81 and 11.02 % compared with the control group without phytase (Fan et al. 2005).

The effect of phytase is dependent on the protein sources and on the dietary amino acids contained in proteins. Since the raw materials used in diet formulations are different, there are different possibilities for the formation of phytic acid and protein complexes with thus different effects of phytase on such substrates.

9.3.4 Regulation of Other Factors on Amino Acid Metabolism

In addition to oligosaccharides, polysaccharides, and enzymes, many other additives, such as probiotics, herbal additives, natural plant extracts, trace elements, amino acid chelators, and active peptides, can exert a regulatory role in amino acid metabolism. They act through different mechanisms that can be divided into the following categories: regulation of the gut microflora, maintenance of intestinal health, improvement of immune functions, regulation of the intestinal metabolism, increase of the activity of digestive enzymes, regulation of expression, and secretion of growth-related factors. These additives, that can directly or indirectly affect the digestion and utilization of amino acids, are considered as effective means to regulate amino acid metabolism.

Some functional amino acids, including arginine, glutamine, and proline, play versatile roles in nutritional metabolism, growth, and immunity. Recently, it has been reported that early weaning stress is associated with marked alterations in amino acid metabolism and some perturbations in population and/or activities of gut microorganisms. These perturbations were manifested in increased levels of organic acids, amino acids, and acetyl-glycoproteins as well as reduced levels of choline metabolism and lipoproteins. Dietary supplementation with arginine could partially counteract the changes of metabolites induced by weaning stress including lipid and amino acid metabolism (He et al. 2011b).

9.4 Summary

In recent years, the nutritional regulation of amino acid metabolism has been extensively studied and has made great progress in areas such as the cloning of genes related to amino acid transporters, classification of the microbial flora in different intestinal segments, and effects of the different structure of the sugar source on metabolism and utilization of amino acids in the PDV tissues (Krehbiel and Matthews 2003; Wang et al. 2008). These have built the theoretical and technological foundations used for the study of the regulation of amino acid metabolism.

However, digestion of proteins, amino acid absorption, and metabolism of amino acids in tissues are very complex. There are many areas that need further studies, such as intestinal metabolism of amino acids, relationship between amino acids and glucose metabolism, regulation of the supply of dietary amino acid and carbohydrate on the balance, and pattern of nutrients and metabolites in the portal vein. Finally, the establishment of the quality of a given dietary protein quality should take into account as an additional parameter the pattern of the amino acids in portal vein. Better understanding of the regulatory mechanisms involved in amino acid metabolism will lead to more rational and thus better use of amino acids in proteins and supplements for animal production, thus allowing to obtain the optimal economic and social benefits.

References

Chen L, Yin Y-L, Jobgen WS, Jobgen SC, Knabe DA, Hu W-X, Wu G (2007) In vitro oxidation of essential amino acids by jejunal mucosal cells of growing pigs. Livest Sci 109:19–23

Chen LX, Li P, Wang JJ, Li XL, Gao HJ, Yin YL, Hou YQ, Wu GY (2009) Catabolism of nutritionally essential amino acids in developing porcine enterocytes. Amino Acids 37: 143–152

Dai QZ, Wang KN, Yin YL, Bin SY (2005) Improvement of study on intestinal amino acid metabolism of growing pig. J Domest Anim Eco 26:63–69

De Filippo C, Cavalieri D, Di Paola M, Ramazzotti M, Poullet JB, Massart S, Collini S, Pieraccini G, Lionetti P (2010) Impact of diet in shaping gut microbiota revealed by a comparative study in children from Europe and rural Africa. Proc Natl Acad Sci U S A 107:14691–14696

Deng D, Huang RL, Li TJ, Wu GY, Xie MY, Tang ZR, Kang P, Zhang YM, Fan MZ, Kong XF, Ruan Z, Xiong H, Deng ZY, Yin YL (2007a) Nitrogen balance in barrows fed low-protein diets supplemented with essential amino acids. Livest Sci 109:220–223

Deng D, Li AK, Chu WY, Huang RL, Li TJ, Kong XF, Liu ZJ, Wu GY, Zhang YM, Yin YL (2007b) Growth performance and metabolic responses in barrows fed low-protein diets supplemented with essential amino acids. Livest Sci 109:224–227

Deng ZY, Zhang JW, Li J, Fan YW, Cao SW, Huang RL, Yin YL, Zhong HY, Li TJ (2007c) Effect of polysaccharides of cassiae seeds on the intestinal microflora of piglets. Asia Pac J Clin Nutr 16(Suppl 1):143–147

Deng ZY, Zhang JW, Wu GY, Yin YL, Ruan Z, Li TJ, Chu WY, Kong XF, Zhang YM, Fan YW, Liu R, Huang RL (2007d) Dietary supplementation with polysaccharides from Semen cassiae enhances immunoglobulin production and interleukin gene expression in early-weaned piglets. J Sci Food Agric 87:1868–1873

Deng D, Yao K, Chu WY, Li TJ, Huang RL, Yin YL, Liu HQ, Zhang JS, Wu GY (2009) Impaired translation initiation activation and reduced protein synthesis in weaned piglets fed a low-protein diet. J Nutr Biochem 20:544–552

Deng J, Wu X, Bin S, Li TJ, Huang R, Liu Z, Liu Y, Ruan Z, Deng Z, Hou Y, Yin YL (2010) Dietary amylose and amylopectin ratio and resistant starch content affects plasma glucose, lactic acid, hormone levels1and protein synthesis in splanchnic tissues. Z Tierphysiol Tierernahr Futtermittelkd 94:220–226

Deutz NE, Bruins MJ, Soeters PB (1998) Infusion of soy and casein protein meals affects interorgan amino acid metabolism and urea kinetics differently in pigs. J Nutr 128:2435–2445

Dudley MA, Wykes LJ, Dudley AW Jr, Burrin DG, Nichols BL, Rosenberger J, Jahoor F, Heird WC, Reeds PJ (1998) Parenteral nutrition selectively decreases protein synthesis in the small intestine. Am J Physiol 274:G131–G137

Everts H, Dekker RA, Smits B, Cone JW (1996) The digestion of maize and native pea starch in the small intestine of pigs. Proc Nutr Soc 55:59A

Fan MZ, Li TJ, Yin YL, Fang RJ, Tang ZY, Hou ZP, Huang RL, Deng ZY, Zhong HY, Zhang YG, Zhang J, Wang B, Schulze H (2005) Effect of phytase supplementation with two levels of phosphorus diets on ileal and faecal digestibilities of nutrients and phosphorus, calcium, nitrogen and energy balances in growing pigs. Anim Sci 81:67–75

Fang Z, Yao K, Zhang X, Zhao S, Sun Z, Tian G, Yu B, Lin Y, Zhu B, Jia G, Zhang K, Chen D, Wu D (2010) Nutrition and health relevant regulation of intestinal sulfur amino acid metabolism. Amino Acids 39:633–640

Garry BP, Fogarty M, Curran TP, O'Connell MJ, O'Doherty JV (2007) The effect of cereal type and enzyme addition on pig performance, intestinal microflora, and ammonia and odour emissions. Animal 1:751–757

He Q, Ren P, Kong X, Xu W, Tang H, Yin Y, Wang Y (2011a) Intrauterine growth restriction alters the metabonome of the serum and jejunum in piglets. Mol Biosyst 7:2147–2155

He Q, Tang H, Ren P, Kong X, Wu G, Yin Y, Wang Y (2011b) Dietary supplementation with l-arginine partially counteracts serum metabonome induced by weaning stress in piglets. J Proteome Res 10:5214–5221

Heijnen ML, Beynen AC (1997) Consumption of retrograded (RS3) but not uncooked (RS2) resistant starch shifts nitrogen excretion from urine to feces in cannulated piglets. J Nutr 127: 1828–1832

Hoerr RA, Matthews DE, Bier DM, Young VR (1993) Effects of protein restriction and acute refeeding on leucine and lysine kinetics in young men. Am J Physiol 264:E567–E575

Hou ZP, Yin YL, Jeaurond EA, Namkung HH, de Lange CFM (2005) Effect of feeding galacto-mannan-oligosaccharide on growth perfoemance, immune response and microflora in early-weaned piglets. J Anim Sci 83:M101

Huang SX, Sauer WC, Marty B (2001) Ileal digestibilities of neutral detergent fiber, crude protein, and amino acids associated with neutral detergent fiber in wheat shorts for growing pigs. J Anim Sci 79:2388–2396

Huang RL, Yin YL, Wu GY, Zhang YG, Li TJ, Li LL, Li MX, Tang ZR, Zhang J, Wang B, He JH, Nie XZ (2005) Effect of dietary oligochitosan supplementation on ileal digestibility of nutrients and performance in broilers. Poult Sci 84:1383–1388

Huang RL, Dai QZ, Li TJ, Bin SY, Hou ZP, Zhan P (2006) Study on the relationship of average daily gain with net portal absorption of nutrients by portal vein—drained organs in growing pigs. Acta Vet Zootech Sin 37(262):269

Huang RL, Yin YL, Li MX, Wu GY, Li TJ, Li LL, Yang CB, Zhang J, Wang B, Deng ZY, Zhang YG, Tang ZR, Kang P, Guo YM (2007) Dietary oligochitosan supplementation enhances immune status of broilers. J Sci Food Agric 87:153–159

Jansman AJ, Verstegen MW, Huisman J, van den Berg JW (1995) Effects of hulls of faba beans (Vicia faba L.) with a low or high content of condensed tannins on the apparent ileal and fecal digestibility of nutrients and the excretion of endogenous protein in ileal digesta and feces of pigs. J Anim Sci 73:118–127

Kong XF, Yin YL, Wu GY, Hou YQ, Liu HJ, Yin FG, Li TJ, Huang RL, Ruan Z, Xiong H, Deng ZY, Xie MY, Liao YY, Chen LX, Kim SW (2007) Dietary supplementation with acanthopanax senticosus extract modulates cellular and humoral immunities in weaned piglets. Asian-Aust J Anim Sci 20:1453–1461

Kong XF, Zhang YZ, Wu X, Yin YL, Tan ZL, Feng Y, Yan FY, Bo MJ, Huang RL, Li TJ (2009) Fermentation characterization of Chinese yam polysaccharide and its effects on the gut microbiota of rats. Int J Microbiol 9:598152. doi:10.1155/2009/598152

Krehbiel CR, Matthews JC (2003) Absorption of amino acids and peptides. In: Mello JPF (ed) Amino acids in animal nutrition. CABI, Wallingford, pp 41–70

Laplace JP, Aumaitre A, Rerat A (2001) Forty years of achievement in French research on digestive physiology in the pig. Reprod Nutr Dev 41:129–151

Li S, Sauer WC, Mosenthin R, Kerr B (1996) Effect of β-glucanase supplementation of cereal-based diets for starter pigs on the apparent digestibility of dry matter, crude protein and energy. Anim Feed Sci Technol 59:223–231

Li TJ, Huang RL, Wu GY, Lin YC, Jiang ZY, Kong XF, Chu WY, Zhang YM, Kang P, Hou ZP, Fan MZ, Liao YP, Yin YL (2007) Growth performance and nitrogen metabolism in weaned pigs fed diets containing different sources of starch. Livest Sci 109:73–76

Li TJ, Dai QZ, Yin YL, Zhang J, Huang RL, Ruan Z, Deng Z, Xie M (2008) Dietary starch sources affect net portal appearance of amino acids and glucose in growing pigs. Animal 2:723–729

Libao-Mercado AJ, Zhu CL, Cant JP, Lapierre H, Thibault JN, Seve B, Fuller MF, de Lange CF (2009) Dietary and endogenous amino acids are the main contributors to microbial protein in the upper gut of normally nourished pigs. J Nutr 139:1088–1094

Low A, Rainbird A (1984) Effect of guar gum on nitrogen secretion into isolated loops of jejunum in conscious growing pigs. Br J Nutr 52:499–505

MacRae JC, Bruce LA, Brown DS, Calder AG (1997a) Amino acid use by the gastrointestinal tract of sheep given lucerne forage. Am J Physiol 273:G1200–G1207

MacRae JC, Bruce LA, Brown DS, Farningham DA, Franklin M (1997b) Absorption of amino acids from the intestine and their net flux across the mesenteric- and portal-drained viscera of lambs. J Anim Sci 75:3307–3314

NRC (1998) Nutrient requirements of swine, 10 Revised Editionth edn. National Academic Press, Washington, DC

Stoll B, Henry J, Reeds PJ, Yu H, Jahoor F, Burrin DG (1998) Catabolism dominates the first-pass intestinal metabolism of dietary essential amino acids in milk protein-fed piglets. J Nutr 128:606–614

Tang ZR, Yin YL, Nyachoti CMM, Zhong HY, Huang RL, Yang CB, Li TJ, Yang XJ, Gong JJ, Xing JJ, Sun ZH, Fan MZ (2005) Effect of dietary supplementation of Chitosan and galacto-mannan-oligosaccharide on serum parameters and the insulin-like growth factor-1 mRNA expression in early-weaned piglets. Domest Anim Endocrinol 28:430–441

Turnbaugh PJ, Ley RE, Mahowald MA, Magrini V, Mardis ER, Gordon JI (2006) An obesity-associated gut microbiome with increased capacity for energy harvest. Nature 444:1027–1031

van der Meulen J, Bakker JG, Smits B, de Visser H (1997) Effects of source of starch on net portal flux of glucose, lactate, volatile fatty acids and amino acids in the pig. Br J Nutr 78:533–544

Velagapudi VR, Hezaveh R, Reigstad CS, Gopalacharyulu P, Yetukuri L, Islam S, Felin J, Perkins R, Boren J, Oresic M, Backhed F (2010) The gut microbiota modulates host energy and lipid metabolism in mice. J Lipid Res 51:1101–1112

Wang J, Chen L, Li P, Li X, Zhou H, Wang F, Li D, Yin Y, Wu G (2008) Gene expression is altered in piglet small intestine by weaning and dietary glutamine supplementation. J Nutr 138: 1025–1032

Wang WW, Qiao SY, Li DF (2009) Amino acids and gut function. Amino Acids 37:105–110

Wang W, Zeng X, Mao X, Wu G, Qiao S (2010) Optimal dietary true ileal digestible threonine for supporting the mucosal barrier in small intestine of weanling pigs. J Nutr 140:981–986
Wang S, Francois B, Feng Z, Yulong Y (2011) Intestinal microbiota: development, metabolism and functions. J Food Agric Environ 9:121–129
Xu JH, Tan XW, Yin YL, Li TJ (2005) Effects of gum on the energy and nitrogen utilization in pigs. Am J Chin Clin Med 7:49–55
Yin YL, Tan BE (2010) Manipulation of dietary nitrogen, amino acids and phosphorus to reduce environmental impact of swine production and enhance animal health. J Food Agric Environ 8: 447–462
Yin YL, McEvoy JDG, Schulze H, Hennig U, Souffranr WB, McCraken KJ (2000a) Apparent digestibility(ileal and overall) of nutrients and endogenous nitrogen losses in growing pigs fed wheat or its by-products without or with xylanase supplementation. Live Prod Sci 62:119–132
Yin YL, McEvoy J, Schulze H, McCracken KJ (2000b) Studies on cannulation method and alternative indigestible markers and the effects of food enzyme supplementation in barley-based diets on ileal and overall apparent digestibility in growing pigs. Anim Sci 70:63–72
Yin YL, McEvoy J, Souffrant WB, Schulze H, McCracken KJ (2000c) Apparent digestibility (ileal and overall) of nutrients as evaluated with PVTC-cannulated or ileo-rectal anastomised pigs fed diets containing two indigestible markers. Livest Prod Sci 62:133–141
Yin YL, Baidoo SK, Boychul JJJ, Simmins H (2001a) Performance and carcass characteristics of pigs and broiler fed diets containing micronized barley, ground barley, wheat and maize. J Sci Food Agric 81:1487–1497
Yin YL, Baidoo SK, Jin ZL, Liu YG, Schulze H, Simmins PH (2001b) The effect of different carbohydrase and protease supplementation on apparent (ileal and overall) digestibility of nutrient of five hulless barley varieties in young pigs. Livest Prod Sci 71:109–120
Yin YL, Baidoo SK, Schulze H, Simmins PH (2001c) Effect of supplementing diets containing hulless barley varieties having different levels of non-starch polysaccharides with β-glucanase and xylanase on the physiological status of gastrointestinal tract and nutrient digestibility of weaned pigs. Livest Prod Sci 71:97–107
Yin YL, McEvoy J, Schulze H, McCracken KJ (2001d) Effects of xylanase and antibiotic addition on ileal and overall apparent digestibility and evaluating HCL-insoluble ash as an indigestible marker in growing pigs. Anim Sci 72:95–103
Yin YL, Huang RL, Yen JT, Libao-Mercado AJ, de Lange CFM (2003) Net portal appearance of amino acids and glucose, and energy expenditure by portal vein-drained organs in growing pigs fed casein and cornstarch based diets or a wheat shorts based diet. 9th International symposium on digestive physiology in pigs, Banff, AB, Canada
Yin YL, Deng ZY, Huang HL, Zhong HY, Hou ZP, Gong J, Liu Q (2004) Nutritional and health functions of carbohydrate for pigs. J Anim Feed Sci 13:523–538
Yin YL, Huang RL, Li TJ, Ruan Z, Xie MY, Deng ZY, Hou YQ, Wu GY (2010a) Amino acid metabolism in the portal-drained viscera of young pigs: effects of dietary supplementation with chitosan and pea hull. Amino Acids 39:1581–1587
Yin FG, Zhang ZZ, Huang J, Yin YL (2010b) Digestion rate of dietary starch affects systemic circulation of amino acids in weaned pigs. Br J Nutr 103:1404–1412
Yin YL, Tang ZR, Sun ZH, Liu ZQ, Li TJ, Huang RL, Ruan Z, Deng ZY, Gao B, Chen LX, Wu GY, Kim SW (2008) Effect of galacto-mannan-oligosaccharides or chitosan supplementation on cytoimmunity and humoral immunity response in early-weaned piglets. Asian-Aust J Anim Sci 21:723–731
Yin F, Yin Y, Zhang Z, Xie M, Huang J, Huang R, Li T (2011) Digestion rate of dietary starch affects the systemic circulation of lipid profiles and lipid metabolism-related gene expression in weaned pigs. Br J Nutr 106:369–377
Zhang YG, Yin YL, Fang J, Wang Q (2012) Pig production in subtropical agriculture. J Sci Food Agric 92:1016–1024

Chapter 10
Amino Acids and Hormone Secretion in Pigs

Bie Tan, Yulong Yin, Yizhen Wang, De Wu, Yingcai Lin, Xuemei Ding, Yuyun Mu, and Jirong Lv

Contents

10.1 Introduction

Amino acids have been implicated in the regulation of hormone synthesis and secretion (Kuhara et al. 1991; Newsholme et al. 2005; Zeng et al. 2012). Amino acids stimulate or inhibit the release of some pituitary, pancreatic, and gastrointestinal hormones by oral or intravenous administration. For example, tyrosine (or

B. Tan • Y. Yin (✉)
Research Center of Healthy Breeding of Livestock and Poultry and Key Laboratory of Agro-ecological Processes in Subtropical Region, Institute of Subtropical Agriculture The Chinese Academy of Sciences, Changsha, Hunan 410125, China
e-mail: yinyulong@isa.ac.cn

Y. Wang
College of Animal Science, Zhejiang University, The Key Laboratory of Molecular Animal Nutrition of Ministry of Education, Hangzhou, P.R. China

D. Wu • X. Ding
Institute of animal Nutrition, Sichuan Agricultural University, Sichuan, China

Y. Lin
Institute of Animal Science, Guangdong Academy of Agricultural Science, Guangdong, China

Y. Mu
Shanghai Xinnong Feed Co. Ltd, Songjiang, China

J. Lv
Chengdu Dadhank Biotech Corp, Chengdu, Sichuan, China

F. Blachier et al. (eds.), *Nutritional and Physiological Functions of Amino Acids in Pigs*,
DOI 10.1007/978-3-7091-1328-8_10, © Springer-Verlag Wien 2013

phenylalanine) is the precursor for the synthesis of epinephrine, norepinephrine, and thyroid hormones. Arginine stimulates the secretion of insulin, growth hormone (GH), prolactin, glucagon, and placental lactogen (Flynn et al. 2002). Glutamine and leucine also increase insulin release from the pancreatic E-cells (Newsholme et al. 2005). Amino acids also induce the secretion of GI hormones such as glucagon-like peptide-1 (GLP-1), cholecystokinin (CCK), and ghrelin (Nakamura et al. 2011). The mechanism of hormones release induced by amino acids may involve membrane depolarization and certain receptors and transporters. It is becoming increasingly recognized that gastrointestinal chemosensing involves additional mechanisms. In this chapter, the regulations and mechanisms of amino acids on GH–IGF-I axis hormones release, pancreatic secretion, and gastrointestinal hormone secretion are discussed.

10.2 Amino Acids and GH/IGF-I Axis

GH and insulin-like growth factor-I (IGF-I) control growth in swine and the GH/IGF-I axis plays an essential and central role in the regulation of body growth and composition, as well as cellular proliferation and differentiation processes (Li et al. 2002; Lucy 2008). Specific amino acids appear to be important in the GH–IGF-I axis. But under normal growing conditions, the GH–IGF-I axis is probably not limited by amino acid availability. However, in conditions of low intake and some pathological states, these signals may become crucial.

The growth-inhibitory effect of amino acid deficiencies might be caused in part by impaired action of GH–IGF-I axis. Bajoria et al. (2002) demonstrated the link between the reduction in certain essential and nonessential amino acids and alterations in fetal circulating levels of IGFBP-I, in growth-restricted twins. Single depletion of arginine, proline, threonine, tryptophan, or Val caused a block of GH-stimulated IGF-I gene expression in cultured pig hepatocytes (Brameld et al. 1999). The GH, IGF-I, and IGFBPs in plasma are sensitive to changes of nutritional status. Many studies have indicated that growth retardation induced by undernutrition is usually followed by an increase of GH and a decrease of IGF-I in plasma of pigs (Vance et al. 1992; Guay and Trottier 2006). Katsumata et al. (2002, 2004) found that a low-lysine or a low-threonine diet significantly reduced plasma IGF-I and IGFBP3 concentrations. The reduction in plasma IGF-I caused by reduced dietary lysine or threonine might have been partly due to suppression of posttranscriptional events of IGF-I expression because low-lysine or low-threonine diet did not affect IGF-I mRNA expression in the liver. An approximately 0.30–0.40 proportional shortage in a single amino acid in a pig diet compared with its requirement does not seems to be severe enough to suppress IGF-I mRNA expression in the liver. Increased clearance rate of circulating IGF-I, indicated by lower IGFBP3 level, may be one reason for the lower plasma IGF-I caused by restriction of a single amino acid in a pig diet. In the study by Wiltafskya et al. (2010), the growth retardation resulting form excess dietary leucine was observed. This might be partially caused by decreased activity of the GH–IGF-I axis as a result of leucine-induced valine deficiency.

Aspartate and arginine are most effective in stimulating GH secretion (Bequette 2003). It has been proposed that excitatory amino acids (EAA) play an important role in regulating secretion of pituitary hormones including GH (Estienne and Barb 2002). Intravenous administration of 100 and 150 mg/kg BW of aspartate or glutamate to gilts increased GH secretion, and aspartate was found to be a more potent secretagogue of GH secretion than was glutamate (Barb et al. 1996). Elevated plasma levels of arginine have been found to correlate with increased secretion of GH and IGF-I. Maximum GH level and the area under the GH curve (AUC) were increased in a dose-dependent manner in response to arginine (0.5–1 g/kg BW) treatment via the duodenum in pigs. However, GH response to the combined arginine and aspartate treatment was delayed compared to arginine alone and was not dose dependent (Cochard et al. 1998). Saleri et al. (2003) demonstrated that arginine also significantly stimulated GH secretion in pig pituitary cells. But in growing-finishing pigs, dietary 1 % arginine supplementation did not affect serum concentrations of growth hormone and insulin-like growth factor-I (Tan et al. 2009).

GH release has a complex regulation with two antagonistic hypothalamic hormones, growth hormone releasing hormone (GHRH) and somatostatin, as well as the liver-derived hormone IGF-I. It has been reported that the mechanism of regulation of GH secretion by amino acid administration is provided by promotion of secretion of a GHRH and by suppression of somatostatin secretion (Casanueva and Diequez 2004). Arginine stimulates GH secretion probably via a direct cholinergic effect, membrane depolarization, and subsequent calcium influx, and by inhibiting hypothalamic somatostatin release. NO production also appears to be involved in the control of GH release by arginine in pig pituitary cells (Saleri et al. 2003). GH secretion is under the negative feedback control of IGF-I, which was observed in GH response to arginine by Nass et al. (2002). This effect was present before the start of the arginine infusion, and this likely accounts for the lower peak GH response to arginine (Nass et al. 2002). The actions of EAAs on GH secretion are mediated through interaction with different postsynaptic receptors, and the *N*-methyl-D,L-aspartic acid (NMA) has attracted attention (Aguilar et al. 2005; Tena-Sempere et al. 2000). NMA treatment increased circulating GH concentrations in prepubertal gilts (Estienne et al. 1995) and boars (Estienne et al. 2000). Xi et al. (2002) reported that feeding

NMA at a dose of 50 mg/kg to finishing pig significantly also increased the IGF-I level in the serum, liver, and muscle and postulated that this increase is primarily caused by the stimulation of GH release in pituitary.

10.3 Amino Acids and Release of Glucagon and Insulin

In normal physiology, protein meals, with their attendant amino acid loads, stimulate glucagon and insulin secretion (Charlton et al. 1996). Secretion of insulin and glucagon by specific cells in the pancreas is regulated usually in the opposite direction; but a striking exception is that insulin and glucagon secretion are stimulated by amino acids or a protein meal (Muller et al. 1971). The balance between insulin and glucagon regulates the metabolism of glucose, fat, and protein, thereby promoting a stable inner metabolic milieu; and this action is done via pathways involving cAMP signaling (McKnight et al. 2010). Glucagon activates adenylyl cyclase to generate cAMP, which stimulates protein kinase A (PKA) (Mersmann and Smith 2005). Insulin has opposing effects to glucagon, causing uptake of glucose and amino acid from the blood to various tissues and stimulation of glycogen, protein, and lipid synthesis.

Unlike glucose, which inhibits glucagon secretion and stimulates insulin release, amino acids stimulate release of both hormones. Specific amino acids are known to acutely and chronically regulate insulin secretion from pancreatic β-cells (Newsholme et al. 2005), and leucine, arginine, and lysine are considered as the most potent stimulators of insulin secretion. The accumulation of positively charged amino acids including L-arginine, L-homoarginine, and L-ornithine in islet cells represents an essential determinant of their secretory action (Blachier et al. 1989b). In the presence of D-glucose, L-lysine stimulates insulin secretion to the same extent as L-arginine or L-ornithine, but the hormonal release is not further enhanced by combinations of these cationic amino acids (Blachier et al. 1989a; Sener et al. 1989a). Arginine, alanine, and glutamine are potent stimulators of glucagon secretion from pancreatic α-cells (Kuhara et al. 1991; Dumonteil et al. 2000; Quesada et al. 2008). Östenson and Grebing (1985) showed that glutamine significantly enhanced glucagon release from 297 to 528 pg/μg DNA/h in normal islets and from 553 to 806 pg/μg DNA/h in α-cell rich islets in guinea pigs. Arginine is the amino acid known to stimulate the highest production of both insulin and glucagon. A great deal of studies have showed that arginine stimulates insulin secretion in pigs (Kim and Wu 2004; Yao et al. 2008) and humans (Eaton and Schade 1974), diabetic subjects (Stingl et al. 2002), and obese subjects (Copinschi et al. 1967). In growing-finishing pigs, dietary 1 % arginine supplementation increased serum concentrations of glucagon by 36 %, compared with the control group (Tan et al. 2009). However, intracellular glutamate or glutamate-derived metabolites contribute to fuel-induced hormone secretion in β- but not α-cells (Feldmann et al. 2011). A few amino acids such as isoleucine can inhibit α-cell secretion while leucine has a dual effect depending on its concentrations (Quesada et al. 2008).

Glucose plays opposite roles in the mechanisms leading to amino acid-induced hormone release from the α- and β-cells, functioning as an inhibitor in the first case and as a permissive agent in the second (Pagliara et al. 1974). There are suggestions that arginine stimulates insulin release by modulating the insulinogenic signal evoked by glucose (Efendic et al. 1971). The regulation of glucagon secretion by glutamine appears to be reciprocally related to factors affecting glucose metabolism and ATP-levels in the α-cell (Östenson and Grebing 1985). Certain amino acids also can directly stimulate insulin secretion primarily by increasing [Ca^{2+}] (Flatt et al. 1991; Maechler et al. 2002). The release of insulin evoked by cationic amino acids could be due to depolarization of the plasma membrane with subsequent gating of voltage-sensitive Ca^{2+} channels and/or to some other biophysical effect (Blachier et al. 1989b; Smith et al. 1997). L-Lysine stimulated insulin release involving the similar biophysical mechanism in pancreatic β-cells and also act as a fuel in islet cells (Sener et al. 1989a). In addition, endogenously formed polyamines and NO may play a role in the secretory response of islet cells to amino acids (Henningsson and Lundquist 1998; Sener et al. 1989b). NO is a negative modulator of insulin release and a positive modulator of glucagon release induced by L-arginine or L-homoarginine. These NO effects are also mainly exerted independently of membrane depolarization events (Henningsson and Lundquist 1998). Recent studies have showed that G-protein-coupled receptors (GPCRs) are involved in the regulation of amino acids-induced insulin and glucagon secretion, which provided potential new treatments for diabetes (Layden et al. 2010; Winzell and Ahrén 2007).

10.4 Amino Acids and Gastrointestinal Hormone Secretion

It has been reported that free amino acids, liberated from protein during digestion, induce the secretion of GI hormones such as GLP-1, CCK, and ghrelin (Nakamura et al. 2011). Ingestion of an amino acid mixture stimulates GLP-1 release in humans (Herrmann et al. 1995). GLUTag cells secrete GLP-1 in response to a range of individual amino acids, including alanine, serine, glutamine, asparagine, and glycine. And glutamine was a more effective stimulus of GLP-1 release than the other amino acids tested (Gameiro et al. 2005; Reimann et al. 2004). Glutamine has also been found to trigger GLP-1 secretion in normal weight and obese human subjects (Greenfield et al. 2009). But in dogs, basal concentrations of CCK and gastrin were not affected by intravenous infusion of amino acids (Fried et al. 1982). Of the amino acids that stimulate CCK secretion, the aromatic amino acids, L-phenylalanine and L-tryptophan, are the most effective (Liddle 1994). With the opposite effect on appetite compared with CCK, ghrelin is increased in plasma in response to oral essential amino acids mixture supplementation in humans (Knerr et al. 2003). High plasma levels of amino acids stimulated ghrelin secretion in horse (Sugino et al. 2010). In weanling pigs, oral tryptophan ingestion also increased plasma ghrelin level and ghrelin expression in gastric fundus (Zhang et al. 2007).

The mechanism for amino acids induced increase of the secretion of GI hormones may involve the membrane depolarization and Ca^{2+}-sensing receptor (CaSR) (Reimann et al. 2006; Wang et al. 2011). The stimuli of certain amino acids result in membrane depolarization and Ca^{2+} entry through voltage-gated Ca^{2+} channels. Alanine and glycine showed great capacity for generating membrane depolarization and increasing intracellular Ca^{2+} in GLUTag cells (Gameiro et al. 2005). Several lines of evidence indicate that intracellular Ca^{2+} concentrations ($[Ca^{2+}]_i$) is important in stimulating CCK release and that amino acids regulate this signal through CaSR (Cancela and Petersen 2002; Mangel et al. 1995). In a CCK-secreting cell line, phenylalanine was shown to stimulate CCK release and to increase $[Ca^{2+}]_i$ through a mechanism that was dependent on an L-type Ca^{2+}

channel (Mangel et al. 1995). In isolated CCK-secreting cells loaded with a Ca^{2+}-sensitive dye, phenylalanine and tryptophan, but not nonaromatic amino acids, caused an increase in $[Ca^{2+}]_i$. The increase in $[Ca^{2+}]_i$ was blocked by the CaSR inhibitor Calhex 231 (Wang et al. 2011). Glutamine triggers depolarization by Na^+-coupled uptake, together with voltage-gated Ca^{2+} entry, while a second pathway involves elevation of intracellular cAMP levels (Diakogiannaki et al. 2011; Tolhurst et al. 2011). Glycine and γ-amino butyric acid (GABA) receptors open Cl^- channels on the surface membrane resulting in Cl^- efflux and subsequent membrane depolarization (Reimann et al. 2006; Tolhurst et al. 2011).

The GI hormones secretion increased by amino acids can be viewed as the effects of chemical "messengers," which signal the gastrointestinal tract and accessory organs to perform different actions (Jahan-Mihan et al. 2011). Chemoreceptors located in the small intestine detect luminal amino acids and trigger the release of gut hormones from mucosal enteroendocrine cells (EECs). These hormones act on gut vagal receptors or are released into the blood and reach the central nervous system. Absorbed amino acids can also induce metabolic signal produced in the periphery or directly in some specific brain area (Tomé et al. 2009). In addition to CaSR described above, metabotropic glutamate receptor subtypes (mGluRs), taste receptor type 1 (T1R1), and G-protein-coupled receptor family C group 6 subtype A (GPCR6A) also partially mediated the chemical sensing between dietary amino acids and EECs (Raybould 2010). The amino acid selectivity profiles for CaSR, GPRC6A, and the T1R1/T1R3 heterodimer indicate that aromatic amino acids such as phenylalanine, tryptophan, and histidine are CaSR specific. Cysteine is both acting on CaSR- and GPRC6A. In contrast, lysine seems to act on GPRC6A (Nakamura et al. 2011). GPR93 has been proposed to act like a taste receptor, on the basis that its expression and activation promote CCK secretion, at least in a CCK-secreting EEC line (Choi et al. 2007). Amino acid transporters also may function as an electrogenic and metabolic transceptors. For example, glutamine triggered GLP-1 release in a murine EEC line upon binding to the Na^+-dependent amino acid transporter SLC38A2 (Reimann et al. 2004). However, the involvement of amino acid metabolic transceptors in mammals in intestinal amino acid sensing has not been fully investigated (Miguel-Aliaga 2012).

References

Aguilar E, Tena-Sempere M, Pinilla L (2005) Role of excitatory amino acids in the control of growth hormone secretion. Endocrine 28:295–301

Bajoria R, Sooranna SR, Ward S, Hancock M (2002) Placenta as a link between amino acids, insulin-IGF axis, and low birth weight: evidence from twin studies. J Clin Endocrinol Metab 87:308–315

Barb CR, Campbell RM, Armstrong JD, Cox NM (1996) Aspartate and glutamate modulation of growth hormone secretion in the pig: possible site of action. Domest Anim Endocrinol 13(1):81–90

Bequette BJ (2003) Amino acid metabolism in animals: an overview. In: D'Mello JPF (ed) Amino acids in animal nutrition. CABI Publishing, Wallingford, pp 87–101

Blachier F, Leclercq-Meyer V, Marchand J, Woussen-Colle MC, Mathias PC, Sener A, Malaisse WJ (1989a) Stimulus-secretion coupling of arginine-induced insulin release. Functional response of islets to L-arginine and L-ornithine. Biochim Biophys Acta 1013:144–151

Blachier F, Mourtada A, Sener A, Malaisse WJ (1989b) Stimulus-secretion coupling of arginine-induced insulin release. Uptake of metabolized and nonmetabolized cationic amino acids by pancreatic islets. Endocrinology 124(1):134–141

Brameld JM, Gilmour RS, Buttery PJ (1999) Glucose and amino acids interact with hormones to control expression of insulin-like growth factor-1 and growth hormone receptor mRNA in cultured pig hepatocytes. J Nutr 129:1298–1306

Cancela JM, Petersen OH (2002) Regulation of intracellular Ca^{2+} stores by multiple Ca^{2+}-releasing messengers. Diabetes 51(3):S349–S357

Casanueva FF, Diequez C (2004) Ghrelin a new hormone implicated in the regulation of growth hormone secretion and body energy homeostasis. Growth Genet Horm 20:1–8

Charlton MR, Adey DB, Nair KS (1996) Evidence for a catabolic role of glucagon during an amino acid load. J Clin Invest 98(1):90–99

Choi S, Lee M, Shiu AL, Yo SJ, Hallden G, Aponte GW (2007) GPR93 activation by protein hydrolysate induces CCK transcription and secretion in STC-1 cells. Am J Physiol Gastrointest Liver Physiol 292:G1366–G1375

Cochard A, Guilhermet R, Bonneau M (1998) Plasma growth hormone (GH), insulin and amino acid responses to arginine with or without aspartic acid in pigs. Effect of the dose. Reprod Nutr Dev 38(3):331–343

Copinschi G, Wegienka LC, Hane S, Forsham PH (1967) Effect of arginine on serum levels of insulin and growth hormone in obese subjects. Metabolism 16:485–491

Diakogiannaki E, Gribble FM, Reimann F (2011) Nutrient detection by incretin hormone secreting cells. Physiol Behav 106(3):387–393. doi:10.1016/j.physbeh.2011.12.001

Dumonteil E, Magnan C, Ritz-Laser B, Ktorza A, Meda P, Philippe J (2000) Glucose regulates proinsulin and prosomatostatin but not proglucagon messenger ribonucleic acid levels in rat pancreatic islets. Endocrinology 141:174–180

Eaton RP, Schade DS (1974) Effect of clofibrate on arginine-stimulated glucagon and insulin secretion in man. Metabolism 23:445–454

Efendic S, Cerasi E, Luft R (1971) Role of glucose in arginine-induced insulin release in man. Metabolism 20(6):568–579

Estienne MJ, Barb CR (2002) Modulation of growth hormone, luteinizing hormone, and testosterone secretion by excitatory amino acids in boars. Reprod Biol 2(1):13–24

Estienne MJ, Harter-Dennis JM, Barb CR, Hartsock TG (1995) Luteinizing hormone and growth hormone concentrations in serum of prepubertal gilts treated with n-methyl-d, l-aspartate. Domest Anim Endocrinol 12:207–213

Estienne MJ, Broughton DS, Barb CR (2000) Serum concentrations of luteinizing hormone, growth hormone, testosterone, estradiol, and leptin in boars treated with n-methyl-d, l-aspartate. J Anim Sci 78:365–370

Feldmann N, del Rio RM, Gjinovci A, Tamarit-Rodriguez J, Wollheim CB, Wiederkehr A (2011) Reduction of plasma membrane glutamate transport potentiates insulin but not glucagon secretion in pancreatic islet cells. Mol Cell Endocrinol 338(1–2):46–57

Flatt PR, Barnett CR, Shibier O, Swanston-Flatt SK (1991) Direct and indirect actions of nutrients in the regulation of insulin secretion from the pancreatic β cells. Proc Nutr Soc 50(3):559–566

Flynn NE, Meininger CJ, Haynes TE, Wu G (2002) The metabolic basis of arginine nutrition and pharmacotherapy. Biomed Pharmacother 56:427–438

Fried GM, Ogden WD, Rhea A, Greeley G, Thompson JC (1982) Pancreatic protein secretion and gastrointestinal hormone release in response to parenteral amino acids and lipid in dogs. Surgery 92(5):902–905

Gameiro A, Reimann F, Habib AM, O'Malley D, Williams L, Simpson AK, Gribble FM (2005) The neurotransmitters glycine and GABA stimulate glucagon-like peptide-1 release from the GLUTag cell line. J Physiol 569:761–772
Greenfield JR, Farooqi IS, Keogh JM, Henning E, Habib AM, Blackwood A, Reimann F, Holst JJ, Gribble FM (2009) Oral glutamine increases circulating glucagon-like peptide 1, glucagon, and insulin concentrations in lean, obese, and type 2 diabetic subjects. Am J Clin Nutr 89:106–113
Guay F, Trottier NL (2006) Muscle growth and plasma concentrations of amino acids, insulin-like growth factor-I, and insulin in growing pigs fed reduced-protein diets. J Anim Sci 84:3010–3019
Henningsson R, Lundquist I (1998) Arginine-induced insulin release is decreased and glucagon increased in parallel with islet NO production. Am J Physiol Endocrinol Metab 275: E500–E506
Herrmann C, Göke R, Richter G, Fehmann HC, Arnold R, Göke B (1995) Glucagon-like peptide-1 and glucose-dependent insulin-releasing polypeptide plasma levels in response to nutrients. Digestion 56:117–126
Jahan-Mihan A, Luhovyy BL, El Khoury D, Anderson GH (2011) Dietary proteins as determinants of metabolic and physiologic functions of the gastrointestinal tract. Nutrients 3(5):574–603
Katsumata M, Kawakami S, Kaji Y, Takada R, Dauncey MJ (2002) Differential regulation of porcine hepatic IGFI mRNA expression and plasma IGF-I concentration by a low lysine diet. J Nutr 132:688–692
Katsumata M, Kawakami S, Kaji Y, Takada R (2004) Circulating levels of insulin-like growth factor-1 and associated binding proteins in plasma and mRNA expression in tissues of growing pigs on a low threonine diet. Anim Sci 79:85–92
Kim SW, Wu G (2004) Dietary arginine supplementation enhances the growth of milk-fed young pigs. J Nutr 134:625–630
Knerr I, Groschl M, Rascher W, Rauh M (2003) Endocrine effects of food intake: insulin, ghrelin, and leptin responses to a single bolus of essential aminoacids in humans. Ann Nutr Metab 47:312–318
Kuhara T, Ikeda S, Ohneda A, Sasaki Y (1991) Effects of intravenous infusion of 17 amino acids on the secretion of GH, glucagon, and insulin in sheep. Am J Physiol Endocrinol Metabol 260: E21–E26
Layden BT, Durai V, Lowe WL Jr (2010) G-protein-coupled receptors, pancreatic islets, and diabetes. Nat Educ 3(9):13
Li J, Forhead AJ, Dauncey MJ, Gilmour RS, Fowden AL (2002) Control of growth hormone receptor and insulin-like growth factor-I expression by cortisol in ovine fetal skeletal muscle. J Physiol 541:581–589
Liddle RA (1994) Cholecystokinin. In: Walsh JH, Dockray GJ (eds) Gut peptides. Raven, New York, pp 175–216
Lucy MC (2008) Functional differences in the growth hormone and insulin-like growth factor axis in cattle and pigs: implications for post-partum nutrition and reproduction. Reprod Domest Anim 43(2):S31–S39
Maechler P, Gjinovci A, Wollheim CB (2002) Implication of glutamate in the kinetics of insulin secretion in rat and mouse perfused pancreas. Diabetes 51(1):S99–S102
Mangel AW, Prpic V, Wong H, Basavappa S, Hurst LJ, Scott L, Garman RL, Hayes JS, Sharara AI, Snow ND, Walsh JH, Liddle RA (1995) Phenylalanine-stimulated secretion of cholecystokinin is calcium dependent. Am J Physiol Gastrointest Liver Physiol 268:G90–G94
McKnight JR, Satterfield MC, Jobgen WS, Smith SB, Spencer TE, Meininger CJ, McNeal CJ, Wu G (2010) Beneficial effects of L-arginine on reducing obesity: potential mechanisms and important implications for human health. Amino Acids 39:349–357
Mersmann HJ, Smith SB (2005) Development of white adipose tissue lipid metabolism. In: Burrin DG, Mersmann HJ (eds) Biology of metabolism in growing animals. Elsevier, Oxford, pp 275–302
Miguel-Aliaga I (2012) Nerveless and gutsy: intestinal nutrient sensing from invertebrates to humans. Semin Cell Dev Biol 23(6):614–620. doi:10.1016/j.semcdb.2012.01.002
Muller WA, Faloona GR, Unger RH (1971) The influence of the antecedent diet upon glucagon and insulin secretion. N Engl J Med 285(26):1450–1454

Nakamura E, Hasumura M, Uneyama H, Torii K (2011) Luminal amino acid-sensing cells in gastric mucosa. Digestion 83:S13–S18

Nass R, Pezzoli SS, Chapman IM, Patrie J, Hintz RL, Hartman ML, Thorner MO (2002) IGF-I does not affect the net increase in GH release in response to arginine. Am J Physiol Endocrinol Metab 283:E702–E710

Newsholme P, Brennan L, Rubi B, Maechler P (2005) New insights into amino acid metabolism, beta-cell function and diabetes. Clin Sci (Lond) 108(3):185–194

Östenson CG, Grebing C (1985) Evidence for metabolic regulation of pancreatic glucagon secretion by l-glutamine. Acta Endocrinol 108:386–391

Pagliara AS, Stillings SN, Hover B, Martin DM, Matschinsky FM (1974) Glucose modulation of amino acid-induced glucagon and insulin release in the isolated perfused rat pancreas. J Clin Invest 54(4):819–832

Quesada I, Tudurí E, Ripoll C, Nadal Á (2008) Physiology of the pancreatic α-cell and glucagon secretion: role in glucose homeostasis and diabetes. J Endocrinol 199:5–19

Raybould HE (2010) Gut chemosensing: interactions between gut endocrine cells and visceral afferents. Auton Neurosci 153:41–46

Reimann F, Williams L, da Silva XG, Rutter GA, Gribble FM (2004) Glutamine potently stimulates glucagon-like peptide-1 secretion from GLUTag cells. Diabetologia 47:1592–1601

Reimann F, Ward PS, Gribble FM (2006) Signaling mechanisms underlying the release of glucagon-like peptide 1. Diabetes 55(2):S78–S85

Saleri R, Cavalli V, Tamanini C (2003) GH and nitric oxide production by pig pituitary cells in different culture conditions. Endocrine Abstr 6:P21

Sener A, Blachier F, Rasschaert J, Mourtada A, Malaisse-Lagae F, Malaisse WJ (1989a) Stimulus-secretion coupling of arginine-induced insulin release: comparison with lysine-induced insulin secretion. Endocrinology 124(5):2558–2567

Sener A, Lebrun P, Blachier F, Malaisse WJ (1989b) Stimulus-secretion coupling of arginine-induced insulin release. Insulinotropic action of agmatine. Biochem Pharmacol 38:327–330

Smith PA, Sakura H, Coles B, Gummerson N, Proks P, Ashcroft FM (1997) Electrogenic arginine transport mediates stimulus-secretion coupling in mouse pancreatic b-cells. J Physiol 499:625–635

Stingl H, Raffesberg W, Nowotny P, Waldhäusl W, Roden M (2002) Reduction of plasma leptin concentrations by arginine but not lipid infusion in humans. Obes Res 10:1111–1119

Sugino T, Kawakita Y, Fukumori R, Hasegawa Y, Kojima M, Kangawa K, Obitsu T, Taniguchi K (2010) Effects of glucose and amino acids on ghrelin secretion in sheep. Anim Sci J 81 (2):199–204

Tan BE, Yin YL, Liu ZQ, Li XG, Xu HJ, Kong XF, Huang RL, Tang WJ, Shinzato I, Smith SB, Wu GY (2009) Dietary L-arginine supplementation increases muscle gain and reduces body fat mass in growing-finishing pigs. Amino Acids 37:169–175

Tena-Sempere M, Pinilla L, González LC, Aguilar E (2000) Regulation of growth hormone (GH) secretion by different glutamate receptor subtypes in the rat. Amino Acids 18:1–6

Tolhurst G, Zheng Y, Parker HE, Habib AM, Reimann F, Gribble FM (2011) Glutamine triggers and potentiates glucagon-like peptide-1 secretion by raising cytosolic Ca^{2+} and cAMP. Endocrinology 152:405–413

Tomé D, Schwarz J, Darcel N, Fromentin G (2009) Protein, amino acids, vagus nerve signaling, and the brain. Am J Clin Nutr 90(3):S838–S843

Vance ML, Hartman ML, Thorner MO (1992) Growth hormone and nutrition. Horm Res 38(1): S85–S88

Wang Y, Chandra R, Samsa LA, Gooch B, Fee BE, Cook JM, Vigna SR, Grant AO, Liddle RA (2011) Amino acids stimulate cholecystokinin release through the Ca^{2+}-sensing receptor. Am J Physiol Gastrointest Liver Physiol 300(4):G528–G537

Wiltafskya MK, Pfaffla MW, Rotha FX (2010) The effects of branched-chain amino acid interactions on growth performance, blood metabolites, enzyme kinetics and transcriptomics in weaned pigs. Br J Nutr 103:964–976

Winzell MS, Ahrén B (2007) G-protein-coupled receptors and islet function—implications for treatment of type 2 diabetes. Pharmacol Ther 116(3):437–448

Xi G, Xu ZR, Xiao P (2002) Growth associated hormones response and fat metabolism change in finishing pigs fed with n-methyl-d, L-aspartate. Asian-Australas J Anim Sci 15(7):1026–1030

Yao K, Yin YL, Chu W, Liu Z, Deng D, Li T, Huang R, Zhang J, Tan B, Wang W, Wu G (2008) Dietary arginine supplementation increases mTOR signaling activity in skeletal muscle of neonatal pigs. J Nutr 138:867–872

Zhang HW, Yin JD, Li DF, Zhou X, Li XL (2007) Tryptophan enhances ghrelin expression and secretion associated with increased food intake and weight gain in weanling pigs. Domest Anim Endocrinol 33(1):47–61

Zeng L, Tan B, Xiao H, Yin Y, Lu X, Fang J (2012) Amino acid sensing signaling induced by amino acid transporters. Scientia Sinca Vitae 42: 699–708.

Chapter 11
Amino Acids, Gene Expression, and Cell Signaling in the Pig Intestine

Qiwen Fan, Xianghua Yan, Yulong Yin, and Zhen Liu

Contents

Abbreviations

4E-BP1	The eIF4E-binding proteins
5-HT	5-Hydroxytryptamine
AADR	Amino acid deprivation responses
ASN	L-Asparagine

Q. Fan • X. Yan
Animal Science College, Huazhong Agricultural University, Wuhan 430033, China

Y. Yin (✉)
Research Center for Healthy Breeding of Livestock and Poultry, Hunan Engineering and Research Center of Animal and Poultry Science and Key Laboratory for Agro-ecological Processes in Subtropical Region, Institute of Subtropical Agriculture, The Chinese Academy of Sciences, Changsha, Hunan 410125, People's Republic of China
e-mail: yinyulong@isa.ac.cn

Z. Liu
Department of Biotechnology and Environmental Science, Changsha University, Changsha, China

F. Blachier et al. (eds.), *Nutritional and Physiological Functions of Amino Acids in Pigs*, DOI 10.1007/978-3-7091-1328-8_11, © Springer-Verlag Wien 2013

ASNase	Asparaginase
ASNS	Asparagine synthase
ASRGL1	Asparaginase-like protein 1
CCBL1	Computational formalization of cognitive binary logic
CHOP	CCAAT/enhancer-binding protein-homologous protein
DEPTOR	DEP domain containing mTOR-interacting protein
EAA	Essential amino acids
EGF	Epidermal growth factor
eIF2	Eukaryotic initiation factor 2
GCN2	General control non-depressible 2
GLN	L-Glutamine
GPx	Glutathione peroxidase
GR	Glutathione reductase
GST	Glutathione-s-transferase
HA	Histamine
HisDC	L-Histidine decarboxylase
HRG	Histidine-rich glycoprotein
IL-8	Interleukin-8
LPS	Lipopolysaccharide
MAPK	Mitogen-activated protein kinase
MCTs	Monocarboxylate transporters
mLST8	Mammalian lethal with sec-13 protein 8
MOCS3	Multi-objective clustering selection operator section 3
mSin1	Mammalian stress-activated map kinase-interacting protein 1
mTOR	The mammalian target of rapamycin
mTORC1/2	mTOR complexes 1/2
NAC	*N*-acetylcysteine
NEAA	Nonessential amino acids
NMDAR	*N*-methyl-D-aspartate receptor
NO	Nitric oxide
ODC	Ornithine decarboxylase
PHD	Phenylalanine deficiency
PheH	Phenylalanine hydroxylase
PKU	Phenylketonuria
PRAS40	Proline-rich Akt substrate 40 kDa
protor1/2	Protein observed with rictor 1 and 2
raptor	Regulatory-associated protein of mammalian target of rapamycin
rictor	Rapamycin-insensitive companion of Mtor
S6K	p70 S6 ribosomal kinase
SAA	Sulfur-containing amino acid
SCFAs	Short-chain fatty acids
SOD	Superoxide dismutase
SREBP-2	Sterol regulatory element-binding protein-2

TH	Tyrosine hydroxylase
TLR4	Toll-like receptor-4
TNF-α	Tumor necrosis factor-alpha

11.1 Introduction

As the component of proteins amino acids serve as precursors for nucleic acids, hormones, vitamins, and other important molecules. Amino acids are incorporated into mammalian protein mostly as α-amino acids (Yin 2008). The only exception is proline, which is an α-imino acid. So the general structure of amino acids is made of a carboxyl group, an amino nitrogen group, and a side chain attached to a central α-carbon. Functional differences among the amino acids lie in the structure of their side chains. In addition to differences in size, these side groups carry different charges at physiological pH (e.g., nonpolar, uncharged but polar, negatively charged, positively charged). Some amino acids are hydrophobic (e.g., branched-chain and aromatic amino acids) and some are hydrophilic (most others).

Older views of the nutritional classification of amino acids categorized them into two groups: essential and nonessential. The nine essential amino acids (Table 11.1) are those that have carbon skeletons that cannot be synthesized to meet body needs from simpler molecules in animals, and therefore must be provided in the diet. The small intestine is the primary organ responsible for terminal digestion and absorption of dietary nutrients including protein and amino acids (Yin 2008; Wu 1998). There is a growing awareness that the small intestine may be an important site for the metabolism of drugs and other foreign compounds, particularly orally administered ones because these must traverse mucosal epithelial cells of the intestine before entering the bloodstream and body tissues (Windmueller and Spaeth 1981).

For a long time, it has been assumed that all amino acids can enter the blood when absorbed by intestine epithelial cells. But more recently, it has been found that the small intestine can metabolize to a large extent some amino acids like dietary glutamate and most glutamine in addition to circulating glutamine derived from other tissues (Windmueller and Spaeth 1975). Glutamine (66 %), glutamate (98 %), and asparagine (99 %) are metabolized by the jejunal mucosa after uptake (Windmueller and Spaeth 1975, 1976). Recent in vivo studies have demonstrated extensive catabolism of many amino acids in the pig small intestine in the first pass (Yin et al. 2010). The biochemical bases for the metabolism of nonessential amino acids (NEAA) in pig enterocytes are now well established (Yin 2008). And metabolism of essential amino acids (EAA) is believed to play an important role in regulating the efficiency of utilization of dietary protein and amino acids (Le Floc'h et al. 1997).

Table 11.1 Indispensable, dispensable, and conditionally indispensable amino acids in the human diet (Laidlaw and Kopple 1987)

Indispensable	Dispensable	Conditionally indispensable[a]	Precursors of conditionally indispensable
Histidine[b]	Alanine	Arginine	Glutamine/glutamate, aspartate
Isoleucine	Aspartic acid	Cysteine	Methionine, serine
Leucine	Asparagine	Glutamine	Glutamic acid/ammonia
Lysine	Glutamic acid	Glycine	Serine, choline
Methionine	Serine	Proline	Glutamate
Phenylalanine		Tyrosine	Phenylalanine
Threonine			
Tryptophan			
Valine			

[a]Conditionally indispensable is defined as requiring a dietary source when endogenous synthesis cannot meet metabolic need

[b]Although histidine is considered indispensable, unlike the other eight indispensable amino acids, it does not fulfill the criteria used in this report of reducing protein deposition and inducing negative nitrogen balance promptly upon removal from the diet

Although the classification of the EAA and their assignment into a single category have been maintained in this report, the definition of NEAA has become blurred as more information on the intermediary metabolism and nutritional characteristics of these compounds has accumulated (Nutrition Board and Institute of Medicine 2005). Laidlaw and Kopple (1987) divided dispensable amino acids into two classes: truly dispensable and conditionally indispensable. Five of the amino acids in Table 11.1 are termed dispensable as they can be synthesized in the body from other amino acids.

11.2 The Function of Amino Acids

11.2.1 Metabolism of L-Amino Acid

Although the free amino acids present in the body fluids represent only a very small proportion of the body's total mass of amino acids, they are very important for the nutritional and metabolic control of the body's proteins.

The complex metabolism of amino acids can adapt to the physiological functions of intestinal tract and play important roles in the maintenance of intestinal structure and regulation of intestinal function.

11.2.1.1 Asparagine

L-Asparagine (ASN) is one of the NEAA, but it plays an important role in the metabolism. It can inhibit the function of monocarboxylate transporters (MCTs), which are sodium-independent transporters that mediate the update of aromatic acid and function as a net efflux pathway for aromatic amino acids in the basolateral epithelial cells (Kim et al. 2001) and regulate carbohydrate transport and metabolism in the body. Through the action of the asparagine synthetase, asparaginase-like protein 1 (ASRGL1) protein is involved in amino acid transport and metabolism (Fresquet et al. 2004; Al Sarraj et al. 2005; Overington et al. 2006). In addition, ASN is involved in the adjustment of energy production and conversion (Oppedisano et al. 2004; Dun et al. 2007).

In the pig intestine, ASN is a more stable amino acid in solution than glutamine but with similar structure. This amino acid can stimulate enterocyte cell proliferation (Rhoads et al. 1995). Asparagine synthetase is increased in cultured mouse spleen lymphocytes after stimulation by phytohemagglutinin, lignins, and lipopolysaccharide (LPS) (Hongo et al. 1989; Suzuki et al. 2002). Those studies suggest the possible link between the stimulated asparagine production and macrophage, lymphocyte activation.

11.2.1.2 Cysteine

Sulfur-containing amino acid (SAA) metabolism has been receiving increased attention because of the link to chronic diseases such as cardiovascular disease, Alzheimer's disease, and diabetes. SAA have been regarded as involved in cardiovascular risk factors (O'Callaghan et al. 2005; Ozkan et al. 2002). In order to get a sulfur atom for subsequent sulfuration reactions, cysteine is first desulfurated by Nfs1 which transfers it onto a cysteine of multi-objective clustering selection operator section 3 (MOCS3), yielding a protein persulfide (Marelja et al. 2008).

L-Cysteine use by gut epithelial cells may play an important role for maintenance of glutathione synthesis and cellular redox function (Bauchart-Thevret et al. 2011). Glutathione synthesis is limited by L-cysteine availability, and there are many studies which have shown the potential for dietary L-cysteine intake to affect overall antioxidant status (Griffith 1999; Boebel and Baker 1983; Burgunder et al. 1989). Research on cysteine in the pig intestine has been often focused in relationship with *N*-acetylcysteine (NAC). This latter compound is the precursor for cysteine and glutathione production, therefore playing an important role in the protection of cells against oxidative stress (Wu et al. 2004). NAC reduces inflammation in the small intestine by regulating redox, epidermal growth factor (EGF), and Toll-like receptor-4 (TLR4) signaling (Hou et al. 2012a), and alleviates the mucosal damage and improves the absorptive function of the small intestine in LPS-challenged piglets (Hou et al. 2012b). When injury occurred in intestinal mucosa, the activities of antioxidant NAC, superoxide dismutase (SOD), catalase, and glutathione peroxidase (GPx) were enhanced. These enhancements might be related to the elimination of the superoxide radical and of H_2O_2, and are accompanied by a fall in glutathione-s-transferase (GST) and glutathione reductase (GR) activities (Cuzzocrea et al. 2000).

11.2.1.3 Histidine

L-Histidine is one of the EAA for many animals. L-histidine decarboxylase (HisDC) is the enzyme catalyzing the formation of histamine from L-histidine (Mamune-Sato et al. 1992). Recent report shows that histidine-rich glycoprotein (HRG) is an abundant plasma glycoprotein that has been shown to regulate a number of important biological processes, such as angiogenesis, cell adhesion and migration, fibrinolysis and coagulation, complement activation, immune complex clearance, and phagocytosis of apoptotic cells (Jones et al. 2005).

L-Histidine's function can be related to histamine (HA) synthesis. Several reports suggest that HA is involved in the regulation of arousal state (Lin et al. 1990), locomotor activity (Clapham and Kilpatrick 1994), cardiovascular control (Imamura et al. 1996), water intake (Lecklin et al. 1998), and food intake (Leurs et al. 1998). Histamine is present in the mucosa of the gastrointestinal tract (Lorenz et al. 1973) and the intestine (Lee and Silverberg 1976; Linaker et al. 1981; Cooke et al. 1984; McCabe and Smith 1984). Histamine is the main mediator for inflammatory response; it can regulate immune function by stimulating receptors on target cells, these latter including hematopoietic progenitors, macrophages, platelets, dendritic cells, and T cells (Dy and Schneider 2004).

11.2.1.4 Tryptophan

L-Tryptophan is catabolized in seven steps to yield aminomuconate. Intermediates in this process are used for the synthesis of serotonin and kynurenine (Peters 1991). The serotonin 5-hydroxytryptamine (5-HT) neurotransmitter system contributes to various physiological and pathological conditions. 5-HT is the first neurotransmitter for which a developmental role was suspected (Nakamura and Hasegawa 2007).

L-Tryptophan is readily absorbed from the gastrointestinal tract (Yao et al. 2011). This amino acid is extensively bound to serum albumin. It is metabolized to serotonin and other metabolites and excreted in the urine. Pyridoxine and ascorbic acid appear to be related to the metabolism of tryptophan.

11.2.1.5 Tyrosine

L-Tyrosine is an NEAA, but it plays a key role in signaling transduction. Many signaling pathways do so by altering the phosphorylation state of tyrosine, serine, or threonine residues in target proteins (Pawson and Scott 1997). Tyrosine hydroxylase (TH) is the first enzyme in catecholamine biosynthesis, as well as being the rate-limiting enzyme in that process. TH requires tetrahydrobiopterin and uses iron as a cofactor in the 3,4-hydroxylation of tyrosine to produce dopa (Lewis et al. 1993; Flatmark 2000). Four isoforms of TH are expressed in the human brain and all of them possess the enzymatic activity (Lewis et al. 1993; Flatmark 2000).

L-Tyrosine can regulate absorption in small intestine (Musch et al. 1987). This amino acid is the substrate for the biosynthesis of the thyroid hormone thyroxine, the neurotransmitter dopamine, the adrenal hormones, and the pigment molecule melanin (Fitzpatrick 2003). A recent study has shown that acetaldehyde induces tyrosine phosphorylation and disrupts tight junction and adherent junction in human colonic mucosa. These latter effects can be prevented by EGF and glutamine (Basuroy et al. 2005).

11.2.1.6 Lysine

The function of L-lysine has been related to intestinal protection in pigs. L-Lysine may be a partial 5-HT 4 receptor antagonist and suppresses 5-HT 4 receptor-mediated intestinal pathologies (Smriga and Torii 2003). An increase in L-lysine intake might be a useful tool in treating stress-induced anxiety and 5-HT-related diarrhea-type intestinal dysfunctions (Smriga and Torii 2003). Recently the research on L-lysine has focused on L-lysine methylation and L-lysine-specific demethylase (Eliazer et al. 2011; DiTacchio et al. 2011). L-Lysine effect on the proteins of the ubiquitin–proteasome pathway has been described (Kensche et al. 2012; Van der Veen et al. 2011).

11.2.1.7 Methionine

L-Methionine is an EAA necessary for normal growth and development in mammals (Finkelstein 1990). L-Methionine is the precursor for cellular methylation reactions and for the synthesis of cysteine and can thus replace part of the cysteine requirement (Finkelstein et al. 1988; Finkelstein and Martin 1984; Mackay et al. 2012). Methionine salvage is important in humans for the recycling of sulfur that has to be assimilated in energy-consuming reactions (Pirkov et al. 2008; Albers 2009).

It is concluded that the transmethylation–transsulfuration sequence is the major pathway for L-methionine metabolism in the mammalian liver (Finkelstein 1990). L-Methionine is one of the EAA and it has very important functions in the intestine. SAA deficiency up-regulates intestinal L-methionine cycle activity and suppresses epithelial growth in neonatal pigs (Bauchart-Thevret et al. 2009). In proteins, one of the amino acids most easily oxidized is L-methionine, which is converted to L-methionine sulfoxide (Weissbach et al. 2002).

11.2.1.8 Phenylalanine

L-Phenylalanine and tyrosine constitute the two initial steps in the biosynthesis of dopamine, which, in turn, is the metabolic precursor of noradrenaline and adrenaline (Lou 1994). The extracellular L-phenylalanine concentration influences brain function in phenylalanine deficiency (PHD) by decreasing dopamine synthesis (Lou 1994). Phenylalanine hydroxylase (PheH) is an iron(II)-dependent enzyme that catalyzes the hydroxylation of aromatic amino acid L-phenylalanine to L-tyrosine (Zhang et al. 2011). Adrenaline, dopamine, and norepinephrine can be used as neurotransmitters and involved in the regulation of brain function.

L-Phenylalanine transport is depressed by L-tryptophan and L-methionine (Spencer and Samiy 1961). In the intestine, phenylalanine represents a competitive inhibitor involved in the regulation of the absorption of amino acids. There are also interactions which have been described between phenylalanine and the anionic amino acids, aspartate and glutamate (Philip et al. 2011). Phenylketonuria (PKU) is characterized by phenylalanine accumulation and progressive mental retardation. Phenylalanine assembly can form amyloid-like deposits, suggesting a new amyloidosis-like etiology for PKU (Adler-Abramovich et al. 2012).

11.2.1.9 Threonine

L-Threonine is also an EAA. This amino acid is the second or third limiting amino acid in swine or poultry diets behind the total SAA and lysine. So this amino acid can influence the balance of amino acids in vivo. This nutrient plays a critical role in the maintenance of intestinal mucosal integrity and barrier function (Mao et al. 2011).

Intestinal mucin synthesis is sensitive to dietary threonine supply, which suggests that the gut's requirement for threonine may comprise a significant proportion of the whole body requirement (Nichols and Bertolo 2008). L-Threonine is of great importance for the maintenance of intestinal health. It is an EAA that is abundantly present in glycoproteins produced by the intestine (Sophie RD et al. 2007). The high requirement of the gut for L-threonine has often been ascribed to the synthesis of mucins, which are secreted glycoproteins protecting the intestinal epithelium from injury (Mao et al. 2011; Remond et al. 2009). This requirement could be even greater during intestinal inflammation, when mucin synthesis is enhanced and there is no oxidation of L-threonine by enterocytes (Remond et al. 2009). Because mucin proteins are little digested and reused, intestinal mucin secretion represents a net loss of L-threonine for the body (McCabe and Smith 1984).

11.2.1.10 Alanine

Among the 20 amino acids, L-alanine is the principal amino acid released by muscle, in accordance with the fact that alanine is the principal amino acid extracted by liver for gluconeogenesis (Felig et al. 1970). L-Alanine plays a major role in the transfer of nitrogen from peripheral tissue to the liver. This amino acid is involved in the metabolism of amino acids and derivatives (Yang et al. 2002), vitamins and cofactors (Marelja et al. 2008), nucleotides (Lee et al. 1995; Tamaki et al. 1990), L-phenylalanine and L-tyrosine catabolism (Wu 1998), (Han et al. 2009; Rossi et al. 2004), L-tryptophan catabolism (Perry et al. 1995), and glyoxylate metabolism (Rodionov et al. 2010). L-alanine is protective against the buildup of toxic substances that are released into muscle cells when muscle proteins are broken down quickly to meet energy needs, as it happens with aerobic exercise (Chang and Goldberg 1978). L-Alanine strengthens the immune system, boosts muscular anaerobic endurance and muscular strength, and increases aerobic endurance (Chang and Goldberg 1978).

11.2.1.11 Arginine

L-Arginine functions in the body are related to its utilization as a free amino acid, as a component of most proteins, and as the substrate for several nonprotein/nitrogen-containing compounds (Nieves and Langkamp-Henken 2002). L-Arginine participates in multiple pathways with important nutritional and physiological functions, including the synthesis of protein, nitric oxide (NO), creatine, proline, glutamate, and polyamines as well as the secretion of hormones (Tan et al. 2009a, b, 2011; He et al. 2009; Li et al. 2007; Yao et al. 2008). L-Arginine which enhances the immune system stimulates cell division, increases wound healing and hormone secretion, and regulates the vascular tone and many other various physiological processes (Nieves and Langkamp-Henken 2002). L-Arginine is the main substance of the urea cycle, which yields urea as the major form in which excess nitrogen is excreted from the human body (Brusilow and Horwich 2001). L-Arginine also is the main substance involved in the metabolism of polyamines (Moinard et al. 2005).

NO, a multifunctional second messenger, is implicated in numerous physiological functions in mammals that range from immune response and potentiation of synaptic transmission to dilation of blood vessels and muscle relaxation (Pacher et al. 2007). NO is also a regulator of intestinal electrolyte transport (MacNaughton 1993), and it has a dual excitatory and inhibitory effect on intestinal motility (Holzer et al. 1997; Tan et al. 2012a, b; Kong et al. 2012; Liu et al. 2012; Wu et al. 2012). L-Arginine is the only substrate for the synthesis of NO. As the precursor of NO, arginine possesses many functions related to effects on the immune system.

11.2.1.12 Aspartate

Aspartate is an NEAA in mammals, being produced from oxaloacetate by transamination. Oxaloacetate is one of several 2-oxoacids that function efficiently in the computational formalization of cognitive binary logic (CCBL1)-mediated transamination (Han and Li 2004) and oxaloacetate is an important substance involved in gluconeogenesis. The active form of the enzyme asparagine synthase (ASNS) is a dimer which can catalyze the reaction of aspartate, glutamine, and ATP to form asparagine, glutamate, AMP, and pyrophosphate (Van Heeke and Schuster 1989). Moreover, aspartate donates one nitrogen atom in the biosynthesis of inosine, the precursor to the purine bases (Chen et al. 2005). The glutamate *N*-methyl-D-aspartate receptor (NMDAR) plays a critical role in learning and memory (Burgdorf et al. 2011). Aspartate also stimulates NMDAR, though not as strongly as the amino acid neurotransmitter glutamate does (Chen et al. 2005). In the intestine, aspartate can affect the absorption of nutrients and, as the dietary supplement, can modify the intestinal ecosystem.

11.2.1.13 Citrulline

Citrulline is a key intermediate in the urea cycle, the pathway by which mammals excrete ammonia. Citrulline is made from ornithine and carbamoyl phosphate in one of the central reactions in the urea cycle (Rabier and Kamoun 1995). In cells, NO was synthesized from citrulline (via arginine) as well as from arginine, indicating the operation of the citrulline–NO cycle involved in arginine recycling (Mori 2007). Although citrulline is not coded for by DNA directly, several proteins are known to contain citrulline as a result of a posttranslational modification. Circulating citrulline is emerging as an innovating biomarker candidate for assessment of intestinal function (Crenn et al. 2011). Citrulline is a precursor for the production of arginine by the kidney (Crenn et al. 2011).

Glutamine, as arginine is a precursor of ornithine, which can be converted to citrulline by the intestine (van de Poll et al. 2007). Hepatic citrulline uptake limits the amount of gut-derived citrulline reaching the kidney and this may have implications for interventions aimed at increasing systemic arginine concentrations (van de Poll et al. 2007).

11.2.1.14 Glutamate

Monosodium glutamate is the crystalline salt of glutamate often used as a flavor enhancer. In neuroscience, glutamate is the most abundant excitatory neurotransmitter that plays a key role in long-term potentiation as well as in learning and memory (Burgdorf et al. 2011; McEntee and Crook 1993). Glutamate is a key molecule in cellular metabolism and can be involved in the synthesis of some amino acids.

Glutamate is also a major intestinal oxidative fuel and may represent a useful dietary supplement for the infant gut metabolism (Janeczko et al. 2007). Extensive gut metabolism limits the absorption of supplemental dietary glutamate even at excessive intakes (Janeczko et al. 2007).

11.2.1.15 Glutamine

The amino acid glutamine plays a central role in nitrogen transport within the body and is a fuel for rapidly dividing cells, such as in the gut and the immune system (Novak et al. 2002). It has been reported that glutamine has the ability to regulate gene expression in a number of physiological processes (Curi et al. 2007). Glutamine is also involved in amino acid synthesis and interconversion (transamination). These reactions mediate the synthesis of aspartate, asparagine, glutamate, and glutamine from ammonia or the intermediate of glycolysis, and allow the utilization of the carbon atoms from these four amino acids for glucose synthesis under fasting conditions.

Ko et al. (1993) show that L-glutamine, when added in amounts greater than required to prevent apoptosis, is essential for the EGF stimulation of cell proliferation as well as DNA, RNA, and protein synthesis in IEC-6 cells. This in vitro study further extends our knowledge on the important role of GLN in intestinal cell growth and function, and is consistent with a large number of in vivo studies demonstrating gut-trophic effects of this amino acid (Burke et al. 1989; Souba et al. 1992).

11.2.1.16 Ornithine

Ornithine has several common characteristics with citrulline. Both of them play a key role in the urea cycle. Ornithine is also not coded by DNA, and, in that sense, is not involved in protein synthesis (Weber and Miller 1981). Ornithine can realize the reversible transfer of an acetyl-group to an acidic (glutamate) amino acid by employing a common mechanism involving an acetyl-enzyme intermediate, but using different side chain binding modes, which is catalyzed by ornithine acetyltransferases (Iqbal et al. 2011).

Ornithine decarboxylase (ODC) is the first enzyme of the polyamine biosynthetic pathway. ODC is induced by androgens in the mouse kidney. ODC in the small intestine is stimulated three- to tenfold by refeeding or administration of insulin (Maudsley et al. 1976). The upregulation of proglucagon and ODC gene expression may be the mechanism by which short-chain fatty acids (SCFAs) facilitate intestinal adaptation (Tappenden et al. 1996). Systemic administration of SCFAs has been shown to facilitate adaptation to small bowel resection (Tappenden et al. 1996). It has been shown that ODC activity in macroscopically normal tissue of patients with tumors of stomach and large intestine increased as the disease progressed. In addition, there is an inverse relationship between ODC activity in adenocarcinomas and differentiation (Berdinskikh et al. 1991). Thus in the intestine, ODC is indeed an important regulator and plays a role in cell signaling.

11.3 Amino Acids, Gene Expression, and Cell Signaling

Many studies have shown that glutamine modulates expression of genes that are crucial for intestinal metabolism and function. Dietary glutamine supplementation increased intestinal expression (120–124 %) of genes that are necessary for cell growth and removal of oxidants while reducing (34–75 %) expression of genes that promote oxidative stress and immune activation (Wang et al. 2008). Previous studies have shown that glutamine regulates the mitogen-activated protein kinase (MAPK) activation and c-Jun signaling in enterocytes (Rhoads 1999). Son et al. (2005) found that histidine significantly inhibited both hydrogen peroxide and tumor necrosis factor-alpha (TNF-α) induced interleukin-8 (IL-8) secretion as well as mRNA expression in intestinal epithelial-like Caco-2 cells and HT-29 cells. Moreover, other amino acids also contribute to regulate the intestinal metabolism since treatment of human intestinal cell line with branched-chain amino acids (leucine, isoleucine, or valine) down regulated the expression of the sterol regulatory element-binding protein-2 (SREBP-2), a transcription factor involved in cholesterol metabolism (Chen and Reimer 2009). Finally, amino acids such as branched-chain amino acids, glycine, lysine, threonine, and cysteine may also play important regulatory roles in the intestine, particularly in improving the immune function (Li et al. 2007; Calder 2006; Burrin and Stoll 2007; Newsholme et al. 2011; Wang et al. 2009).

It seems that the functions, in which amino acids are implicated as gene expression regulators, involve the activation of different signaling pathways and transcription factors. In fact, the regulation to gene expression is editing the structure of RNA, including deadenylation-dependent mRNA decay, microRNA-mediated RNA cleavage, regulation of mRNA stability, and tRNA aminoacylation. The two main processes by which amino acids regulate gene expression are control of tRNA aminoacylation and regulation of mRNA through signaling pathway.

11.3.1 Amino Acids Regulate Aminoacylation of tRNA

tRNAs are directly involved in the translation process of gene expression. In that way, they can influence the gene expression. tRNA synthetases catalyze the ligation of tRNAs to their cognate amino acids in an ATP-dependent manner. The reaction is shown in Fig. 11.1. The rapid hydrolysis of pyrophosphate makes these reactions essentially irreversible under physiological conditions (Fersht and Kaethner 1976a). Specificity of the tRNA charging reactions is achieved both by specific recognition of amino acid and tRNA substrates, by the synthetase, and by an editing process in which incorrect aminoacyladenylate molecules (e.g., valyladenylate associated with isoleucyl tRNA synthetase) are hydrolyzed rather than conjugated to tRNAs in the second step of the reaction (Baldwin and Berg 1966a, b; Fersht and Kaethner 1976b).

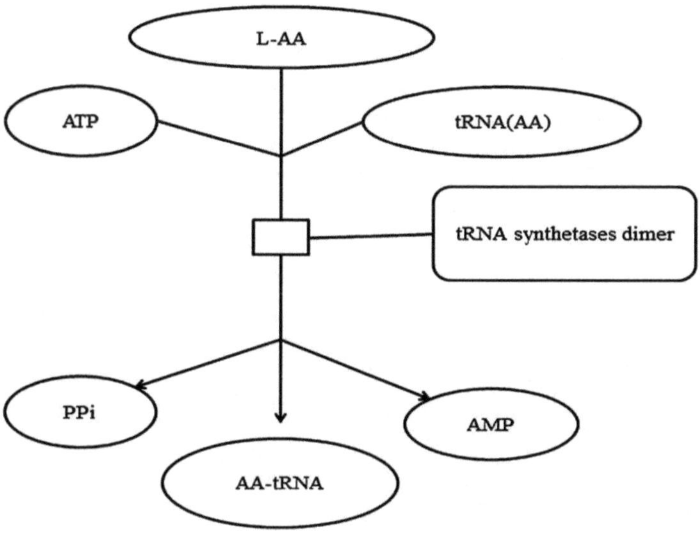

Fig. 11.1 L-AA tRNA aminoacylation. The reaction proceeds in two steps. First, amino acid and ATP form an aminoacyladenylate molecule, releasing pyrophosphate. The aminoacyladenylate remains associated with the synthetase enzyme where in the second step it reacts with tRNA to form aminoacyl tRNA and AMP

The tRNA synthetases can be divided into two structural classes based on conserved amino acid sequence features (Burbaum and Schimmel 1991).

A number of tRNA synthetases are known to have functions distinct from tRNA charging (Park et al. 2005). Additionally, mutations in several of the tRNA synthetases, often affecting protein domains that are dispensable in vitro for aminoacyl tRNA synthesis, are associated with a diverse array of neurological and other diseases (Park et al. 2005, 2008; Antonellis and Green 2008). These findings raise interest into the role of these enzymes in human development and disease.

11.3.2 Amino Acids Regulate Gene Expression Through Cell Signaling Pathway

The best-characterized example of amino acid-induced regulation of a signal transduction pathway is one involving a protein kinase referred to as the mammalian target of rapamycin (mTOR), which acts to modulate the function of proteins engaged in both global mRNA translation and the selection of specific mRNAs for translation (Kimball and Jefferson 2004).

mTOR in most organisms has evolved to be able to make a transition between anabolic and catabolic states. This would allow these organisms to survive and

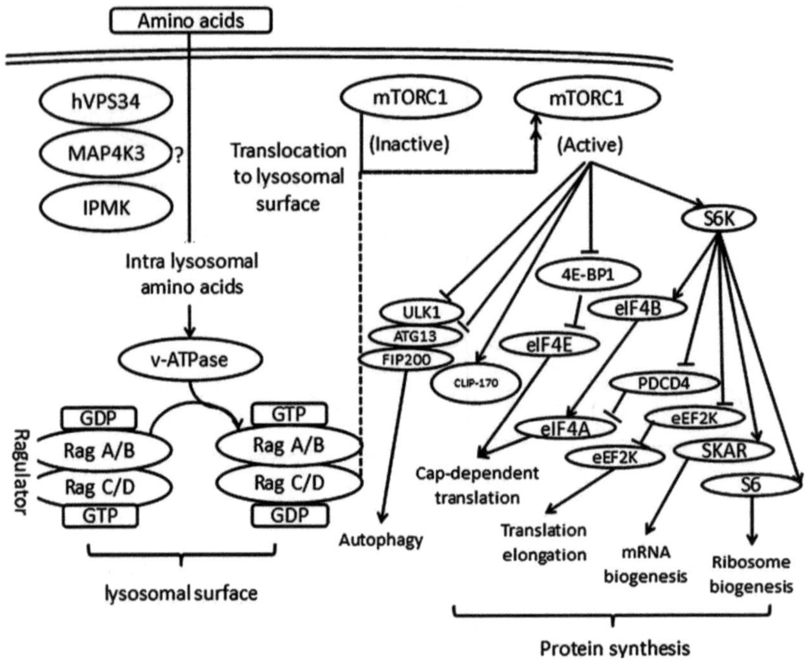

Fig. 11.2 The molecular mechanisms involved in the amino acid regulation of gene expression through the mTORC1 signaling pathway

grow in environments in which nutrient availability is variable (Laplante and Sabatini 2012). There are two mTOR complexes and both mTOR complexes are large. mTORC1 is known to have six and mTORC2 seven known protein components. They share the catalytic mTOR subunit, and also mLST8 (also known as GbL) (Jacinto et al. 2006; Kim et al. 2003), DEP domain containing mTOR-interacting protein (DEPTOR) (Peterson et al. 2009), and the Tti1/Tel2 complex (Kaizuka et al. 2010). In contrast, regulatory-associated protein of mammalian target of rapamycin (raptor) (Hara et al. 2002; Kim et al. 2002) and proline-rich Akt substrate 40 kDa (PRAS40) (Sancak et al. 2007; Thedieck et al. 2007; Wang et al. 2007) are specific to mTORC1, whereas rapamycin-insensitive companion of Mtor (rictor) (Jacinto et al. 2004; Sarbassov et al. 2004), mammalian stress-activated map kinase-interacting protein 1 (mSin1) (Jacinto et al. 2006; Frias et al. 2006), and protein observed with rictor 1 and 2 (protor1/2) (Thedieck et al. 2007; Pearce et al. 2007, 2011) are only part of mTORC2. Figure 11.2 describes the known molecular functions of amino acid-induced regulation of the mTOR complex.

Amino acids are the nutrition signal which can induce activated mTORC1. When activated, mTORC1 promotes protein synthesis mainly by phosphorylating the kinase p70 S6 ribosomal kinase (S6K) and the translation regulator of the eIF4E-binding proteins (4E-BP1) (Ma and Blenis 2009). This complex also induces lipid biogenesis by activating SREBP1 and PPARg transcription factors (Laplante and Sabatini 2009) and inhibits catabolism by blocking autophagy through the phosphorylation of the ULK1–Atg13–FIP200 complex (Ganley et al. 2009; Hosokawa et al. 2009).

Besides the mTOR pathway, many amino acids show their own way to reflect the cell signaling. In that sense, asparagine and glutamine are good examples. Asparagine starvation by asparaginase (ASNase) enhances phosphorylation of the eukaryotic initiation factor 2 (eIF2) by general control non-depressible 2 (GCN2) kinase, leading to reduced global mRNA translation rates. This conserves energy and allows cells time to reprogram stress-related gene expression to alleviate cell injury (Bunpo et al. 2010). Asparaginase depletes circulating asparagine and glutamine, activating amino acid deprivation responses (AADR) such as phosphorylation of eukaryotic initiation factor 2 (p-eIF2) leading to increased mRNA levels of asparagine synthetase and CCAAT/enhancer-binding protein-homologous protein (CHOP) and decreased mTORC1 signaling (Bunpo et al. 2009). NAC inhibits activation of c-Jun N-terminal kinase, p38 MAP kinase, and redox-sensitive activating protein-1 and nuclear factor kappa B transcription factor activities regulating expression of numerous genes (Zafarullah et al. 2003).

Particular attention has been paid to the effects of glutamine at appropriate concentrations for its capacity to enhance a great number of cell functions via the activation of various transcription factors (Brasse-Lagnel et al. 2009). Glutamine is known to play a regulatory role at the gene and protein level in several cell-specific processes including metabolism (e.g., oxidative fuel, gluconeogenesis precursor, and lipogenic precursor), cell integrity (survival, cell proliferation), protein synthesis and degradation, redox potential, respiratory burst, insulin resistance, insulin secretion, and extracellular matrix synthesis (Curi et al. 2007). Glutamine has been shown to regulate the expression of many genes related to metabolism, signal transduction, and cell defense and repair and to activate intracellular signaling pathways (Curi et al. 2007).

Figure 11.3 is a schematic view of the regulatory functions of glutamate. Both the diversity of the studies into the effect of glutamine and the variety of transcription factors involved in its action can be seen in this figure. Although glutamine deprivation was also able to stimulate the expression of ASNS (Hutson and Kilberg 1994) and CHOP (Huang et al. 1999) genes in different kinds of mammalian cells, the involvement of the NSRE and AARE sequences in these effects was not studied. Moreover, none of these responsive elements were identified in the other target genes studied.

11.4 Summary and Perspectives

EAA cannot be synthesized in vivo, so their functions are often placed at the center of nutrition and animal science. NEAA can be synthesized in the body and used for conversion into other amino acids. In addition, the so-called NEAA can become

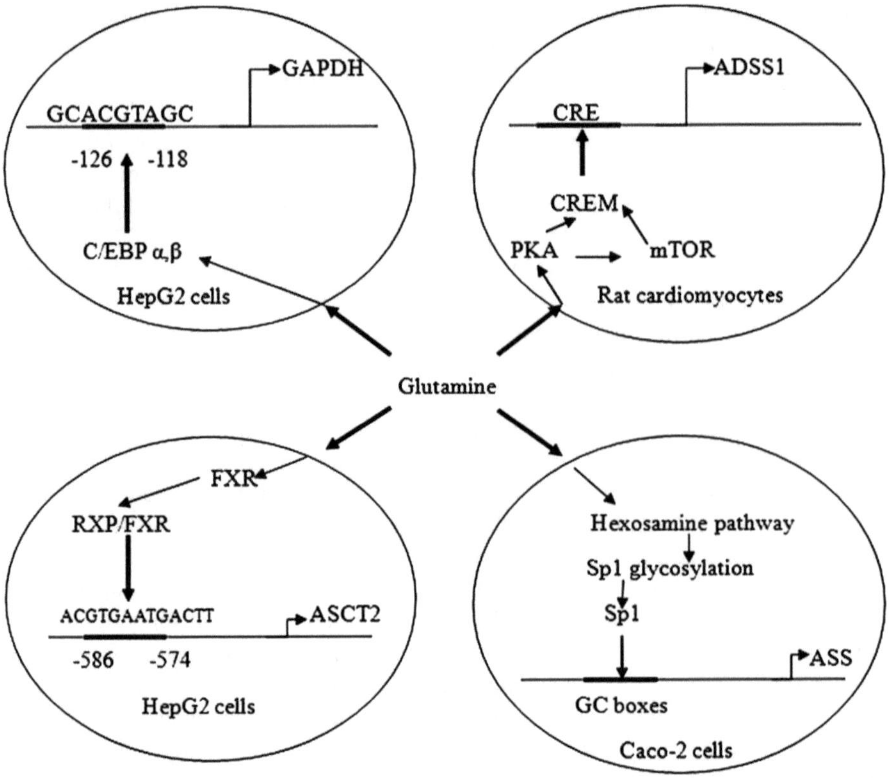

Fig. 11.3 Schematic representation of the influence of glutamine on the transcription of genes involved in intermediary metabolism (Brasse-Lagnel et al. 2009)

essential at some special physiological conditions. The functional amino acids (Wu et al. 2007, 2009), it can be considered that one given functional amino acid can be either a "nonessential" or an "essential" amino acid.

The precise pathways from the entrance of the amino acid into the cell to the activation of gene transcription remain elusive and the molecular mechanisms involved represent an important perspective for future research.

References

Adler-Abramovich L, Vaks L, Carny O, Trudler D, Magno A, Caflisch A, Frenkel D, Gazit E (2012) Phenylalanine assembly into toxic fibrils suggests amyloid etiology in phenylketonuria. Nat Chem Biol 8(8):701–706

Al Sarraj J, Vinson C, Thiel G (2005) Regulation of asparagine synthetase gene transcription by the basic region leucine zipper transcription factors ATF5 and CHOP. Biol Chem 386(9): 873–879

Albers E (2009) Metabolic characteristics and importance of the universal methionine salvage pathway recycling methionine from 5′-methylthioadenosine. IUBMB Life 61(12):1132–1142
Antonellis A, Green ED (2008) The role of aminoacyl-tRNA synthetases in genetic diseases. Annu Rev Genomics Hum Genet 9:87–107
Baldwin AN, Berg P (1966a) Transfer ribonucleic acid-induced hydrolysis of valyladenylate bound to isoleucyl ribonucleic acid synthetase. J Biol Chem 241(4):839–845
Baldwin AN, Berg P (1966b) Purification and properties of isoleucyl ribonucleic acid synthetase from Escherichia coli. J Biol Chem 241(4):831–838
Basuroy S, Sheth P, Mansbach CM, Rao RK (2005) Acetaldehyde disrupts tight junctions and adherens junctions in human colonic mucosa: protection by EGF and L-glutamine. Am J Physiol Gastrointest Liver Physiol 289(2):G367–G375
Bauchart-Thevret C, Stoll B, Burrin DG (2009) Intestinal metabolism of sulfur amino acids. Nutr Res Rev 22(2):175–187
Bauchart-Thevret C, Cottrell J, Stoll B, Burrin DG (2011) First-pass splanchnic metabolism of dietary cysteine in weanling pigs. J Anim Sci 89(12):4093–4099
Berdinskikh NK, Ignatenko NA, Zaletok SP, Ganina KP, Chorniy VA (1991) Ornithine decarboxylase activity and polyamine content in adenocarcinomas of human stomach and large intestine. Int J Cancer 47(4):496–498
Boebel KP, Baker DH (1983) Blood and liver concentrations of glutathione, and plasma concentrations of sulfur-containing amino acids in chicks fed deficient, adequate, or excess levels of dietary cysteine. Proc Soc Exp Biol Med 172(4):498–501
Brasse-Lagnel C, Lavoinne A, Husson A (2009) Control of mammalian gene expression by amino acids, especially glutamine. FEBS J 276(7):1826–1844
Brusilow SW, Horwich AL (2001) Urea cycle enzymes. In: Scriver CR, Sly WS, Valle D (eds) Childs B, Kinzler KW, Vogelstein B (assoc eds) The metabolic and molecular bases of inherited disease. McGraw-Hill, New York
Bunpo P, Dudley A, Cundiff JK, Cavener DR, Wek RC, Anthony TG (2009) GCN2 protein kinase is required to activate amino acid deprivation responses in mice treated with the anti-cancer agent L-asparaginase. J Biol Chem 284(47):32742–32749
Bunpo P, Cundiff JK, Reinert RB, Wek RC, Aldrich CJ, Anthony TG (2010) The eIF2 kinase GCN2 is essential for the murine immune system to adapt to amino acid deprivation by asparaginase. J Nutr 140(11):2020–2027
Burbaum JJ, Schimmel P (1991) Structural relationships and the classification of aminoacyl-tRNA synthetases. J Biol Chem 266(26):16965–16968
Burgdorf J, Zhang XL, Weiss C, Matthews E, Disterhoft JF, Stanton PK, Moskal JR (2011) The N-methyl-D-aspartate receptor modulator GLYX-13 enhances learning and memory, in young adult and learning impaired aging rats. Neurobiol Aging 32(4):698–706
Burgunder JM, Varriale A, Lauterburg BH (1989) Effect of N-acetylcysteine on plasma cysteine and glutathione following paracetamol administration. Eur J Clin Pharmacol 36(2):127–131
Burke DJ, Alverdy JC, Aoys E, Moss GS (1989) Glutamine-supplemented total parenteral nutrition improves gut immune function. Arch Surg 124(12):1396–1399
Burrin DG, Stoll B (2007) Emerging aspects of gut sulfur amino acid metabolism. Curr Opin Clin Nutr Metab Care 10(1):63–68
Calder PC (2006) Branched-chain amino acids and immunity. J Nutr 136(1 Suppl):288S–293S
Chang TW, Goldberg AL (1978) The origin of alanine produced in skeletal muscle. J Biol Chem 253(10):3677–3684
Chen Q, Reimer RA (2009) Dairy protein and leucine alter GLP-1 release and mRNA of genes involved in intestinal lipid metabolism in vitro. Nutrition 25(3):340–349
Chen PE, Geballe MT, Stansfeld PJ, Johnston AR, Yuan H, Jacob AL, Snyder JP, Traynelis SF, Wyllie DJ (2005) Structural features of the glutamate binding site in recombinant NR1/NR2A N-methyl-D-aspartate receptors determined by site-directed mutagenesis and molecular modeling. Mol Pharmacol 67(5):1470–1484

Clapham J, Kilpatrick GJ (1994) Thioperamide, the selective histamine H3 receptor antagonist, attenuates stimulant-induced locomotor activity in the mouse. Eur J Pharmacol 259(2): 107–114

Cooke HJ, Nemeth PR, Wood JD (1984) Histamine action on guinea pig ileal mucosa. Am J Physiol 246(4 Pt 1):G372–G377

Crenn P, Hanachi M, Neveux N, Cynober L (2011) Circulating citrulline levels: a biomarker for intestinal functionality assessment. Ann Biol Clin 69(5):513–521

Curi R, Newsholme P, Procopio J, Lagranha C, Gorjao R, Pithon-Curi TC (2007) Glutamine, gene expression, and cell function. Front Biosci 12:344–357

Cuzzocrea S, Mazzon E, Costantino G, Serraino I, De Sarro A, Caputi AP (2000) Effects of n-acetylcysteine in a rat model of ischemia and reperfusion injury. Cardiovasc Res 47(3): 537–548

DiTacchio L, Le HD, Vollmers C, Hatori M, Witcher M, Secombe J, Panda S (2011) Histone lysine demethylase JARID1a activates CLOCK-BMAL1 and influences the circadian clock. Science 333(6051):1881–1885

Dun Y, Mysona B, Itagaki S, Martin-Studdard A, Ganapathy V, Smith SB (2007) Functional and molecular analysis of D-serine transport in retinal Muller cells. Exp Eye Res 84(1):191–199

Dy M, Schneider E (2004) Histamine-cytokine connection in immunity and hematopoiesis. Cytokine Growth Factor Rev 15(5):393–410

Eliazer S, Shalaby NA, Buszczak M (2011) Loss of lysine-specific demethylase 1 non-autonomously causes stem cell tumors in the Drosophila ovary. Proc Natl Acad Sci USA 108(17):7064–7069

Felig P, Pozefsky T, Marliss E, Cahill GF Jr (1970) Alanine: key role in gluconeogenesis. Science 167(3920):1003–1004

Fersht AR, Kaethner MM (1976a) Mechanism of aminoacylation of tRNA. Proof of the aminoacyl adenylate pathway for the isoleucyl- and tyrosyl-tRNA synthetases from Escherichia coli K12. Biochemistry 15(4):818–823

Fersht AR, Kaethner MM (1976b) Enzyme hyperspecificity. Rejection of threonine by the valyl-tRNA synthetase by misacylation and hydrolytic editing. Biochemistry 15(15): 3342–3346

Finkelstein JD (1990) Methionine metabolism in mammals. J Nutr Biochem 1(5):228–237

Finkelstein JD, Martin JJ (1984) Methionine metabolism in mammals. Distribution of homocysteine between competing pathways. J Biol Chem 259(15):9508–9513

Finkelstein JD, Martin JJ, Harris BJ (1988) Methionine metabolism in mammals. The methionine-sparing effect of cystine. J Biol Chem 263(24):11750–11754

Fitzpatrick PF (2003) Mechanism of aromatic amino acid hydroxylation. Biochemistry 42(48): 14083–14091

Flatmark T (2000) Catecholamine biosynthesis and physiological regulation in neuroendocrine cells. Acta Physiol Scand 168(1):1–17

Fresquet V, Thoden JB, Holden HM, Raushel FM (2004) Kinetic mechanism of asparagine synthetase from Vibrio cholerae. Bioorg Chem 32(2):63–75

Frias MA, Thoreen CC, Jaffe JD, Schroder W, Sculley T, Carr SA, Sabatini DM (2006) mSin1 is necessary for Akt/PKB phosphorylation, and its isoforms define three distinct mTORC2s. Curr Biol 16(18):1865–1870

Ganley IG, du Lam H, Wang J, Ding X, Chen S, Jiang X (2009) ULK1.ATG13.FIP200 complex mediates mTOR signaling and is essential for autophagy. J Biol Chem 284(18):12297–12305

Griffith OW (1999) Biologic and pharmacologic regulation of mammalian glutathione synthesis. Free Radic Biol Med 27(9–10):922–935

Han Q, Li J (2004) pH dependence, substrate specificity and inhibition of human kynurenine aminotransferase I. Eur J Biochem 271(23–24):4804–4814

Han Q, Robinson H, Cai T, Tagle DA, Li J (2009) Structural insight into the inhibition of human kynurenine aminotransferase I/glutamine transaminase K. J Med Chem 52(9):2786–2793

Hara K, Maruki Y, Long X, Yoshino K, Oshiro N, Hidayat S, Tokunaga C, Avruch J, Yonezawa K (2002) Raptor, a binding partner of target of rapamycin (TOR), mediates TOR action. Cell 110 (2):177–189

He Q, Kong X, Wu G, Ren P, Tang H, Hao F, Huang R, Li T, Tan B, Li P, Tang Z, Yin Y, Wu Y (2009) Metabolomic analysis of the response of growing pigs to dietary L-arginine supplementation. Amino Acids 37(1):199–208

Holzer P, Lippe IT, Tabrizi AL, Lenard L Jr, Bartho L (1997) Dual excitatory and inhibitory effect of nitric oxide on peristalsis in the guinea pig intestine. J Pharmacol Exp Ther 280(1):154–161

Hongo S, Takeda M, Sato T (1989) Induction of asparagine synthetase during lymphocyte activation by phytohemagglutinin. Biochem Int 18(3):661–666

Hosokawa N, Hara T, Kaizuka T, Kishi C, Takamura A, Miura Y, Iemura S, Natsume T, Takehana K, Yamada N, Guan JL, Oshiro N, Mizushima N (2009) Nutrient-dependent mTORC1 association with the ULK1-Atg13-FIP200 complex required for autophagy. Mol Biol Cell 20(7):1981–1991

Hou Y, Wang L, Yi D, Ding B, Yang Z, Li J, Chen X, Qiu Y, Wu G (2012a) N-acetylcysteine reduces inflammation in the small intestine by regulating redox, EGF and TLR4 signaling. Amino Acids. doi:10.1007/s00726-012-1295-x

Hou Y, Wang L, Zhang W, Yang Z, Ding B, Zhu H, Liu Y, Qiu Y, Yin Y, Wu G (2012b) Protective effects of N-acetylcysteine on intestinal functions of piglets challenged with lipopolysaccharide. Amino Acids 43(3):1233–1242

Huang Q, Lau SS, Monks TJ (1999) Induction of gadd153 mRNA by nutrient deprivation is overcome by glutamine. Biochem J 341(Pt 1):225–231

Hutson RG, Kilberg MS (1994) Cloning of rat asparagine synthetase and specificity of the amino acid-dependent control of its mRNA content. Biochem J 304(Pt 3):745–750

Imamura M, Smith NC, Garbarg M, Levi R (1996) Histamine H3-receptor-mediated inhibition of calcitonin gene-related peptide release from cardiac C fibers. A regulatory negative-feedback loop. Circ Res 78(5):863–869

Iqbal A, Clifton IJ, Chowdhury R, Ivison D, Domene C, Schofield CJ (2011) Structural and biochemical analyses reveal how ornithine acetyl transferase binds acidic and basic amino acid substrates. Org Biomol Chem 9(18):6219–6225

Jacinto E, Loewith R, Schmidt A, Lin S, Ruegg MA, Hall A, Hall MN (2004) Mammalian TOR complex 2 controls the actin cytoskeleton and is rapamycin insensitive. Nat Cell Biol 6(11): 1122–1128

Jacinto E, Facchinetti V, Liu D, Soto N, Wei S, Jung SY, Huang Q, Qin J, Su B (2006) SIN1/MIP1 maintains rictor-mTOR complex integrity and regulates Akt phosphorylation and substrate specificity. Cell 127(1):125–137

Janeczko MJ, Stoll B, Chang X, Guan X, Burrin DG (2007) Extensive gut metabolism limits the intestinal absorption of excessive supplemental dietary glutamate loads in infant pigs. J Nutr 137(11):2384–2390

Jones AL, Hulett MD, Parish CR (2005) Histidine-rich glycoprotein: a novel adaptor protein in plasma that modulates the immune, vascular and coagulation systems. Immunol Cell Biol 83 (2):106–118

Kaizuka T, Hara T, Oshiro N, Kikkawa U, Yonezawa K, Takehana K, Iemura S, Natsume T, Mizushima N (2010) Tti1 and Tel2 are critical factors in mammalian target of rapamycin complex assembly. J Biol Chem 285(26):20109–20116

Kensche T, Tokunaga F, Ikeda F, Goto E, Iwai K, Dikic I (2012) Analysis of nuclear factor-kappaB (NF-kappaB) essential modulator (NEMO) binding to linear and lysine-linked ubiquitin chains and its role in the activation of NF-kappaB. J Biol Chem 287(28): 23626–23634

Kim DK, Kanai Y, Chairoungdua A, Matsuo H, Cha SH, Endou H (2001) Expression cloning of a Na+-independent aromatic amino acid transporter with structural similarity to H+/ monocarboxylate transporters. J Biol Chem 276(20):17221–17228

Kim DH, Sarbassov DD, Ali SM, King JE, Latek RR, Erdjument-Bromage H, Tempst P, Sabatini DM (2002) mTOR interacts with raptor to form a nutrient-sensitive complex that signals to the cell growth machinery. Cell 110(2):163–175

Kim DH, Sarbassov DD, Ali SM, Latek RR, Guntur KV, Erdjument-Bromage H, Tempst P, Sabatini DM (2003) GbetaL, a positive regulator of the rapamycin-sensitive pathway required for the nutrient-sensitive interaction between raptor and mTOR. Mol Cell 11(4):895–904

Kimball SR, Jefferson LS (2004) Amino acids as regulators of gene expression. Nutr Metab 1(1):3

Ko TC, Beauchamp RD, Townsend CM Jr, Thompson JC (1993) Glutamine is essential for epidermal growth factor-stimulated intestinal cell proliferation. Surgery 114(2):147–153, discussion 153–154

Kong XF, Yin YL, Wu GY (2012) Arginine stimulates the mTOR signaling pathway and protein synthesis in porcine trophectoderm cells. J Nutr Biochem 23:1178–1183

Laidlaw SA, Kopple JD (1987) Newer concepts of the indispensable amino acids. Am J Clin Nutr 46(4):593–605

Laplante M, Sabatini DM (2009) An emerging role of mTOR in lipid biosynthesis. Curr Biol 19 (22):R1046–R1052

Laplante M, Sabatini DM (2012) mTOR signaling in growth control and disease. Cell 149(2): 274–293

Le Floc'h N, Thibault JN, Seve B (1997) Tissue localization of threonine oxidation in pigs. Br J Nutr 77(4):593–603

Lecklin A, Etu-Seppala P, Stark H, Tuomisto L (1998) Effects of intracerebroventricularly infused histamine and selective H1, H2 and H3 agonists on food and water intake and urine flow in Wistar rats. Brain Res 793(1–2):279–288

Lee JS, Silverberg JW (1976) Effect of histamine on intestinal fluid secretion in the dog. Am J Physiol 231(3):793–798

Lee IS, Muragaki Y, Ideguchi T, Hase T, Tsuji M, Ooshima A, Okuno E, Kido R (1995) Molecular cloning and sequencing of a cDNA encoding alanine-glyoxylate aminotransferase 2 from rat kidney. J Biochem 117(4):856–862

Leurs R, Blandina P, Tedford C, Timmerman H (1998) Therapeutic potential of histamine H3 receptor agonists and antagonists. Trends Pharmacol Sci 19(5):177–183

Lewis DA, Melchitzky DS, Haycock JW (1993) Four isoforms of tyrosine hydroxylase are expressed in human brain. Neuroscience 54(2):477–492

Li P, Yin YL, Li D, Kim SW, Wu G (2007) Amino acids and immune function. Br J Nutr 98(2): 237–252

Lin JS, Sakai K, Vanni-Mercier G, Arrang JM, Garbarg M, Schwartz JC, Jouvet M (1990) Involvement of histaminergic neurons in arousal mechanisms demonstrated with H3-receptor ligands in the cat. Brain Res 523(2):325–330

Linaker BD, McKay JS, Higgs NB, Turnberg LA (1981) Mechanisms of histamine stimulated secretion in rabbit ileal mucosa. Gut 22(11):964–970

Liu XD, Wu X, Yin YL, Liu YQ, Geng MM, Yang HS, Wu GY (2012) Effects of dietary L-arginine or N-carbamylglutamate supplementation during late gestation of sows on the miR-15b/16, miR-221/222, VEGFA and eNOS expression in umbilical vein. Amino Acids 42: 2111–2119

Lorenz W, Matejka E, Schmal A, Seidel W, Reimann HJ, Uhlig R, Mann G (1973) A phylogenetic study on the occurrence and distribution of histamine in the gastro-intestinal tract and other tissues of man and various animals. Comp Gen Pharmacol 4(15):229–250

Lou HC (1994) Dopamine precursors and brain function in phenylalanine hydroxylase deficiency. Acta Paediatr Suppl 407:86–88

Ma XM, Blenis J (2009) Molecular mechanisms of mTOR-mediated translational control. Nat Rev Mol Cell Biol 10(5):307–318

Mackay DS, Brophy JD, McBreairty LE, McGowan RA, Bertolo RF (2012) Intrauterine growth restriction leads to changes in sulfur amino acid metabolism, but not global DNA methylation, in Yucatan miniature piglets. J Nutr Biochem 23(9):1121–1127

MacNaughton WK (1993) Nitric oxide-donating compounds stimulate electrolyte transport in the guinea pig intestine in vitro. Life Sci 53(7):585–593

Mamune-Sato R, Yamauchi K, Tanno Y, Ohkawara Y, Ohtsu H, Katayose D, Maeyama K, Watanabe T, Shibahara S, Takishima T (1992) Functional analysis of alternatively spliced transcripts of the human histidine decarboxylase gene and its expression in human tissues and basophilic leukemia cells. Eur J Biochem 209(2):533–539

Mao X, Zeng X, Qiao S, Wu G, Li D (2011) Specific roles of threonine in intestinal mucosal integrity and barrier function. Front Biosci 3:1192–1200

Marelja Z, Stocklein W, Nimtz M, Leimkuhler S (2008) A novel role for human Nfs1 in the cytoplasm: Nfs1 acts as a sulfur donor for MOCS3, a protein involved in molybdenum cofactor biosynthesis. J Biol Chem 283(37):25178–25185

Maudsley DV, Leif J, Kobayashi Y (1976) Ornithine decarboxylase in rat small intestine: stimulation with food or insulin. Am J Physiol 231(5 Pt 1):1557–1561

McCabe RD, Smith PL (1984) Effects of histamine and histamine receptor antagonists on ion transport in rabbit descending colon. Am J Physiol 247(4 Pt 1):G411–G418

McEntee WJ, Crook TH (1993) Glutamate: its role in learning, memory, and the aging brain. Psychopharmacology 111(4):391–401

Moinard C, Cynober L, de Bandt JP (2005) Polyamines: metabolism and implications in human diseases. Clin Nutr 24(2):184–197

Mori M (2007) Regulation of nitric oxide synthesis and apoptosis by arginase and arginine recycling. J Nutr 137(6 Suppl 2):1616S–1620S

Musch MW, McConnell FM, Goldstein L, Field M (1987) Tyrosine transport in winter flounder intestine: interaction with Na+–K+–2Cl- cotransport. Am J Physiol 253(2 Pt 2):R264–R269

Nakamura K, Hasegawa H (2007) Developmental role of tryptophan hydroxylase in the nervous system. Mol Neurobiol 35(1):45–54

Newsholme P, Abdulkader F, Rebelato E, Romanatto T, Pinheiro CH, Vitzel KF, Silva EP, Bazotte RB, Procopio J, Curi R, Gorjao R, Pithon-Curi TC (2011) Amino acids and diabetes: implications for endocrine, metabolic and immune function. Front Biosci 16:315–339

Nichols NL, Bertolo RF (2008) Luminal threonine concentration acutely affects intestinal mucosal protein and mucin synthesis in piglets. J Nutr 138(7):1298–1303

Nieves C Jr, Langkamp-Henken B (2002) Arginine and immunity: a unique perspective. Biomed Pharmacother 56(10):471–482

Novak F, Heyland DK, Avenell A, Drover JW, Su X (2002) Glutamine supplementation in serious illness: a systematic review of the evidence. Crit Care Med 30(9):2022–2029

Food and Nutrition Board, Institute of Medicine (2005) A report of the Panel on Macronutrients, and the Standing Committee on the scientific evaluation of dietary reference intakes protein and amino acids. In: Spears GE (ed) Reference intakes for energy, carbohydrate, fiber, fat, fatty acids, cholesterol, protein, and amino acids (macronutrients). The National Academies, Washington, DC

O'Callaghan M, Glare TR, Burgess EP, Malone LA (2005) Effects of plants genetically modified for insect resistance on nontarget organisms. Annu Rev Entomol 50:271–292

Oppedisano F, Pochini L, Galluccio M, Cavarelli M, Indiveri C (2004) Reconstitution into liposomes of the glutamine/amino acid transporter from renal cell plasma membrane: functional characterization, kinetics and activation by nucleotides. Biochim Biophys Acta 1667(2): 122–131

Overington JP, Al-Lazikani B, Hopkins AL (2006) How many drug targets are there? Nat Rev Drug Discov 5(12):993–996

Ozkan Y, Ozkan E, Simsek B (2002) Plasma total homocysteine and cysteine levels as cardiovascular risk factors in coronary heart disease. Int J Cardiol 82(3):269–277

Pacher P, Beckman JS, Liaudet L (2007) Nitric oxide and peroxynitrite in health and disease. Physiol Rev 87(1):315–424

Park SG, Ewalt KL, Kim S (2005) Functional expansion of aminoacyl-tRNA synthetases and their interacting factors: new perspectives on housekeepers. Trends Biochem Sci 30(10):569–574

Park SG, Schimmel P, Kim S (2008) Aminoacyl tRNA synthetases and their connections to disease. Proc Natl Acad Sci USA 105(32):11043–11049

Pawson T, Scott JD (1997) Signaling through scaffold, anchoring, and adaptor proteins. Science 278(5346):2075–2080

Pearce LR, Huang X, Boudeau J, Pawlowski R, Wullschleger S, Deak M, Ibrahim AF, Gourlay R, Magnuson MA, Alessi DR (2007) Identification of Protor as a novel Rictor-binding component of mTOR complex-2. Biochem J 405(3):513–522

Pearce LR, Sommer EM, Sakamoto K, Wullschleger S, Alessi DR (2011) Protor-1 is required for efficient mTORC2-mediated activation of SGK1 in the kidney. Biochem J 436(1):169–179

Perry S, Harries H, Scholfield C, Lock T, King L, Gibson G, Goldfarb P (1995) Molecular cloning and expression of a cDNA for human kidney cysteine conjugate beta-lyase. FEBS Lett 360 (3):277–280

Peters JC (1991) Tryptophan nutrition and metabolism: an overview. Adv Exp Med Biol 294: 345–358

Peterson TR, Laplante M, Thoreen CC, Sancak Y, Kang SA, Kuehl WM, Gray NS, Sabatini DM (2009) DEPTOR is an mTOR inhibitor frequently overexpressed in multiple myeloma cells and required for their survival. Cell 137(5):873–886

Philip V, Harris J, Adams R, Nguyen D, Spiers J, Baudry J, Howell EE, Hinde RJ (2011) A survey of aspartate-phenylalanine and glutamate-phenylalanine interactions in the protein data bank: searching for anion-pi pairs. Biochemistry 50(14):2939–2950

Pirkov I, Norbeck J, Gustafsson L, Albers E (2008) A complete inventory of all enzymes in the eukaryotic methionine salvage pathway. FEBS J 275(16):4111–4120

Rabier D, Kamoun P (1995) Metabolism of citrulline in man. Amino Acids 9:299–316

Remond D, Buffiere C, Godin JP, Mirand PP, Obled C, Papet I, Dardevet D, Williamson G, Breuille D, Faure M (2009) Intestinal inflammation increases gastrointestinal threonine uptake and mucin synthesis in enterally fed minipigs. J Nutr 139(4):720–726

Ren W, Yin Y, Liu G, Yu X, Li Y, Yang G, Li T, Wu G (2012) Effect of dietary arginine supplementation on reproductive performance of mice with porcine circovirus type 2 (PCV2) infection. Amino Acids 42:2089–2094

Rhoads M (1999) Glutamine signaling in intestinal cells. J Parenter Enteral Nutr 23(5 Suppl): S38–S40

Rhoads JM, Argenzio RA, Chen W, Gomez GG (1995) Asparagine stimulates piglet intestinal Cl- secretion by a mechanism requiring a submucosal glutamate receptor and nitric oxide. J Pharmacol Exp Ther 274(1):404–412

Rodionov RN, Murry DJ, Vaulman SF, Stevens JW, Lentz SR (2010) Human alanine-glyoxylate aminotransferase 2 lowers asymmetric dimethylarginine and protects from inhibition of nitric oxide production. J Biol Chem 285(8):5385–5391

Rossi F, Han Q, Li J, Rizzi M (2004) Crystal structure of human kynurenine aminotransferase I. J Biol Chem 279(48):50214–50220

Sancak Y, Thoreen CC, Peterson TR, Lindquist RA, Kang SA, Spooner E, Carr SA, Sabatini DM (2007) PRAS40 is an insulin-regulated inhibitor of the mTORC1 protein kinase. Mol Cell 25 (6):903–915

Sarbassov DD, Ali SM, Kim DH, Guertin DA, Latek RR, Erdjument-Bromage H, Tempst P, Sabatini DM (2004) Rictor, a novel binding partner of mTOR, defines a rapamycin-insensitive and raptor-independent pathway that regulates the cytoskeleton. Curr Biol 14(14):1296–1302

Smriga M, Torii K (2003) L-Lysine acts like a partial serotonin receptor 4 antagonist and inhibits serotonin-mediated intestinal pathologies and anxiety in rats. Proc Natl Acad Sci USA 100(26): 15370–15375

Son DO, Satsu H, Shimizu M (2005) Histidine inhibits oxidative stress- and TNF-alpha-induced interleukin-8 secretion in intestinal epithelial cells. FEBS Lett 579(21):4671–4677

Sophie RD DL, Jan H, Andras V, Johannes B (2007) The gut takes nearly all: threonine kinetics in infants. Am J Clin Nutr 86(4):1132–1138

Souba WW, Pan M, Stevens BR (1992) Kinetics of the sodium-dependent glutamine transporter in human intestinal cell confluent monolayers. Biochem Biophys Res Commun 188(2):746–753

Spencer RP, Samiy AH (1961) Intestinal absorption of L-phenylalanine in vitro. Am Physiol Soc 200(3):501–504

Suzuki F, Okayasu H, Tashiro M, Hashimoto K, Yokote Y, Akahane K, Hongo S, Sakagami H (2002) Effect of lignins and their precursors on nitric oxide, citrulline and asparagine production by mouse macrophage-like Raw 264.7 cells. Anticancer Res 22(5):2719–2724

Tamaki N, Kaneko M, Mizota C, Kikugawa M, Fujimoto S (1990) Purification, characterization and inhibition of D-3-aminoisobutyrate aminotransferase from the rat liver. Eur J Biochem 189(1):39–45

Tan B, Li X, Yin Y, Wu Z, Liu C, Tekwe CD, Wu G (2012) Regulatory roles for L-arginine in reducing white adipose tissue. Front Biosci 17:2237–2246

Tan BE, Li XG, Wu GY, Yulong Y (2012) Dynamic changes in blood flow and oxygen consumption in the portal-drained viscera of growing pigs receiving acute administration of L-arginine. Amino Acids 43:2481-2489

Tan B, Li XG, Kong X, Huang R, Ruan Z, Yao K, Deng Z, Xie M, Shinzato I, Yin Y, Wu G (2009a) Dietary L-arginine supplementation enhances the immune status in early-weaned piglets. Amino Acids 37(2):323–331

Tan B, Yin Y, Liu Z, Li X, Xu H, Kong X, Huang R, Tang W, Shinzato I, Smith SB, Wu G (2009b) Dietary L-arginine supplementation increases muscle gain and reduces body fat mass in growing-finishing pigs. Amino Acids 37(1):169–175

Tan B, Yin Y, Liu Z, Tang W, Xu H, Kong X, Li X, Yao K, Gu W, Smith SB, Wu G (2011) Dietary L-arginine supplementation differentially regulates expression of lipid-metabolic genes in porcine adipose tissue and skeletal muscle. J Nutr Biochem 22(5):441–445

Tappenden KA, Thomson AB, Wild GE, McBurney MI (1996) Short-chain fatty acids increase proglucagon and ornithine decarboxylase messenger RNAs after intestinal resection in rats. J Parenter Enteral Nutr 20(5):357–362

Thedieck K, Polak P, Kim ML, Molle KD, Cohen A, Jeno P, Arrieumerlou C, Hall MN (2007) PRAS40 and PRR5-like protein are new mTOR interactors that regulate apoptosis. PLoS One 2 (11):e1217

van de Poll MC, Siroen MP, van Leeuwen PA, Soeters PB, Melis GC, Boelens PG, Deutz NE, Dejong CH (2007) Interorgan amino acid exchange in humans: consequences for arginine and citrulline metabolism. Am J Clin Nutr 85(1):167–172

Van der Veen AG, Schorpp K, Schlieker C, Buti L, Damon JR, Spooner E, Ploegh HL, Jentsch S (2011) Role of the ubiquitin-like protein Urm1 as a noncanonical lysine-directed protein modifier. Proc Natl Acad Sci USA 108(5):1763–1770

Van Heeke G, Schuster SM (1989) Expression of human asparagine synthetase in Escherichia coli. J Biol Chem 264(10):5503–5509

Wang L, Harris TE, Roth RA, Lawrence JC Jr (2007) PRAS40 regulates mTORC1 kinase activity by functioning as a direct inhibitor of substrate binding. J Biol Chem 282(27):20036–20044

Wang J, Chen L, Li P, Li X, Zhou H, Wang F, Li D, Yin Y, Wu G (2008) Gene expression is altered in piglet small intestine by weaning and dietary glutamine supplementation. J Nutr 138 (6):1025–1032

Wang WW, Qiao SY, Li DF (2009) Amino acids and gut function. Amino Acids 37(1):105–110

Weber AL, Miller SL (1981) Reasons for the occurrence of the twenty coded protein amino acids. J Mol Evol 17(5):273–284

Weissbach H, Etienne F, Hoshi T, Heinemann SH, Lowther WT, Matthews B, St John G, Nathan C, Brot N (2002) Peptide methionine sulfoxide reductase: structure, mechanism of action, and biological function. Arch Biochem Biophys 397(2):172–178

Windmueller HG, Spaeth AE (1975) Intestinal metabolism of glutamine and glutamate from the lumen as compared to glutamine from blood. Arch Biochem Biophys 171(2):662–672

Windmueller HG, Spaeth AE (1976) Metabolism of absorbed aspartate, asparagine, and arginine by rat small intestine in vivo. Arch Biochem Biophys 175(2):670–676

Windmueller HG, Spaeth AE (1981) Source and fate of circulating citrulline. Am J Physiol 241(6): E473–E480

Wu G (1998) Intestinal mucosal amino acid catabolism. J Nutr 128(8):1249–1252
Wu G, Fang YZ, Yang S, Lupton JR, Turner ND (2004) Glutathione metabolism and its implications for health. J Nutr 134(3):489–492
Wu GY, Bazer FW, Davis TA, Johnson GA, Kim SW, Knabe DA, Spencer TE, Yin YL (2007) Important roles for arginine-family amino acids in swine nutrition and production. Livestock Sci 122:8–22
Wu GY, Bazer FW, Davis TA, Kim SW, Li P, Rhoads JM, Satterfield MC, Smith SB, Spencer TE, Yin YL (2009) Arginine metabolism and nutrition in growth, health and disease. Amino Acids 37:169–175
Wu X, Yin YL, Liu YQ, Liu XD, Liu ZQ, Li TJ, Huang RL, Ruan Z, Deng Z (2012) Effect of dietary arginine and N-carbamoylglutamate supplementation reproduction and gene expression of eNOS, VEGFA and PlGF1 in on in late pregnancy of sows placenta. Anim Reprod Sci 132:187–192
Yang RZ, Blaileanu G, Hansen BC, Shuldiner AR, Gong DW (2002) cDNA cloning, genomic structure, chromosomal mapping, and functional expression of a novel human alanine aminotransferase. Genomics 79(3):445–450
Yao K, Yin YL, Chu W, Liu Z, Deng D, Li T, Huang R, Zhang J, Tan B, Wang W, Wu G (2008) Dietary arginine supplementation increases mTOR signaling activity in skeletal muscle of neonatal pigs. J Nutr 138(5):867–872
Yao K, Fang J, Yin YL, Feng ZM, Tang ZR, Wu GY (2011) Tryptophan metabolism in animals: important roles in nutrition and health. Front Biosci S3:286–297
Yin Y (2008) Amino acids nutrition and metabolism in swine. Science China, Beijing
Yin YL, Huang RL, Li TJ, Ruan Z, Xie MY, Deng ZY, Hou YQ, Wu GY (2010) Amino acid metabolism in the portal-drained viscera of young pigs: effects of dietary supplementation with chitosan and pea hull. Amino Acids 39:1581–1587
Zafarullah M, Li WQ, Sylvester J, Ahmad M (2003) Molecular mechanisms of N-acetylcysteine actions. Cell Mol Life Sci 60(1):6–20
Zhang W, Ames BD, Walsh CT (2011) Identification of phenylalanine 3-hydroxylase for meta-tyrosine biosynthesis. Biochemistry 50(24):5401–5403

Chapter 12
Amino Acids and Immune Functions

Bie Tan, Mingyong Xie, and Yulong Yin

Contents

12.1 Introduction

In modern, high-density production systems, swine are challenged by pathogenic microorganisms—bacteria, viruses, and parasites that can cause infectious disease or pathology, especially for neonatal and weaned piglets (Zhang et al. 2012; Ren et al. 2012). In these latter, immune system is not well developed in the first 4 weeks of life (Yang and Schultz 1986). Amino acids have been demonstrated to play important roles in immune responses by regulating (1) the activation of T lymphocytes, B lymphocytes, natural killer cells, and macrophages; (2) cellular redox state, gene expression, and lymphocyte proliferation; and (3) the production of antibodies, cytokines, and other cytotoxic substances (Li et al. 2007b; Kim et al. 2007). A number of studies have showed that dietary specific amino acids

B. Tan • Y. Yin (✉)
Research Center for Healthy Breeding of Livestock and Poultry, Hunan Engineering and Research Center of Animal and Poultry Science and Key Laboratory of Agro-ecological Processes in Subtropical Region, Institute of Subtropical Agriculture, The Chinese Academy of Sciences, Changsha, Hunan 410125, People's Republic of China
e-mail: yinyulong@isa.ac.cn

M. Xie
The Key Laboratory of Food Science of Ministry of Education and Department of Food Science and Engineering, Nanchang University, Nanchang 330047, P.R. China

F. Blachier et al. (eds.), *Nutritional and Physiological Functions of Amino Acids in Pigs*, DOI 10.1007/978-3-7091-1328-8_12,

supplementation to pigs with malnutrition and infectious diseases enhance the immune status, thereby reducing morbidity and mortality (Ewaschuk et al. 2011; Johnsona et al. 2006; Liu et al. 2008; Han et al. 2008; Tan et al. 2009; Ren et al. 2012a, b). In this chapter, functions of amino acids in regulating the immune system and the amino acids requirements of immune system are described, in the hope of providing great promise in improving health and preventing infectious diseases in animals and humans.

12.2 Porcine Immune System

The two primary immune responses to microorganisms and their antigens are generated by the innate and acquired or adaptive immune systems. The two immune systems interact intimately and fulfill the different needs of the host to control microorganisms. Innate immunity of pigs consists of the similar components described for other mammals. The effector functions are realized through two major mechanisms: (a) the recruitment and activation of cellular components, including macrophages, neutrophils, natural killer (NK) cells, and dendritic cells (DCs) and (b) the release of a broad spectrum of extracellular mediators such as cytokines, chemokines, complement, and antimicrobial peptides (AMPs). The cells of the immune system include the myeloid cells, monocytes, macrophages, dendritic cells, neutrophils, eosinophils, basophils and lymphocytes, of which there are two major types, the B and T lymphocytes. B cells and T cells are responsible for humoral and cell-mediated immunity, respectively, which sum to form adaptive immunity. The porcine immune system differs in many aspects from that of humans and mice, including morphological differences in the lymphatic system, and phenotypic differences in immune cells as well as functional differences in immune cell populations (Rothkötter et al. 2002). Unlike other most species, the lymphocytes enter into the lymphoid organs through the lymphatic vessels and exit directly into the blood in the pig. These differences might contribute to the predisposition to and outcomes of bacterial infections such as Salmonella serovars (Scharek and Tedin 2007).

The gastrointestinal tracts (GITs), one of the largest immunological organs of the body, contain greater than 10^{12} lymphocytes and has a greater concentration of antibodies than any other site in the body (Mayer 2000). With respect to immune function within the GITs, it may be equally important to achieve a homeostatic balance between immune tolerance and immune responsiveness (Artis 2008). The mucosal immune system is adequately equipped to generate a protective immune response directed at harmful pathogens, but it also has the capability to be tolerant of the ubiquitous dietary antigens and normal microbial flora while maintaining the ability to permit the absorption of nutrients. Therefore, the development of the gastrointestinal immune system is important for establishing an effective immunological response to a diverse milieu of dietary and microbial antigenic components (Brandtzaeg and Pabst 2004; Burkey et al. 2009).

The neonatal piglets have little active mucosal immune system with the low level of peripheral lymphocytes, underdeveloped lymphoid nodes, rudimentary jejunal Peyers patches, and the low numbers of effector/memory immune cells (Blecha 2001). Changes in activation of immune system by antigenic stimuli by commensal microbial flora, pathogens, and environmental influences result in the appearance of conventional, activated T and B cells and the influx and expansion of mucosal immune cells and those in the peripheral lymphoid pool (Sinkora et al. 1998, 2005). Piglets receive the passive immunity via colostral immunoglobulins intake. As time progresses up to weaning, the active mucosal system gradually gathers the ability to generate its own antibody molecules in the gut wall. In the first 2 weeks of life, the intestine rapidly becomes colonized with lymphoid cells. Between 14 and 28 days of age, the intestinal mucosa becomes colonized with $CD4^+$ cells, and $CD8^+$ cells begin to appear at 35 days age. The immunological architecture of piglet cannot be considered fully mature until 7 weeks of age (Sinkora and Butler 2009). In commercial practice, piglets are weaned at 3–4 weeks of age, and this results in an increased susceptibility to bacterial infection and acute diarrhea and high mortality rates. The obvious increase of the numbers of $CD2^+$ leucocytes was observed in the intestine (Dréau et al. 1995), and piglets showed reduced ability to react to the lymphocyte mitogen phytohemagglutinin after weaning. With the reduction of IL-2 secreted from systemic T cells, production of specific antibody was also reduced (Butler et al. 2002; Sinkora et al. 2000).

12.3 Amino Acids Metabolism by Cells of the Immune System

The utilization of amino acids by immune cells plays an important role in the function of the immune system (Curi et al. 1997). Glutamine utilization has been linked to functional activities of immune cell function such as cytokine production, nitric oxide production, superoxide production, and phagocytosis. Many of cells of the immune system including lymphocytes, macrophages, and neutrophils utilize glutamine at high rates, which is related to the specific function of these cells in the inflammatory response (Curi et al. 1997, 1999; Wu et al. 2007; Wang et al. 2008; Newsholme et al. 1999; Ren et al. 2013). Therefore, glutamine had to be present at 10- to 100-fold in excess of any other amino acid in culture and cannot be replaced by glutamic acid or glucose. Although the activity of the first enzyme responsible for the metabolism of glutamine, glutaminase, is high in these cells, the rate of oxidation is low. Much of the glutamine is converted to glutamate, aspartate, lactate, and in appropriate conditions to CO_2 (Newsholme 2001). Koch et al. (1990) demonstrated that glutamine provides N- and C-atoms required for the synthesis of macromolecules and energy while leucine provides more precursors for incorporation into protein in peripheral lymphocytes.

Cell culture studies have showed that BCAA are absolutely essential for lymphocytes to synthesize protein, RNA, and DNA and to divide in response to stimulation (Calder 2006). Immune cells are able to incorporate BCAA into proteins and oxidize BCAA (Calder 2006; Burns 1975). Walrand et al. (2004) reported that leucine incorporation into immune cell proteins was linear over time with a comparable slope. Incorporation of isoleucine into proteins by lymphocytes is greatest, followed by eosinophils, and then by neutrophils (Burns 1975). Immune cells express branched chain alpha keto acid dehydrogenase and decarboxylase activities and so can oxidize BCAA. Lymphocytes take up and oxidize leucine and isoleucine in vitro (Calder 2006). Mitogen stimulation of lymphocytes increases leucine transport by 270 %, leucine transamination by 195 %, and leucine oxidation by 122 % (Koch et al. 1990). Isoleucine is oxidized by neutrophils and lymphocytes through the Krebs cycle after decarboxylation and lymphocytes are able to oxidize isoleucine eight times more rapidly than neutrophils (Burns 1975). In a B cell line, the pattern of uptake of all three BCAAs through the cell cycle is the same, and the order of the rate of uptake is leucine, isoleucine, and valine (Glassy and Fur long 1981).

Another important immuno-enhancing amino acid, arginine, is metabolized either by inducible nitric oxide synthases (iNOS) or by arginase 1 in immune cells. These enzymes are stimulated by T helper 1 or 2 cytokines, respectively. In the absence of immune stimulation, little arginine is used by immune cells due to a lack of expression of major arginine metabolizing enzymes, iNOS and arginase 1 (Bernard et al. 2001). Myeloid cells expressing arginase 1 are described in a growing number of disease processes, prominently in cancer, autoimmune diseases, and in graft vs. host disease (Bronte and Zanovello 2005; Rodriguez et al. 2005; Serafini et al. 2006; Popovic et al. 2007). The accumulation of arginase 1-expressing myeloid cells in spleens in mice after surgical trauma has also been observed (Makarenkova et al. 2006). Myeloid suppressor cells (MSC) efficiently deplete arginine and generate ornithine. Arginase 1 expression is also detected in mononuclear cells after trauma or surgery. Unlike arginase 1, iNOS is regulated by opposing stimuli. It has been shown that inflammatory stimuli induce the expression of iNOS in myeloid cells and other cell types (Hibbs 1991; Popovic et al. 2007; Johansson et al. 2002, 2010). iNOS can be induced in response to various cytokines such as IL-1, TNFα, and IFNγ, or bacterial products such as LPS. In vivo, IFNγ is the most potent and prevailing inducer of iNOS (Johansson et al. 2002). iNOS exerts a regulatory effect on arginase activity through the production of hydroxy-L-arginine, an intermediate product in the generation of NO. Arginase 1 in turn regulates NO through depletion of arginine availability (Morris 2004). Neither iNOS nor arginase 1 is induced in T lymphocytes, which represents a marked difference between these cells and myeloid cells. T lymphocytes depend on arginine for proliferation, ζ-chain peptide, and T cell receptor complex expression.

12.4 Impact of Amino Acids on Immune Function: Pig Studies

Amino acids affect immune system function usually through actions at several levels in the gastrointestinal tract, thymus, spleen, regional lymph nodes, and immune cells of the circulating blood (Cunningham-Rundles 2002). The regulation of amino acids on the immune function can be considered from two perspectives, namely, the enhancement of the immune response that protects individuals from infections and malignant neoplasms and the reduction of over-responses such as inflammation and autoimmunity (Yoneda et al. 2009). The roles of glutamine, arginine, threonine, methionine, cysteine, and tryptophan in enhancing the immune function in pigs have been well established (Johnsona et al. 2006; Ewaschuk et al. 2011; Tan et al. 2008; Li et al. 2007b; Wang et al. 2006; Grimble 2006; Le Floc'h and Sève 2007).

Glutamine is required to support optimal lymphocyte proliferation and production of cytokines by lymphocytes and macrophages (Wu 1996; Yoo et al. 1997; Yu et al. 2002; Calder and Yaqoob 1999). Adding glutamine to the weaning diet of pigs significantly modified immune cells in the mesenteric lymph nodes, in a potentially beneficial manner by preventing an increase in antigen naive CD4+ cells, increasing the proliferative response to pokeweed mitogen, and supporting a Th-1 type cytokine response after T cell stimulation (Johnsona et al. 2006). Glutamine is preferentially metabolized by the intestinal mucosa and by lymphocytes. As a precursor for glutathione (GSH) it helps maintain the antioxidant status of cells, improving the gut barrier function against bacterial infection. So glutamine supplementation is useful in reducing early steps in weaning-related gastrointestinal infections (Ewaschuk et al. 2011; Liu et al. 2002). Intestinal tissue from control, but not from Gln-supplemented, pigs responded to *Escherichia coli* with a significant increase in mucosal cytokine mRNA (IL-1β, IL-6, transforming growth factor-β, and IL-10) and a decrease in tight-junction protein expression (claudin-1 and occludin) (Ewaschuk et al. 2011).

As a precursor for nitric oxide and polyamine synthesis, arginine profoundly influences immune function (Kelly et al. 1995; Wu and Meininger 2002). Arginine has been demonstrated to exert beneficial effects on pregnant sows (Kim et al. 2006) and weaned pigs (Tan et al. 2009), on LPS-immunostimulated pigs (Liu et al. 2008), and on cyclophosphamide-immunosuppressed pigs (Han et al. 2009). This amino acid reduces morbidity and mortality in response to infectious pathogens. Administration of arginine increased thymus size, weight and cellularity, enhanced lymphocyte proliferation in response to mitogen and alloantigen, augmented macrophage and natural killer (NK)-cell-induced lysis of tumor cells, and increased IL-2 production by lymphocytes and receptor activity (Han et al. 2009; Kim et al. 2006; Tan et al. 2009). Arginine supplementation improves the development of digestive tract, prevents intestinal villous atrophy, and decreases the expression of intestinal pro-inflammatory cytokines, thereby enhancing the mucosal immune status in early-weaned piglets and alleviating mucosal injury of LPS-challenged pigs (Liu et al. 2008; Tan et al. 2008).

Threonine plays an important role in the production of antibodies and in providing overall immune system support. A significant part of the threonine intake is utilized by the gut itself and is used for the synthesis of endogenous secretions, particularly mucus, which is important to maintain the gut barrier. Threonine supplementation can regulate the innate immune function of IPEC-J2 cells infected with Pseudorabies Virus at molecular level, and can inhibit expression of genes corresponding to IL-1β,

TNF-α, and TGF-β1, but enhance expression of gene corresponding to IL-6 and IL-15, while the effects were found to vary with time (Han et al. 2012). Dietary supplementation with threonine also has been demonstrated to promote serum levels of IgG in sows (Cuaron et al. 1984) and increase antibody production, serum IgG levels, and jejunal mucosal concentrations of IgG and IgA, while decreasing jejunal mucosal concentrations of IL-6 in young pigs challenged with *E. coli* (Li et al. 1999; Wang et al. 2006).

Sulfur amino acids, methionine and cysteine, have indeed been shown to be beneficial for the immune system, aside from their role in protein synthesis (Grimble 2006). Additional dietary intake of methionine plus cysteine can reduce the adverse effects of immune system stimulation on whole body protein deposition in growing pigs, and probably accelerates the recovery (Litvak et al. 2011). Dietary supplementation with *N*-acetylcysteine (a stable precursor of cysteine) is highly effective in enhancing immune functions under various disease states (Geudens et al. 2008; Grimble 2001). However, improvement of immune function in pigs challenged with aflatoxin was not observed with supplementation of 0, 0.15, 0.30, or 0.45 % methionine to a basal diet containing 0.33 % methionine (Van Heugten et al. 1994). Methionine and cysteine are precursors of important molecules and important for intestinal mucosal function. We recently demonstrated that a sulfur amino acid-free diet administered enterally to piglets for 7 days led to a reduced intestinal mucosal growth associated with villus atrophy, reduced epithelial cell proliferation, lower goblet cell number, and diminished small intestinal redox capacity (Bauchart-Thevret et al. 2009).

Tryptophan plays an important role in the defense of the body and immune response modulation (Moffet and Namboodiri 2003; Le Floc'h and Sève 2007), in relation with the kynurenine pathway. In pigs, Melchior et al. (2004, 2005) showed that this metabolic pathway is involved in tryptophan metabolism disturbances associated with an inflammatory response. Pigs suffering from lung inflammation had lower plasma tryptophan concentrations and higher IDO activity in lungs and associated lymph nodes, than pair-fed healthy control piglets (Le Floc'h et al. 2004; Melchior et al. 2004, 2005).

12.5 Amino Acids Requirements to Optimize Immunity

A deficiency of dietary protein or amino acids has long been demonstrated to impair immune function and increase the sensitivity of animals to infectious challenges or stressful conditions (Le Floc'h et al. 2004). For example, deficiency of branched chain amino acids and of arginine + lysine increased splenocyte proliferation, but sulfur amino acid deficiency decreased splenocyte and lymphocyte proliferation (Konashi et al. 2000). The study with piglets has shown that threonine deficiency caused higher nitrogen excretion, blood urea, and lower number of acidic mucin-producing goblet cells in the small intestine (Wang et al. 2007). Dietary supplementation with amino acids beyond their requirements for growth deposition might thus be useful depending on environmental conditions particularly during periods of stress and when the immune system is challenged (Reeds and Jahoor 2001).

It is now clear that immune system stimulation (ISS) can cause morphological and physiological changes in the gastrointestinal tract and impact nutrient utilization in pigs (Le Floc'h et al. 2004; Mani et al. 2012). During immunological stress, amino acids are redistributed away from protein production towards tissues involved in inflammation and immune response (Bruins et al. 2000, 2002; Webster et al. 2002). They are used for the synthesis of inflammatory and immune proteins, to support immune cell proliferation, and for the synthesis of other compounds important for body defense functions (Le Floc'h et al. 2004; Webster et al. 2002). Immune activation appears to alter glutamine and arginine metabolism. Bruins et al. (2000) and Deutz et al. (1992) have shown that surgery and endotoxemia induced by LPS injection results in an increase in glutamine efflux from the hindquarter and intestine and in glutamine uptake by the liver and spleen. During the peak of an immune response, the requirement for those nonessential amino acids (glutamine, arginine, cysteine, and so on) increases 2- to 3-fold (Wilmore and Shabert 1998; Pond and Newsholme 1999), becoming, at least potentially, limiting (Reeds and Jahoor 2001). ISS does not change the apparent ileal digestibility (AID) of amino acids but alters the partitioning of sulfur amino acids in favor of nonprotein body stores in growing pigs (Rakhshandeh et al. 2010). These findings reflect an increased need for dietary sulfur amino acids to support the immune response during immune system stimulation in growing pigs (Rakhshandeh et al. 2010). In pigs injected with turpentine, fibrinogen plasma concentrations increase by 30 % and fibrinogen synthesis increases by 140 % (Jahoor et al. 1999). And in pigs with a lung inflammation induced by intravenous injection of complete Freund's adjuvant, plasma tryptophan concentrations declined for 10 days. Therefore, the increase in protein synthesis may require a great quantity of tyrosine, phenylalanine, and tryptophan (Le Floc'h et al. 2004; Melchior et al. 2004, 2005). Li et al. (1999) reported that although maximum growth rate of 17–31 kg pigs occurred at a dietary threonine level of 6.8 g kg^{-1}, higher threonine levels were needed to maximize humoral antibody production and IgG levels. To optimize immunity of 10–25 kg pigs, 6.6 g per day of true ileal digestible threonine should be fed (Wang et al. 2006). Li et al. (2007a) showed that the ideal amino acid pattern of lysine/methionine/threonine/tryptophan on the digestible basis was 100:27:29:59 for 10 kg pigs under immune stress and 100:30:21:61 for piglets under normal conditions.

References

Artis D (2008) Epithelial-cell recognition of commensal bacteria and maintenance of immune homeostasis in the gut. Nat Rev Immunol 8:411–420

Bauchart-Thevret C, Stoll B, Chacko S, Burrin DG (2009) Sulfur amino acid deficiency upregulates intestinal methionine cycle activity and suppresses epithelial growth in neonatal pigs. Am J Physiol Endocrinol Metab 296:E1239–E1250

Bernard AC, Mistry SK, Morris SM Jr, O'Brien WE, Tsuei BJ, Maley ME, Shirley LA, Kearney PA, Boulanger BR, Ochoa JB (2001) Alterations in arginine metabolic enzymes in trauma. Shock 15:215–219

Blecha F (2001) Immunology. In: Pond WG, Mersmann HJ (eds) Biology of the domestic pig. Cornell University Press, Ithaca, NY, pp 688–711

Brandtzaeg P, Pabst R (2004) Let's go mucosal: communication on slippery ground. Trends Immunol 25:570–577

Bronte V, Zanovello P (2005) Regulation of immune responses by L-arginine metabolism. Nat Rev Immunol 5:641–654

Bruins MJ, Soeters PB, Deutz NE (2000) Endotoxemia affects organ protein metabolism differently during prolonged feeding in pigs. J Nutr 130:3003–3013

Bruins MJ, Soeters PB, Lamers WH, Deutz NE (2002) L-arginine supplementation in pigs decreases liver protein turnover and increases hindquarter protein turnover both during and after endotoxemia. Am J Clin Nutr 75:1031–1044

Burkey TE, Skjolaas KA, Minton JE (2009) BOARD-INVITED REVIEW: Porcine mucosal immunity of the gastrointestinal tract. J Anim Sci 87:1493–1501

Burns CP (1975) Isoleucine metabolism by leukemic and normal human leukocytes in relation to cell maturity and type. Blood 45:643–651

Butler JE, Weber P, Sinkora M, Baker D, Schoenherr A, Mayer B, Francis D (2002) Antibody repertoire development in fetal and neonatal piglets. VIII. Colonization is required for newborn piglets to make serum antibodies to T-dependent and type 2 T-independent antigens. J Immunol 169(12):6822–6830

Calder PC (2006) Branched-chain amino acids and immunity. J Nutr 136:288S–293S

Calder PC, Yaqoob P (1999) Glutamine and the immune system. Amino Acids 17:227–241

Cuaron JA, Chapple RP, Easter RA (1984) Effect of lysine and threonine supplementation of sorghum gestation diets on nitrogen balance and plasma constituents in first-litter gilts. J Anim Sci 58:631–637

Cunningham-Rundles S (2002) Evaluation of the effects of nutrients on immune function. In: Calder PC, Field CJ, Gill HS (eds) Nutrition and immune function. CABI Publishing, Wallingford, UK, pp 57–92

Curi TCP, Demelo MP, Deazevedo RB, Zorn TMT, Curi R (1997) Glutamine utilization by rat neutrophils: presence of phosphate-dependent glutaminase. Am J Physiol 42:Cl124–Cl129

Curi R, Newsholme P, Pithon-Curi TC, Pires-de-Melo M, Garcia C, Homem-de-Bittencourt PI Jr, Guimarães ARP (1999) Metabolic fate of glutamine in lymphocytes, macrophages and neutrophils. Braz J Med Biol 32:15–21

Deutz NE, Reijven PL, Athanasas G, Soeters PB (1992) Post-operative changes in hepatic, intestinal, splenic and muscle fluxes of amino acids and ammonia in pigs. Clin Sci 83:607–614

Dréau D, Lallès JP, Toullec R, Salmon H (1995) B and T lymphocytes are enhanced in the gut of piglets fed heat-treated soyabean proteins. Vet Immunol Immunopathol 47:69–79

Ewaschuk JB, Murdoch GK, Johnson IR, Madsen KL, Field CJ (2011) Glutamine supplementation improves intestinal barrier function in a weaned piglet model of Escherichia coli infection. Br J Nutr 106:870–877

Geudens N, Wuyts WA, Rega FR, Vanaudenaerde BM, Neyrinck AP, Verleden GM, Lerut TE, Van Raemdonck DEM (2008) N-acetyl cysteine attenuates the inflammatory response in warm ischemic pig lungs. J Surg Res 146(2):177–183

Glassy MC, Fur long CE (1981) Neutral amino acid transport during the cell cycle of cultured human lymphocytes. J Cell Physiol 107:69–74

Grimble RF (2001) Nutritional modulation of immune function. Proc Nutr Soc 60:389–397

Grimble RF (2006) The effects of sulfur amino acid intake on immune function in humans. J Nutr 136:1660S–1665S

Han J, Liu YL, Fan W, Chao J, Hou YQ, Yin YL, Zhu HL, Meng GQ, Che ZQ (2009) Dietary L-arginine supplementation alleviates immunosuppression induced by cyclophosphamide in weaned pigs. Amino Acids 37:643–651

Han GQ, Yu B, Chen DW, Xiang ZT, Chen Y, Chen H, Mao Q (2012) Effects of threonine on expression of immune-related genes of porcine intestinal epithelial cell infected with Pseudorabies virus in vitro. Chin J Anim Nutr 3:487–496

Hibbs JB (1991) Synthesis of nitric oxide from L-arginine: a recently discovered pathway induced by cytokines with antitumour and antimicrobial activity. Res Immunol 142:565–569

Jahoor F, Wykes L, del Rosario MP, Frazer ME, Reeds PJ (1999) Chronic protein undernutrition and an acute inflammatory stimulus elicit different protein kinetic responses in plasma but not in muscle of piglets. J Nutr 129:693–699

Johansson AC, Hegardt P, Janelidze S, Visse E, Widegren B, Siesjö P (2002) Enhanced expression of iNOS intra-tumorally and at the immunization site after immunization with IFNγ -secreting rat glioma cells. J Neuroimmunol 123:135–143

Johansson CC, Mougiakakos D, Trocme E, All-Ericsson C, Economou MA, Larsson O, Seregard S, Kiessling R (2010) Expression and prognostic significance of iNOS in uveal melanoma. Int J Cancer 126:2682–2689

Johnsona IR, Ball RO, Baracos VE, Field CJ (2006) Glutamine supplementation influences immune development in the newly weaned piglet. Dev Comp Immunol 30:1191–1202

Kelly E, Morris SM Jr, Billiar TR (1995) Nitric oxide, sepsis, and arginine metabolism. J Parenter Enteral Nutr 19:234–238

Kim SW, Mateo RD, Wu G, Carroll JA, Shinzato I (2006) Dietary L-arginine supplementation affects immune status of pregnant gilts. FASEB J A266.1

Kim SW, Mateo RD, Yin YL, Wu GY (2007) Functional amino acids and fatty acids for enhancing production performance of sows and piglets. Asian-Australas J Anim Sci 20:295–306

Koch B, Schroder MT, Schafer G, Schauder P (1990) Comparison between transport and degradation of leucine and glutamine by peripheral human lymphocytes exposed to concanavalin A. J Cell Physiol 143:94–99

Konashi S, Takahashi K, Akiba Y (2000) Effects of dietary essential amino acid deficiencies on immunological variables in broiler chickens. Br J Nutr 83:449–456

Le Floc'h N, Sève B (2007) Biological roles of tryptophan and its metabolism: potential implications for pig feeding. Livest Sci 112:23–32

Le Floc'h N, Melchiora D, Obledb C (2004) Modifications of protein and amino acid metabolism during inflammation and immune system activation. Livest Prod Sci 87(1):37–45

Li DF, Xiao CH, Qiao SY, Zhang JH, Johnsonb EW, Thacker PA (1999) Effects of dietary threonine on performance, plasma parameters and immune function of growing pigs. Anim Feed Sci Technol 78:179–188

Li JW, Chen DW, Zhang KY, Long DB (2007a) The effect of immune stress on ideal amino acid pattern for piglets. Front Agric China 1(2):210–213

Li P, Yin YL, Li DF, Kim SW, Wu G (2007b) Amino acids and immune function. Br J Nutr 98:237–252

Litvak N, Rakhshandeh A, Htoo J, de Lange CFM (2011) Optimal dietary methionine to methionine plus cysteine ratio is increased during immune system stimulation in growing pigs. In: Proceedings of Mid-West American society of animal science meeting, Des Moines, Iowa

Liu T, Peng J, Xiong Y, Zhou S, Cheng X (2002) Effects of dietary glutamine and glutamate supplementation on small intestinal structure, active absorption and DNA, RNA concentrations in skeletal muscle tissue of weaned piglets during d 28 to 42 of age. Asian-Australas J Anim Sci 15:238–242

Liu YL, Huang JJ, Hou YQ, Zhu H, Zhao S, Ding B, Yin Y, Yi G, Shi J, Fan W (2008) Dietary arginine supplementation alleviates intestinal mucosal disruption induced by Escherichia coli lipopolysaccharide in weaned pigs. Bri J Nutr 100:552–560

Makarenkova VP, Bansal V, Matta BM, Perez LA, Ochoa JB (2006) CD11b+/Gr-1+ myeloid suppressor cells cause T cell dysfunction after traumatic stress. J Immunol 176:2085–2094

Mani V, Weber TE, Baumgard LH, Gabler NK (2012) Endotoxin inflammation and intestinal function in livestock. J Anim Sci. doi:10.2527/jas.2011-4627

Mayer L (2000) Mucosal immunity and gastrointestinal antigen processing. J Pediatr Gastroenterol Nutr 30:S4–S12
Melchior D, Sève B, Le Floc'h N (2004) Chronic lung inflammation affects plasma amino acid concentrations in pigs. J Anim Sci 82:1091–1099
Melchior D, Mézière N, Sève B, Le Floc'h N (2005) Is tryptophan catabolism increased under indoleamine 2,3 dioxygenase activity during chronic lung inflammation in pigs? Reprod Nutr Dev 45:175–183
Moffet JR, Namboodiri MA (2003) Tryptophan and immune response. Immunol Cell Biol 81:247–265
Morris SM Jr (2004) Enzymes of arginine metabolism. J Nutr 134(10):S2743–S2747
Newsholme P (2001) Why is L-glutamine metabolism important to cells of the immune system in health, postinjury, surgery or infection? J Nutr 131:S2515–S2522
Newsholme P, Curi R, Pithon-Curi TC, Murphy CJ, Garcia C, Pires-de-Melo M (1999) Glutamine metabolism by lymphocytes, macrophages and neutrophils. Its importance in health and disease. J Nutr Biochem 10:316–324
Pond CM, Newsholme EA (1999) Coping with metabolic stress in wild and domestic animals. Brit Soc Anim Sci Publ Ser 24:9–20
Popovic PJ, Zeh HJ III, Ochoa JB (2007) Arginine and immunity. J Nutr 137:S1681–S1686
Rakhshandeh A, Htoo JK, de Lange CFM (2010) Immune system stimulation of growing pigs does not alter apparent ileal amino acid digestibility but reduces the ratio between whole body nitrogen and sulfur retention. Livest Sci 134(1):21–23
Reeds PJ, Jahoor F (2001) The amino acid requirement of disease. Clin Nutr 20:15–22
Ren WK, Yin YL, Liu G, Yu XL, Li YH, Yang G, Li TJ, Wu GY (2012a) Effect of dietary arginine supplementation on reproductive performance of mice with porcine circovirus type 2 (PCV2) infection. Amino Acids 42:2089–2094
Ren WK, Luo W, Wu MM, Liu G, Yu XL, Fang J, Li TJ, Yin YL, Wu GY (2012b) Dietary L-glutamine supplementation improves pregnancy outcome in mice infected with type-2 porcine circovirus. Amino Acids. doi:10.1007/s00726-011-1134-5
Ren W, Zou L, Li N, Wang Y, Gang L, Peng Y, Ding J, Yin Y, Wu G (2013) Dietary arginine supplementation enhances immune responses to inactivated Pasteure llamultocid avaccination in mice. Br J Nutr 109:867–872
Rodriguez PC, Hernandez CP, Quiceno D, Dubinett SM, Zabaleta J, Ochoa JB, Gilbert J, Ochoa AC (2005) Arginase I in myeloid suppressor cells is induced by COX-2 in lung carcinoma. J Exp Med 202:931–939
Rothkötter HJ, Sowa E, Pabst R (2002) The pig as a model of developmental immunology. Hum Exp Toxicol 21:533–536
Scharek L, Tedin K (2007) The porcine immune system—differences compared to man and mouse and possible consequences for infections by Salmonella serovars. Berl Munch Tierarztl Wochenschr 120(7–8):347–354
Serafini P, Borrello I, Bronte V (2006) Myeloid suppressor cells in cancer: recruitment, phenotype, properties, and mechanisms of immune suppression. Semin Cancer Biol 16:53–65
Sinkora M, Butler JE (2009) The ontogeny of the porcine immune system. Dev Comp Immunol 33 (3):273–283
Sinkora M, Sinkora J, Rehakova Z, Splichal I, Yang H, Parkhouse RM, Trebichavsky I (1998) Prenatal ontogeny of lymphocyte subpopulations in pigs. Immunology 95(4):595–603
Sinkora M, Sinkora J, Rehakova Z, Butler JE (2000) Early ontogeny of thymoctes in pigs: sequential colonization of the thymus by T cell progenitors. J Immunol 165(4):1832–1839
Sinkora M, Butler JE, Holtmeier W, Sinkorova J (2005) Lymphocyte development in fetal piglets: facts and surprises. Vet Immunol Immunopathol 108(1–2):177–184
Tan BE, Li XG, Kong XF, Yao K, Wu GY, Yin YL (2008) Effect of arginine on development, mucosal morphology and IL-2 gene expression levels of digestive tract in early-weaned piglets. Scientia Agricultura Sinica 41(9):2783–2788
Tan BE, Li XG, Kong XF, Huang RL, Ruan Z, Yao K, Deng ZY, Xie MY, Shinzato I, Yin YL, GY W (2009) Dietary L-arginine supplementation enhances the immune status in early-weaned piglets. Amino Acids 37:323–331

Van Heugten E, Spears JW, Coffey MT, Kegley EB, Qureshi MA (1994) The effect of methionine and aflatoxin on immune function in weanling pigs. J Anim Sci 72:658–664

Walrand S, Guillet C, Gachon P, Rousset P, Giraudet C, Vasson MP, Boirie Y (2004) Protein synthesis rates of human PBMC and PMN can be determined simultaneously in vivo using small blood samples. Am J Physiol Cell Physiol 286:C1474–C1478

Wang X, Qiao SY, Liu M, Ma YX (2006) Effects of graded levels of true ileal digestible threonine on performance, serum parameters and immune function of 10–25 kg pigs. Anim Feed Sci Technol 129:264–278

Wang X, Qiao SY, Yin YL, Yue LY, Wang ZY, Wu GY (2007) A deficiency or excess of dietary threonine reduces protein synthesis in jejunum and skeletal muscle of young pigs. J Nutr 137:1442–1446

Wang JJ, Chen LX, Li P, Li XL, Zhou HJ, Wang FL, Li DF, Yin YL, Wu GY (2008) Gene expression is altered in piglet small intestine by weaning and dietary glutamine supplementation. J Nutr 138:1025–1032

Webster JI, Tonelli L, Sternberg EM (2002) Neuroendocrine regulation of immunity. Annu Rev Immunol 20:125–163

Wilmore DW, Shabert JK (1998) Role of glutamine in immunologic responses. Nutrition 14:618–626

Wu G (1996) Effects of concanavalin A and phorbolmyristate acetate on glutamine metabolism and proliferation of porcine intraepithelial lymphocytes. Comp Biochem Physiol 114:363–368

Wu G, Meininger CJ (2002) Regulation of nitric oxide synthesis by dietary factors. Annu Rev Nutr 22:61–86

Wu GY, Bazer FW, Davis TA, Johnson GA, Kim SW, Knabe DA, Spencer TE, Yin YL (2007) Important roles for arginine-family amino acids in swine nutrition and production. Livest Sci 122:8–22

Yang WC, Schultz RD (1986) Ontogeny of natural killer cell activity and antibody dependent cell-mediated cytotoxicity in pigs. Dev Comp Immunol 10:405–418

Yoneda J, Andou A, Takehana K (2009) Regulatory roles of amino acids in immune response. Curr Rheumatol Rev 5:252–258

Yoo SS, Field CJ, McBurney MI (1997) Glutamine supplementation maintains intramuscular glutamine concentrations and normalizes lymphocyte function in infected early-weaned pigs. J Nutr 127:2253–2259

Yu IT, Wu JF, Yang PC, Liu CY, Lee DN, Yen HT (2002) Roles of glutamine and nucleotides in combination in growth, immune responses and FMD antibody titres of weaned pigs. Anim Sci 75:379–385

Zhang Y, Yin Y, Fang J, Wang Q (2012) Pig production in subtropical agriculture. J Sci Food Agric 92:1016–1024

Chapter 13
The Pig Model for Studying Amino Acid-Related Human Diseases: Amino Acids and Intestinal Diseases in Preterm Infants

Caihong Hu, Yulong Yin, Anshan Shan, Xingguo He, Weifen Li, and Jun Fang

Contents

C. Hu • W. Li
The Key Laboratory of Molecular Animal Nutrition of Ministry of Education, College of Animal Science, Zhejiang University, Hangzhou, P. R. China

Y. Yin (✉)
Research Center for Healthy Breeding of Livestock and Poultry, Hunan Engineering and Research Center of Animal and Poultry Science and Key Laboratory for Agro-ecological Processes in Subtropical Region, Institute of Subtropical Agriculture, The Chinese Academy of Sciences, Changsha, Hunan 410125, P. R. China
e-mail: yinyulong@isa.ac.cn

A. Shan
Institute of Animal Nutrition, Northeast Agricultural University, Harbin 150030, P. R. China

X. He
Changsha Luye Biotechnology Co., Ltd, Changsha, Hunan 410125, P. R. China

J. Fang
College of Bioscience and Biotechnology, Hunan Agricultural University, Changsha, Hunan 410125, P. R. China

F. Blachier et al. (eds.), *Nutritional and Physiological Functions of Amino Acids in Pigs*,
DOI 10.1007/978-3-7091-1328-8_13,

13.1 Preterm Delivery

Preterm delivery in humans, defined as birth before 90 % gestation, is a leading cause of infant morbidity and mortality worldwide, occurring in approximately 10 % of all pregnancies (McIntire et al. 1999). Infants born prematurely account for the majority of all neonatal deaths. Not surprisingly, preterm infants show various signs of organ immaturity and this may make preterm neonates more sensitive to serious feeding-induced gastrointestinal complications (Siggers et al. 2011). The immature gastrointestinal tract is less able to deal with microbiology, immunology, and nutrition-related challenges of postnatal life as a result of deficiencies in intestinal structural integrity, digestive capacity, and intestinal immunity (Neu 2007). Such deficiencies are associated with increased enteric disease susceptibility in preterm versus term neonates. Thus, investigating means of improving these deficiencies will aid in improving the maturation of the preterm gastrointestinal tract, in reducing gut inflammation, and in optimizing nutrition and health in this compromised population.

13.2 Animal Models

However, less detailed information is available from human infants partly due to the difficulties in performing well-controlled studies on this vulnerable population of infants. In addition, it is neither ethical nor practical to conduct these experiments with the human fetus or infant.

Often, the more suitable approach is the use of animal models. While some earlier studies have focused on information derived mainly from rodent models (e.g., rats and mice) (Sodhi et al. 2008), it is difficult to conduct experiments with laboratory rodents because of their small body size and immature organs at birth, and this makes the large farm animals (e.g., pigs, cattle, sheep) more attractive models in this field.

Although no animal model will ever perfectly mimic the human condition, the pig has emerged as a superior non-primate experimental animal model because of similarities in anatomy, development, nutrition, and physiology between the pig and the human (Ball et al. 1996; Clouard et al. 2012). Pigs are also the only widely utilized animal model that is truly omnivorous, and they have strikingly similar nutritional requirements to that of humans (Patterson et al. 2008). The gut in the newborn pig is more mature than in newborn rodents, although less mature than in infants (Sangild 2006). Thus, in pigs, preterm delivery at 90 % gestation is comparable to preterm infants born at approximately 75 % gestation (30 weeks) (Siggers et al. 2011). In contrast to rodent models, the size of the newborn pig easily allows for clinically tissue collections and experimental manipulation of physiologic conditions. Besides, the ontogeny, the physiology of digestion, and associated metabolic processes are very similar between humans and pigs (Patterson et al. 2008; Patrycja and

Barbara 2008), which makes the pig an attractive animal model for further studies on intestinal complications related to preterm birth.

13.3 Determinants in Intestinal Inflammatory Diseases

Preterm infants suffer from numerous devastating intestinal diseases, including necrotizing enterocolitis (NEC). Nutritional, microbial, and immunological dysfunctions may all play a role in disease progression. The lowered digestive and nutrient absorptive function, impaired intestinal epithelial barrier, inappropriate bacterial colonization, and a dysregulated mucosal immune system may add to this increased susceptibility to enteric disease.

13.3.1 Immature Digestion and Nutrient Absorption in Preterm Neonates

Preterm birth is associated with immature motility, digestive capacity, and nutrient absorption, thus leading to nutrient fermentation, bacterial overgrowth, and mucosal inflammation.

Gastrointestinal motility is limited largely to infants less than 34 weeks gestation (Riezzo et al. 2009). The motility is found considerably less organized in premature infants (Neu 2007), probably, because of the intrinsic immaturity of the enteric neurons. Also in preterm piglets, bowel movements are not well developed during the first days of enteral feeding (Sangild et al. 2002a, b, c). The incomplete innervation and poor motility of the immature gut, thereby, may lead to stasis, nutrient fermentation, inappropriate colonization, and further contribute to the development of enteric disease in preterm infants (Neu 2007; Oste et al. 2005).

Studies in both animals and infants indicate that immature brush-border enzyme activities following preterm delivery may result in maldigestion, excessive nutrient fermentation, intestinal distension, and mucosal ischemia in preterm infants. In preterm infants, the intestine is relatively short (Weaver et al. 1991) and may have a reduced absorptive area, consistent with studies in pigs (Sangild et al. 2000, 2002a, b, c). Besides, preterm-delivered pigs also differ from term neonates in their intestinal cell proliferative and apoptotic responses (Burrin et al. 2000; Bittrich et al. 2004). The impaired ability of the immature intestine to increase cell proliferation, decrease apoptosis, and regulate the mesenteric blood flow (Crissinger et al. 1994; Clark et al. 2005; Dyess et al. 1993) may lead to mucosal atrophy, dysfunction, and necrosis in preterm neonates. In addition to ontogenetic immaturity of enterocyte function, the possible hypoxia, hypothermia, altered endocrine and metabolic status may make preterm neonates more sensitive to serious feeding-induced complications (Sangild 2006).

These deficiencies in intestinal structure and function appear to be one of the most critical problems resulting in feeding intolerance, a commonly encountered problem in neonatal care. Formula-fed preterm neonates are thus at increased risk of developing diseases due to their compromised digestive system. Similar results are observed in preterm newborn pigs. Preterm birth affects the intestinal response to enteral nutrition in newborn piglets. Formula feeding in preterm newborn pigs leads to a diminished intestinal trophic responses relative to colostrum (Bjornvad et al. 2005; Oste et al. 2005; Sangild et al. 2006), marked atrophy of the mucosal surface, and increased permeability (Rouwet et al. 2002). Preterm neonates, thus, may require a period of total parenteral nutrition (TPN) before enteral nutrition is administered (Heird and Gomez 1996; Sangild et al. 2002a, b, c). Nevertheless, parenteral nutrition has significant detrimental side effects, including intestinal atrophy, malfunction, and sepsis leading to increased susceptibility to inflammatory stimuli and the development of intestinal inflammation (Siggers et al. 2011). This is consistent with the observation that TPN increases the risk of NEC in prematurely born piglets.

13.3.2 Deficient Host-Associated Defense Mechanisms in Preterm Neonates

Relative to term neonates, preterm neonates show immature intestinal barrier that lacks several key protective mechanisms that normally prevent invasion by luminal bacteria. The premature gastrointestinal tract also has increased intestinal permeability (Neu 1996) commonly observed in preterm neonates (Neu 2007), making the immature intestine more permeable to macromolecules, while little is known about the maturation of tight junction proteins such as occludin and claudins, which constitute the major paracellular barrier of the epithelium (Nusrat et al. 2000). The mucus layer forms a physical barrier between the underlying epithelium and the lumen of the gastrointestinal tract (Atumal et al. 2001), protecting the epithelium against noxious agents and pathogenic bacteria. However, the production of mucous was reported to be immature in preterm infants (Claud and Walker 2001; Omari and Davidson 2003; Sangild 2006), leading to a diminished intestinal barrier function, impaired mucosal repair, and lowered degradation of bacterial toxins.

Additionally, impaired functioning of immune defenses in preterms (Sangild et al. 2002a, b, c; Baxter 2010), as well as lower levels of immunoglobulins (e.g., IgA, IgM, IgG) (Lin 2004), gut B and T lymphocytes, makes premature neonates particularly susceptible to enteric inflammation and injury during the early postnatal period (Claud and Walker 2001; Kuitunen and Savilahti 1995). Coupled with the increased intestinal mucosal permeability in preterms (Rouwet et al. 2002; van Elburg et al. 2003), this impairment leads to transmural translocation of microbes or their toxic products into the immature intestinal mucosal barrier in neonates. This may, in turn, further compromise intestinal defense mechanisms and eventually culminate in an inflammatory cascade, leading to NEC (Berman and Moss 2011). Furthermore, the premature gastrointestinal tract also has decreased regenerative capabilities, and the imbalance between epithelial cell injury and repair usually leads to a vicious cycle of maldigestion, impaired mucosal protection, immune activation and results in a greater potential for tissue damage (Siggers et al. 2011).

13.3.3 Gut Microflora and Inflammatory Responses in Preterm Neonates

Several studies in both human infants and piglets show that assemblages of gut bacteria differ markedly between preterm and term neonates (Schmidt et al. 2008; Cilieborg et al. 2010). While the full-term infants are rapidly colonized with a more diverse microbiota, preterm neonates have a slow bacterial colonization and decreased bacterial diversity present, which may predispose the premature gut to bacterial overgrowth by pathogenic bacteria (Arboleya et al. 2012; Fanaro et al. 2003; Schwiertz et al. 2003), and this difference may be directly related to the degree of intestinal prematurity, the deficient mechanical defense barriers, as well as the environmental factors (Hallstrom et al. 2004). Wang et al (2011) showed that intestinal microbiota had important functions in host energy metabolism, amino acid nutrition, immunity, and health.

Moreover, the nature of the enteral foods may affect initial bacterial colonization patterns in preterm infants (Caicedo et al. 2005; Claud and Walker 2001) and pigs (Shulman 2002; Wang et al. 2011), and the use of parenteral nutrition further delay colonization (Fanaro et al. 2003; Caicedo et al. 2005) in neonates following preterm delivery. Maldigestion and disturbances in intestinal barrier function may lead to bacterial overgrowth and excessive nutrient fermentation, thereby rendering the mucosa more susceptible to bacterial infections and further initiating uncontrolled inflammatory reactions (Siggers et al. 2011).

A disordered enterocyte signaling to bacterial toxins, via the production of various pro-inflammatory cytokines (e.g., IL-1, IL-6, IL-8, TNF-alpha), is thought to be crucial in the development of NEC in the susceptible preterm infants (Hunter et al. 2008). Those pro-inflammatory mediators may further initiate the inflammatory cascade, thus favoring mucosal barrier disruption and adversely affecting mucosal repair. Toll-like receptors are identified among the immunological components of the early mucosal dysfunction. It has been shown that TLRs (2 and 4) are upregulated in intestinal tissue from preterm pigs with NEC (Sangild 2006). In addition, p38 kinase, cyclooxygenase-2, and NF-kB signaling pathways may all be involved in mucosal inflammation (Grishin et al. 2006; Wang et al. 2010). It remains difficult, however, to further understand why the premature newborn is susceptible to NEC as well as other inflammatory bowel diseases, which may need further study in preterm animal models.

13.4 Nutritional Modulation of Mucosal Defense and Immunology in Preterm Neonates

As previously stated, the immature neonates are prone to bacterial infection and exaggerated immune responses, potentially resulting in irreversible tissue damage. Mother's milk and colostrum contains numerous bioactive factors, including growth factors, immunoglobulins, anti-inflammatory components, as well as amino acids

(Møller et al. 2011; Claud et al. 2003), which help to protect against the development of intestinal diseases. Immunonutrition with specific nutrients may also be effective to modulate the activity of the immune system. Some dietary components, such as amino acids, have already been shown to enhance mucosal barrier function and immunologic responses in animal models and in humans (Stechmiller et al. 2004). Kim et al (2007) reported that provision of amino acids and fatty acids with specific functions may enhance the performance of pregnant and lactating sows by modulating key metabolic pathways by which can enhance conception rates, embryogenesis, blood flow, antioxidant activity, appetite, translation initiation for protein synthesis, immune cell proliferation, and intestinal development. Increasing evidence shows that dietary supplementation of specific amino acids to animals and humans with malnutrition and infectious disease enhances the immune status, thereby reducing morbidity and mortality (Li et al. 2007; Yin et al. 2010).

13.4.1 Arginine

The arginine, which is nutritionally essential for neonates, is involved in a number of biological and physiological processes. Arginine is crucial for the synthesis of protein and molecules (e.g., nitric oxide (NO), creatine, and polyamines) with enormous physiological importance (Flynn et al. 2002; Rhoads et al. 2004; He et al. 2009; Kim et al. 2007; Li et al. 2007; Wu et al. 2009). Nitric oxide is a vasodilator involved in intestinal permeability, mucosal integrity, and barrier function (Upperman et al. 2005; Wu et al. 2007). Moderate levels of NO are important for regulation of mesenteric blood flow and protect the mucosa from injury. Polyamines are involved in the regulation of gene expression, DNA and protein synthesis, apoptosis, as well as cellular division (Flynn et al. 2002). Moreover, arginine stimulates the secretion of growth hormone and insulin in preterm infants (Vlaardingerbroek et al. 2011), thereby playing an important role in regulating nutrients metabolism (Liu et al. 2008; Yao et al. 2008, 2011; Yin and Tan 2010; Tan et al. 2009, 2011).

Intestinal amino acid metabolism differs between preterm and term birth. A significant nutritional problem in preterm infants is a severe deficiency of arginine (hypoargininemia), which occurs in more than 50 % of the preterm infant population (Wu et al. 2004). Arginine deficiency may result in hyperammonemia as well as intestinal, immunological, and neurological dysfunction (Flynn et al. 2002) and it is often associated with an increased incidence of NEC in preterm infants (Becker et al. 2000). Thus, the knowledge of arginine metabolism and physiological effects is beneficial for optimizing neonatal survival and health in this compromised population.

Wu et al. (1999) reported that the amino acid composition of the fetal pig was similar to that of human fetus. It has been reported that endogenous synthesis of arginine is important for maintaining arginine homeostasis in the neonatal pigs (Flynn and Wu 1996), and the underdevelopment of intestinal arginine synthesis may be primarily responsible for hypoargininemia in preterm neonates (Dekaney et al. 2003). Glutamine and proline are major substrates for intestinal synthesis of citrulline in pigs (Wu et al. 2000a, b, 2009; Wang et al. 2008). However, synthesis of citrulline is low and there is little conversion of citrulline into arginine in enterocytes of preterm neonates owing to the limited expression of the genes for key enzymes (e.g., pyrroline-5-carboxylate synthase, argininosuccinate synthase, and lyase), thereby contributing to hypoargininemia (Wu et al. 2004). Furthermore, the possible increase in whole-body arginine catabolism, as well as absence of perinatal cortisol surge due to premature delivery, may also be responsible for the limited endogenous synthesis of arginine. In preterm piglets, low rates of intestinal arginine synthesis are associated with low plasma arginine concentrations (Urschel et al. 2007). Plasma levels of citrulline, arginine, and glutamine are lower in premature neonates with NEC compared with healthy infants (Wu et al. 2001; Becker et al. 2000); and provision of exogenous arginine prevents hyperammonemia and reduces NEC (Amin et al. 2002). Thus, an enhancement of endogenous arginine synthesis in preterm neonates may be obtained by the promotion of the maturation of intestinal arginine-synthetic enzymes.

Glucocorticoids play a crucial role in advancing the maturation of intestinal arginine synthesis and possibly decrease the incidence of NEC in preterm infants (Bauer et al. 1984). Administration of cortisol is effective to advance the maturation of intestinal arginine synthesis in preterm neonates (Wu et al. 2004). Another promising candidate is glucagon-like peptide-2 (GLP-2), a nutrient-responsive gut hormone, which may exert multiple effects on intestinal mucosa growth in preterm neonates (Estall and Drucker 2005). In neonates, proline is a dietary precursor for arginine and is dependent on intact gut metabolism (Vlaardingerbroek et al. 2011). Preterm infants receiving PN are unable to synthesize sufficient proline de novo (Miller et al. 1995). Future research is needed to define mechanisms for arginine metabolism and develop strategies for arginine deficiency in preterm infants.

Arginine plays an important role in improving intestinal function and regulating nutrient metabolism, but the underlying mechanisms are largely unknown (Liu et al. 2008). He et al. (2009) conducted metabolomic analysis of the response of growing pigs to dietary L-arginine and found that arginine alters the catabolism of fat and amino acids in the whole body, enhances protein synthesis in skeletal muscle, and modulates intestinal microbial metabolism. Tan et al (2011) indicated that Arg differentially regulates expression of fat-metabolic genes and increases mTOR signaling activity in skeletal muscle (Yao et al. 2008) and white adipose tissue, therefore favoring lipogenesis in muscle but lipolysis in adipose tissue.

13.4.2 Glutamine

Glutamine is the preferred fuel for rapidly proliferating cells including enterocytes (Chauhan et al. 2008). It is essential for many metabolic processes, and supplementation with this amino acid has been demonstrated to improve mucosal integrity and intestinal barrier function in critically ill patients. In vitro, glutamine is required for barrier function (Ewaschuk et al. 2011) and helps recovery from loss of transepithelial resistance and increase of permeability induced by stress in Caco-2 cells (Li et al. 2003). Wang et al (2008) found that early weaning resulted in increased expression of genes related to oxidative stress and immune activation but decreased expression of genes related to macronutrient metabolism and cell proliferation in the gut. Dietary glutamine supplementation increased intestinal expression (120–124 %) of genes that are necessary for cell growth and removal of oxidants, while reducing (34–75 %) expression of genes that promote oxidative stress and immune activation. Functionally, the glutamine treatment enhanced intestinal oxidative-defense capacity, prevented jejunal atrophy, and promoted small intestine growth and body weight gain in weaned piglets. These findings reveal coordinate alterations of gene expression in response to weaning and aid in providing molecular mechanisms for the beneficial effect of dietary glutamine supplementation to improve nutrition status in young mammals.

Glutamine is abundant in mother's milk but present in much lower levels in formula milk (Agostoni et al. 2000). Thus, neonates, prematurely born infants, would benefit from glutamine addition. However, clinical studies of glutamine supplementation remain inconclusive. Parenteral glutamine appears to be well tolerated and safe in preterm neonates and this amino acid reduces the time to achieve full enteral nutrition (Thompson et al. 2003). Provision of glutamine showed some beneficial effects such as inhibition of whole body protein breakdown (Kadrofske et al. 2006) and activation of immune system in preterm infants (Parimi and Kalhan 2007). However, Tubman and Thompson (2001) reported that no additional benefit of the addition of glutamine to preterm infants was observed in their study. More studies are needed to evaluate the efficacy of this amino acid in neonatal nutrition and to understand the mechanism of glutamine dysfunction-related pathology.

13.4.3 Methionine and Cysteine

Methionine, an essential amino acid, is also a source through cysteine production for the synthesis of glutathione (GSH). This latter compound plays a crucial role in reducing intestinal oxidative damage and inflammation (Thomas et al. 2008). The rates of transsulfuration of methionine are high in prematurely born low birth weight infants (Maaike et al. 2007a, b). This may reflect high demands for glutathione (GSH) and methionine in parenteral amino acid mixtures for premature babies. It is reported that general cysteine requirement is less than 18 mg/kg per day and that cysteine is probably not a conditionally essential amino acid in the prematurely born infant (Maaike et al. 2007a, b). Regardless of the adequate GSH and protein synthesis, methionine has been implicated in increased homocysteine concentration (Courtney-Martin et al. 2008; Shoveller et al. 2004) in the neonate. Therefore, a balance between methionine and cysteine should be taken in consideration to provide the adequate total sulfur amino acid (SAA) in neonatal nutrition (Courtney-Martin et al. 2010).

13.4.4 Branched-Chain Amino Acids

The essential branched-chain amino acids (BCAAs), leucine, isoleucine, and valine, are used for incorporation into body protein (Maingay-de Groof et al. 2010; Li et al. 2011), and utilization by intestine is also high. The uptakes of total leucine and valine carbon are relatively large and the oxidation rates of these essential branched-chain amino acids are high in fetuses and neonates (van den Akker et al. 2011). Among the BCAAs, leucine can act as a nutrient signal and stimulates protein synthesis via the activation of translation initiation factors (Vlaardingerbroek et al. 2011). Recent work with young pigs shows that reducing dietary protein intake can improve gut function after weaning but result in inadequate provision of essential amino acids for muscle growth. Yin et al. (2010) reported that supplementing L-leucine to a low-protein diet may maintain the activation of translation initiation factors and adequate protein synthesis in multiple organs of post-weaning pigs. This novel finding provides a molecular basis for designing effective nutritional means to increase the efficiency of nutrient utilization for protein accretion in neonates. Leucine is not only a substrate for protein synthesis of skeletal muscle but also plays as signaling molecules to affect feeding behavior, energy balance, and fuel efficiency (Li et al. 2011). Leucine activates signaling factor of mammalian target of rapamycin (mTOR) to promote protein synthesis in skeletal muscle and in adipose tissue. It is also a major regulator of the mTOR sensitive response of food intake to high protein diet. Meanwhile, leucine regulates blood glucose level by promoting gluconeogenesis and aids in the retention of lean mass in a hypocaloric state. It is beneficial to animal nutrition and clinical application and extrapolation to humans (Li et al. 2011). Studies in newborn pigs suggest that enteral leucine supplementation may have a beneficial effect on neonatal growth as it may enhance protein synthesis in an mTORC1-dependent pathway (Suryawan et al. 2012; Li et al. 2011). The gut has a high demand for leucine and protein synthesis has been found to be limited by deficient leucine intake (Elango et al. 2002). BCAA-enriched parenteral nutrition in preterm neonates might influence functional outcome in the direct postnatal phase. The requirements of the individual BCAAs are almost twice the current recommendations (Maingay-de Groof et al. 2010). To optimize current parenteral and enteral feeding, the optimal BCAA ratio should be determined for both. Further studies are required to better understand the role of BCAAs in the regulation of neonatal growth.

13.4.5 Threonine

Threonine is an indispensable amino acid that must come from dietary sources. It is critical in the production of mucins in the gut (Schaart et al. 2009). Therefore, this amino acid is of presumably vital nutritional importance to maintain the protective mucus layer and thus the intestinal barrier function. Lack of threonine can result in diarrhea and reduced mucin production, indicating the important role of threonine in the structure and function of the gut (Vlaardingerbroek et al. 2011). However, prolonged dietary excess of threonine fed to neonates, may have negative behavioral consequences and may induce serious metabolic disturbances (Chapman et al. 2009). Dietary threonine imbalance is known to reduce the growth of the small intestine, liver, and skeletal muscle in young animals. Using the pig model, Wang et al. (2007) found that either a deficiency or an excess of dietary threonine impairs protein synthesis in these tissues. This finding provides a mechanism for the low growth performance of animals fed a threonine-imbalanced diet. Currently, neonatal amino acid solutions provide intakes of threonine (111–165 mg/kg per day) are greater than an infant's enteral intake from breast milk (76 mg/kg per day) (WHO 2007). Chapman et al. (2009) concluded that current parenteral solutions should be revised to incorporate the population-safe requirements of threonine to promote optimum metabolic and neurologic growth in neonates.

13.4.6 Tryptophan

Tryptophan is an essential amino acid while the concentration is low in plasma and content low in proteins compared with the other essential amino acids (Vlaardingerbroek et al. 2011). Intestinal inflammation, malnutrition, and pro-inflammatory situation may result in tryptophan depletion (Christmas et al. 2011), thus affecting the weight gain and nitrogen balance in neonates. In young pigs, experimental inflammation was associated to a decrease in plasma tryptophan concentrations compared with healthy piglets (Le Floc'h et al. 2008). Furthermore, studies showed that in neonatal piglets, no difference is found in tryptophan requirements when enteral and parenteral feeding are compared (Alegria et al. 1999). Therefore, most parenteral neonatal amino acid solutions contain similar concentrations of tryptophan compared with human breast milk (Cvitkovic et al. 2004). Lastly, the commercially available preterm formula content ranges from 18 to 36 mg/kg per day when infants receive 160 ml/kg per day of milk (Vlaardingerbroek et al. 2011).

13.4.7 Other Amino Acids

Taurine is considered conditionally essential because needs are not met when intake is low (Verner et al. 2007). It has important roles in intestinal absorption, membrane stability, and visual development in preterm infants. Currently, taurine concentrations in modern parenteral amino acid solutions are more than sufficient to meet recommendations (Verner et al. 2007). Tyrosine is also a conditionally essential amino acid resulting from the insufficient enzymatic activity in preterm infants. Due to poor tyrosine solubility in parenteral nutrition (Roberts et al. 2001), hydroxylation of phenylalanine to tyrosine may be a good way to provide tyrosine when the diet is tyrosine-deficient.

13.5 Conclusion

Deficiencies in intestinal integrity, barrier function, digestive capacities, and intestinal immunity, make preterm neonates more susceptible to inflammatory diseases. Using preterm pigs, recent evidence suggests that nutritional modulation has great potential to improve neonatal intestinal development, manipulate the gut microbiota, in parallel with direct effects on the mucosal immune system, to prevent the onset of NEC. Some amino acids have been shown to enhance mucosal barrier function, immunologic responses, and NEC resistance in piglets and in neonates. However, it always remains a challenge to translate data generated from animal models to corresponding conditions in humans. Understanding the unique function of amino acids could eventually play a pivotal role in improving future nutritional strategies for premature infants. Clearly, further studies involving amino acids as compounds for prevention and clinical therapies against devastating intestinal diseases are needed, both in parenterally and enterally fed infants. Such work will greatly advance our knowledge with regard to the "optimal" amino acid pattern and it will also be beneficial for designing the next generation of amino acid supplemental solutions to optimize survival and health in preterm neonates.

References

Agostoni C, Carratù B, Boniglia C, Riva E, Sanzini E (2000) Free amino acid content in standard infant formulas: comparison with human milk. J Am Coll Nutr 19:434–438

Alegria A, Barbera R, Farre R, Lagarda MJ, Lopez JC (1999) Amino acid contents of infant formulas. J Food Compos Anal 12:137–146

Amin HJ, Zamora SA, McMillan DD, Fick GH, Butzner JD, Parsons HG et al (2002) Arginine supplementation prevents necrotizing enterocolitis in the premature infant. J Pediatr 140:425–431

Arboleya S, Binetti A, Salazar N, Fernández N, Solís G, Hernández-Barranco A, Margolles A, de Los Reyes-Gavilán CG, Gueimonde M (2012) Establishment and development of intestinal microbiota in preterm neonates. FEMS Microbiol Ecol 79:763–772

Atumal C, Strugala V, Allen A, Holm L (2001) The adherent gastrointestinal mucus gel layer: thickness and physical state in vivo. Am J Physiol Gastrointest Liver Physiol 280:G922–G929

Ball RO, House JD, Wykes LL, Pencharz PB (1996) A piglet model for neonatal amino acid metabolism during total parenteral nutrition. In: Tumbleson ME, Schook LB (eds) Advances in swine in biomedical research. Plenum Press, New York, pp 713–731

Bauer CR, Morrison JC, Poole WK, Korones SB, Boehm JJ, Rigatto H, Zachman RD (1984) A decreased incidence of necrotizing enterocolitis after prenatal glucocorticoid therapy. Pediatrics 73:682–688

Baxter D (2010) Impaired functioning of immune defenses to infection in premature and term infants and their implications for vaccination. Hum Vaccin 6:494–505

Becker RM, Wu G, Galanko JA, Chen W, Maynor AR, Bose CL, Rhoads JM (2000) Reduced serum amino acid concentrations in infants with necrotizing enterocolitis. J Pediatr 137:785–793

Berman L, Moss RL (2011) Necrotizing enterocolitis: an update. Semin Fetal Neonatal Med 16:145–150

Bittrich S, Philipona C, Hammon HM, Rome V, Guilloteau P, Blum JW (2004) Preterm as compared with full-term neonatal calves are characterized by morphological and functional immaturity of the small intestine. J Dairy Sci 87:1786–1795
Bjornvad CR, Schmidt M, Petersen YM, Jensen SK, Offenberg H, Elnif J, Sangild PT (2005) Preterm birth makes the immature intestine sensitive to feeding-induced intestinal atrophy. Am J Physiol Regul Integr Comp Physiol 289:R1212–R1222
Burrin DG, Stoll B, Jiang R, Petersen Y, Elnif J, Buddington RK, Schmidt M, Holst JJ, Hartmann B, Sangild PT (2000) GLP-2 stimulates intestinal growth in premature TPN-fed pigs by suppressing proteolysis and apoptosis. Am J Physiol Gastrointest Liver Physiol 279: G1249–G1256
Caicedo RA, Schanler RJ, Li N, Neu J (2005) The developing intestinal ecosystem: implications for the neonate. Pediatr Res 58:625–628
Chapman KP, Courtney-Martin G, Moore AM, Ball RO, Pencharz PB (2009) Threonine requirement of parenterally fed postsurgical human neonates. Am J Clin Nutr 89:134–141
Chauhan M, Henderson G, McGuire W (2008) Enteral feeding for very low birth weight infants: reducing the risk of necrotising enterocolitis. Arch Dis Child Fetal Neonatal Ed 93:F162–166
Christmas DM, Potokar J, Davies SJ (2011) A biological pathway linking inflammation and depression: activation of indoleamine 2,3-dioxygenase. Neuropsychiatr Dis Treat 7:431–439
Cilieborg M, Boye M, Thymann T, Mølbak L, Sangild PT (2010) Preterm birth and necrotizing enterocolitis alter gut colonization in pigs. Pediatr Res 69:10–16
Clark JA, Lane RH, Maclennan NK, Holubec H, Dvorakova K, Halpern MD, Williams CS, Payne CM, Dvorak B (2005) Epidermal growth factor reduces intestinal apoptosis in an experimental model of necrotizing enterocolitis. Am J Physiol Gastrointest Liver Physiol 288:G755–G762
Claud EC, Walker WA (2001) Hypothesis: inappropriate colonization of the premature intestine can cause neonatal necrotizing enterocolitis. FASEB J 15:1398–1403
Claud EC, Savidge T, Walker WA (2003) Modulation of human intestinal epithelial cell IL-8 secretion by human milk factors. Pediatr Res 53:419–425
Clouard C, Meunier-Salaün MC, Val-Laillet D (2012) Food preferences and aversions in human health and nutrition: how can pigs help the biomedical research? Animal 6:118–36
Courtney-Martin G, Chapman KP, Moore AM, Kim JH, Ball RO, Pencharz PB (2008) Total sulfur amino acid requirement and metabolism in parenterally fed postsurgical human neonates. Am J Clin Nutr 88:115–124
Courtney-Martin G, Moore AM, Ball RO, Penchar PB (2010) The addition of cysteine to the total sulphur amino acid requirement as methionine does not increase erythrocytes glutathione synthesis in the parenterally fed human neonate. Pediatr Res 67:320–324
Crissinger KD, Burney DL, Velasquez OR, Gonzalez E (1994) An animal model of necrotizing enterocolitis induced by infant formula and ischemia in developing piglets. Gastroenterology 106:1215–1222
Cvitkovic S, Bertolo RF, Brunton JA, Pencharz PB, Ball RO (2004) Enteral tryptophan requirement determined by oxidation of gastrically or intravenously infused phenylalanine is not different from the parenteral requirement in neonatal piglets. Pediatr Res 55:630–636
Dekaney CM, Wu G, Jaeger LA (2003) Gene expression and activity of enzymes in the arginine biosynthetic pathway in porcine fetal small intestine. Pediatr Res 53:274–80
Dyess DL, Peeples GL, Ardell JL, Tacchi EJ, Roberts WS, Ferrara JJ, Powell RW (1993) Indomethacin-induced blood flow distribution in premature and full-term piglets. J Pediatr Surg 28:1396–1400
Elango R, Pencharz PB, Ball RO (2002) The branched-chain amino acid requirement of parenterally fed neonatal piglets is less than the enteral requirement. J Nutr 132:3123–3129
Estall JL, Drucker DJ (2005) Tales beyond the crypt: glucagon-like peptide-2 and cytoprotection in the intestinal mucosa. Endocrinology 146:19–21
Ewaschuk JB, Murdoch GK, Johnson IR, Madsen KL, Field CJ (2011) Glutamine supplementation improves intestinal barrier function in a weaned piglet model of *Escherichia coli* infection. Br J Nutr 106:870–877

Fanaro S, Chierici R, Guerrini P, Vigi V (2003) Intestinal microflora in early infancy: composition and development. Acta Paediatr Suppl 91:48–55

Flynn NE, Wu G (1996) An important role for endogenous synthesis of arginine in maintaining arginine homeostasis in neonatal pigs. Am J Physiol Regul Integr Comp Physiol 271: R1149–R1155

Flynn NE, Meininger CJ, Haynes TE, Wu G (2002) The metabolic basis of arginine nutrition and pharmacotherapy. Biomed Pharmacother 56:427–438

Grishin AV, Wang J, Potoka DA, Hackam DJ, Upperman JS, Boyle P, Zamora R, Ford HR (2006) Lipopolysaccharide induces cyclooxygenase-2 in intestinal epithelium via a noncanonical p38 MAPK pathway. J Immunol 176:580–588

Hallstrom M, Eerola E, Vuento R, Janas M, Tammela O (2004) Effects of mode of delivery and necrotising enterocolitis on the intestinal microflora in preterm infants. Eur J Clin Microbiol Infect Dis 23:463–470

He QH, Kong XF, Wu GY, Ren PP, Tang HR, Hao FH, Huang RL, Li TJ, Tan BE, Li P, Tang ZR, Yin YL, Wu YN (2009) Metabolomic analysis of the response of growing pigs to dietary L-arginine supplementation. Amino Acids 37:199–208

Heird WC, Gomez MR (1996) Parenteral nutrition in low-birth-weight infants. Annu Rev Nutr 16:471–499

Hunter CJ, Upperman JS, Ford HR, Camerini V (2008) Understanding the susceptibility of the premature infant to necrotizing enterocolitis (NEC). Pediatr Res 63:117–123

Kadrofske MM, Parimi PS, Gruca LL, Kalhan SC (2006) Effect of intravenous amino acids on glutamine and protein kinetics in low-birth-weight preterm infants during the immediate neonatal period. Am J Physiol Endocrinol Metab 290:E622–E630

Kim SW, Mateo RD, Yin YL, Wu GY (2007) Functional amino acids and fatty acids for enhancing production performance of sows and piglets. Asian Australas J Anim Sci 20:295–306

Kuitunen M, Savilahti E (1995) Mucosal IgA, mucosal cow's milk antibodies, serum cow's milk antibodies and gastrointestinal permeability in infants. Pediatr Allergy Immunol 6:30–35

Le Floc'h N, Melchior D, Sève B (2008) Dietary tryptophan helps to preserve tryptophan homeostasis in pigs suffering from lung inflammation. J Anim Sci 86:3473–3479

Li N, DeMarco VG, West CM, Neu J (2003) Glutamine supports recovery from loss of transepithelial resistance and increase of permeability induced by media change in Caco-2 cells. J Nutr Biochem 14:401–408

Li P, Yin YL, Li DF, Kim SW, Wu GY (2007) Amino acids and immune function. Br J Nutr 98:237–252

Li FN, Yin YL, Tan B, Kong XF, Wu GY (2011) Leucine nutrition in animals and humans: mTOR signaling 3 and beyond. Amino Acids 41:1185–1193

Lin J (2004) Too much short chain fatty acids cause neonatal necrotizing enterocolitis. Med Hypotheses 62:291–293

Liu YL, Huang JJ, Hou YQ, Zhu HL, Zhao SJ, Ding BY, Yin YL, Yi GF, Shi JX, Fan W (2008) Dietary arginine supplementation alleviates intestinal mucosal disruption induced by *Escherichia coli* lipopolysaccharide in weaned pigs. Br J Nutr 100:552–560

Maaike AR, van Beek RH, Gardi V, de Bie MH, Dassel AC, van Goudoever JB (2007a) Cysteine: a conditionally essential amino acid in low-birth-weight preterm infants? Am J Clin Nutr 86:1120–1125

Maaike AR, Barbara S, Shaji C, Henk S, Agneta LS, Johannes BG, Douglas GB (2007b) Methionine transmethylation and transsulfuration in the piglet gastrointestinal tract. Proc Natl Acad Sci U S A 104:3408–3413

Maingay-de Groof F, Huang L, Voortman GJ, Chao C, Huang Y, Van Goudoeve JB (2010) 148 branched chain amino acid requirements for term neonates. Pediatr Res 68:78

McIntire DD, Bloom SL, Casey BM, Leveno KJ (1999) Birth weight in relation to morbidity and mortality among newborn infants. N Engl J Med 340:1234–1238

Miller RG, Jahoor F, Jaksic T (1995) Decreased cysteine and praline synthesis in parenterally fed, premature infants. J Pediatr Surg 30:953–958

Møller HK, Thymann T, Fink LN, Frokiaer H, Kvistgaard AS, Sangild PT (2011) Bovine colostrum is superior to enriched formulas in stimulating intestinal function and necrotising enterocolitis resistance in preterm pigs. Br J Nutr 105:44–53

Neu J (1996) Necrotizing enterocolitis: the search for a unifying pathogenic theory leading to prevention. Pediatr Clin North Am 43:409–32

Neu J (2007) Gastrointestinal development and meeting the nutritional needs of premature infants. Am J Clin Nutr 85:629S–634S

Nusrat A, Turner JR, Madara JL (2000) Molecular physiology and pathophysiology of tight junctions. IV. Regulation of tight junctions by extracellular stimuli: nutrients, cytokines, and immune cells. Am J Physiol Gastrointest Liver Physiol 279:G851–G857

Omari TI, Davidson GP (2003) Multipoint measurement of intragastric pH in healthy preterm infants. Arch Dis Child Fetal Neonatal Ed 88:F517–F520

Oste M, Van Ginneken C, Van Haver E, Bjornvad CR, Thymann T, Sangild PT (2005) The intestinal trophic response to enteral food is reduced in parenterally fed preterm pigs and is associated with more nitrergic neurons. J Nutr 135:2657–2663

Parimi PS, Kalhan SC (2007) Glutamine supplementation in the newborn infant. Semin Fetal Neonatal Med 12:19–25

Patrycja P, Barbara S (2008) Animal models to study neonatal nutrition in humans. Curr Opin Clin Nutr Metab Care 11:601–606

Patterson JK, Lei XG, Miller DD (2008) The pig as an experimental model for elucidating the mechanisms governing dietary influence on mineral absorption. Exp Biol Med (Maywood) 233:651–64

Rhoads JM, Chen W, Gookin J, Wu GY, Fu Q, Blikslager AT, Rippe RA, Argenzio RA, Cance WG, Weaver EM, Romer LH (2004) Arginine stimulates intestinal cell migration through a focal adhesion kinase dependent mechanism. Gut 53:514–522

Riezzo G, Indrio F, Raimondi F, Montagna O, Salvia G, Massimo B, Polimeno L, Cavallo L, Francavilla R (2009) Maturation of gastric electrical activity, gastric emptying and intestinal permeability in preterm newborns during the first month of life. Ital J Pediatr 35:6

Roberts SA, Ball RO, Moore AM, Filler RM, Pencharz PB (2001) The effect of graded intake of glycyl-L-tyrosine on phenylalanine and tyrosine metabolism in parenterally fed neonates with an estimation of tyrosine requirement. Pediatr Res 49:111–119

Rouwet EV, Heineman E, Buurman WA, ter Riet G, Ramsay G, Blanco CE (2002) Intestinal permeability and carrier-mediated monosaccharide absorption in preterm neonates during the early postnatal period. Pediatr Res 51:64–70

Sangild PT (2006) Gut responses to enteral nutrition in preterm infants and animals. Exp Biol Med 231:1695–1711

Sangild PT, Schmidt M, Jacobsen H, Fowden AL, Forhead A, Avery B, Greve T (2000) Blood chemistry, nutrient uptake and organ weights of fetal and newborn calves derived from in vitro produced embryos. Biol Reprod 62:1495–1504

Sangild PT, Petersen YM, Schmidt M, Elnif J, Petersen TK, Buddington RK, Greisen G, Michaelsen KF, Burrin DG (2002a) Preterm birth affects the intestinal response to parenteral and enteral nutrition in newborn pigs. J Nutr 132:3786–3794

Sangild PT, Petersen YM, Schmidt M, Elnif J, Petersen TK, Buddington RK, Michaelsen KF, Greisen G, Burrin DG (2002b) Preterm birth affects the gastrointestinal responses to parenteral and enteral nutrition in the newborn pig. J Nutr 132:2673–2681

Sangild PT, Schmidt M, Elnif J, Bjornvad CR, Buddington RK (2002c) Prenatal development of the gastrointestinal tract in pigs and the effect of fetal gut obstruction. Pediatr Res 52:416–424

Sangild PT, Siggers RH, Schmidt M, Elnif J, Bjornvad CR, Thymann T, Grondahl ML, Hansen AK, Jensen SK, Boye M, Moelbak L, Buddington RK, Westrom BR, Holst JJ, Burrin DG (2006) Diet- and colonization-dependent intestinal dysfunction predisposes to necrotizing enterocolitis in preterm pigs. Gastroenterology 130:1776–1792

Schaart MW, de Bruijn AC, Schierbeek H, Tibboel D, Renes IB, van Goudoever JB (2009) Small intestinal MUC2 synthesis in human preterm infants. Am J Physiol Gastrointest Liver Physiol 296:G1085–G1090

Schmidt M, Buddington RK, Sangild PT, Siggers RH, Thymann T, Jensen BB, Mølbak L, Heegaard PM (2008) Elective cesarean delivery affects gut maturation and delays microbial colonization but does not increase necrotizing enterocolitis in preterm pigs. Am J Physiol Regul Integr Comp Physiol 294:R929–R938

Schwiertz A, Gruhl B, Löbnitz M, Michel P, Radke M, Blaut M (2003) Development of the intestinal bacterial composition in hospitalized preterm infants in comparison with breast-fed, full-term infants. Pediatr Res 54:393–399

Shoveller AK, House JD, Brunton JA, Pencharz PB, Ball RO (2004) The balance of dietary sulfur amino acids and the route of feeding affect plasma homocysteine concentrations in neonatal piglets. J Nutr 134:609–612

Shulman RJ (2002) Effect of enteral administration of insulin on intestinal development and feeding tolerance in preterm infants: a pilot study. Arch Dis Child Fetal Neonatal 86: F131–F133

Siggers RH, Siggers J, Thymann T, Boye M, Sangild PT (2011) Nutritional modulation of the gut microbiota and immune system in preterm neonates susceptible to necrotizing enterocolitis. J Nutr Biochem 22:511–521

Sodhi C, Richardson W, Gribar S, Hackam DJ (2008) The development of animal models for the study of necrotizing enterocolitis. Dis Model Mech 1:94–98

Stechmiller JK, Childress B, Porter T (2004) Arginine immunonutrition in critically ill patients: a clinical dilemma. Am J Crit Care 13:17–23

Suryawan A, Nguyen HV, Almonaci RD, Davis TA (2012) Differential regulation of protein synthesis in skeletal muscle and liver of neonatal pigs by leucine through an mTORC1-dependent pathway. J Anim Sci Biotechnol 3:3

Tan B, Yin YL, Liu ZQ, Li XG, Xu HJ, Kong XF, Huang RL, Tang WJ, Shinzato I, Smith SB, Wu GY (2009) Dietary L-arginine supplementation increases muscle gain and reduces body fat mass in growing-finishing pigs. Amino Acids 37:169–175

Tan BE, Yin YL, Liu ZQ, Tang WJ, Xu HJ, Konga XF, Li XG, Yao K, Gu WT, Smith SB, Wu GY (2011) Dietary L-arginine supplementation differentially regulates expression of fat-metabolic genes in porcine adipose tissue and skeletal muscle. J Nutr Biochem 22:441–445

Thomas B, Gruca LL, Bennett C, Parimi PS, Hanson RW, Kalhan SC (2008) Metabolism of methionine in the newborn infant: response to the parenteral and enteral administration of nutrients. Pediatr Res 64:381–386

Thompson SW, McClure BG, Tubman TRJ (2003) A randomized, controlled trial of parenteral glutamine in ill, very low birth-weight neonates. J Pediatr Gastroenterol Nutr 37:550–553

Tubman TR, Thompson SW (2001) Glutamine supplementation for prevention of morbidity in preterm infants. Cochrane Database Syst Rev 4:CD001457

Upperman JS, Potoka D, Grishin A, Hackam D, Zamora R, Ford HR (2005) Mechanisms of nitric oxide-mediated intestinal barrier failure in necrotizing enterocolitis. Semin Pediatr Surg 14:159–66

Urschel KL, Evans AR, Wilkinson CW, Pencharz PB, Ball RO (2007) Parenterally fed neonatal piglets have a low rate of endogenous arginine synthesis from circulating proline. J Nutr 137:601–606

van den Akker CH, Schierbeek H, Minderman G, Vermes A, Schoonderwaldt EM, Duvekot JJ, Steegers EA, van Goudoever JB (2011) Amino acid metabolism in the human fetus at term: leucine, valine, and methionine kinetics. Pediatr Res 70:566–71

van Elburg RM, Fetter WP, Bunkers CM, Heymans HS (2003) Intestinal permeability in relation to birth weight and gestational and postnatal age. Arch Dis Child Fetal Neonatal Ed 88: F52–F55

Verner A, Craig S, McGuire W (2007) Effect of taurine supplementation on growth and development in preterm or low birth weight infants. Cochrane Database Syst Rev 17:CD006072

Vlaardingerbroek H, van den Akker CHP, de Groof F, Hogewind-Schoonenboom JE, Huang L, Riedijk MA, van der Schoor SRD, Huang Y, van Goudoever JB (2011) Amino acids for the neonate: search for the ideal dietary composition. NeoRev 12:e506–e516

Wang X, Qiao SY, Yin YL, Yu LG, Wang ZY, Wu GY (2007) A deficiency or excess of dietary threonine reduces protein synthesis in jejunum and skeletal muscle of young pigs. J Nutr 137:1442–1446

Wang JJ, Chen LX, Li P, Li XL, Zhou HJ, Wang FL, Li DF, Yin YL, Wu GY (2008) Gene expression is altered in piglet small intestine by weaning and dietary glutamine supplementation. J Nutr 138:1025–1032

Wang J, Ford HR, Grishin AV (2010) NF-κB-mediated expression of MAPK phosphatase-1 is an early step in desensitization to TLR ligands in enterocytes. Mucosal Immunol 3:523–534

Wang SP, Blachier F, Zhao F, Yin YL (2011) Intestinal microbiota: development, metabolism and functions. J Food Agric Environ 9:121–129

Weaver LT, Austin S, Cole TJ (1991) Small intestinal length: a factor essential for gut adaptation. Gut 32:1321–1323

WHO (2007) Protein and amino acid requirements in human nutrition. WHO Tech Rep Ser 935:1–265

Wu G, Ott TL, Knabe DA, Bazer FW (1999) Amino acid composition of the fetal pig. J Nutr 129:1031–8

Wu G, Meininger CJ, Kelly K, Watford M, Morris SM Jr (2000a) A cortisol surge mediates the enhanced expression of pig intestinal pyrroline-5-carboxylate synthase during weaning. J Nutr 130:1914–1919

Wu G, Meininger CJ, Knabe DA, Bazer FW, Rhoads JM (2000b) Arginine nutrition in development, health and disease. Curr Opin Clin Nutr Metab Care 3:59–66

Wu G, Becker RM, Bose CL, Rhoads JM (2001) Serum amino acid concentrations in preterm infants. J Pediatr 139:334–337

Wu G, Jaeger LA, Bazer FW, Rhoads JM (2004) Arginine deficiency in preterm infants: biochemical mechanisms and nutritional implications. J Nutr Biochem 15:442–451

Wu GY, Bazer FW, Davis TA, Johnson GA, Kim SW, Knabe DA, Spencer TE, Yin YL (2007) Important roles for arginine-family amino acids in swine nutrition and production. Livest Sci 122:8–22

Wu GY, Bazer FW, Davis TA, Kim SW, Li P, Rhoads JM, Satterfield MC, Smith SB, Spencer TE, Yin YL (2009) Arginine metabolism and nutrition in growth, health and disease. Amino Acids 37:169–175

Yao K, Yin YL, Chu WY, Liu ZQ, Dun D, Li TJ, Huang RL, Zhang JS, Bie T, Wang WC, Wu GY (2008) Dietary arginine supplementation increases mTOR signaling activity in skeletal muscle of neonatal pigs. J Nutr 138:867–872

Yao K, Guan S, Li TJ, Huang RL, Wu GY, Ruan Z, Yin YL (2011) Dietary L-arginine supplementation enhances intestinal development and expression of vascular endothelial growth factor in weanling piglets. Br J Nutr 105:703–709

Yin YL, Tan BE (2010) Manipulation of dietary nitrogen, amino acids and phosphorus to reduce environmental impact of swine production and enhance animal health. J Food Agric Environ 8:447–462

Yin YL, Yao K, Liu ZJ, Gong M, Ruan Z, Deng D, Tan BE, Liu ZQ, Wu GY (2010) Supplementing L-leucine to a low-protein diet increases tissue protein synthesis in weanling pigs. Amino Acids 39:1477–1486

Chapter 14
Amino Acids and Obesity, Diabetes, and Dyslipidemia in Pigs and Other Mammals

Wu Xin, Li Fengna, and Yulong Yin

Contents

14.1 Introduction

The incidence of diabetes, obesity, and dyslipidemia dramatically increased in all parts of the world, a fact largely attributed to dietary and lifestyle changes that come with economic development (Nguyen et al. 2008). Obesity is closely linked with diabetes and can lead to endocrine, cerebrovascular diseases. Various complications induced by diabetes and obesity are endangering public health and have become one of the most challenging problems facing the world today (Nguyen et al. 2008).

Regional body fat distribution has an important influence on metabolic and cardiovascular risk factors. Increased abdominal (visceral) fat accumulation is a risk factor for type 2 diabetes, dyslipidemia, hypertension, stroke, and coronary artery disease (CAD) (Franssen et al. 2011; Repas 2011).

The primary dyslipidemia related to obesity is characterized by increased triglycerides, decreased HDL levels, and abnormal LDL composition. The recent emphasis on treatment of the dyslipidemia of the metabolic syndrome

W. Xin • L. Fengna • Y. Yin (✉)
Research Center for Healthy Breeding of Livestock and Poultry, Hunan Engineering and Research Center of Animal and Poultry Science and Key Laboratory for Agro-ecological Processes in Subtropical Region, Institute of Subtropical Agriculture, The Chinese Academy of Sciences, Changsha, Hunan 410125, People's Republic of China
e-mail: yinyulong@isa.ac.cn

F. Blachier et al. (eds.), *Nutritional and Physiological Functions of Amino Acids in Pigs*, DOI 10.1007/978-3-7091-1328-8_14,

(hypertriglyceridemia, reduced high-density lipoprotein, and increased small, dense low-density lipoprotein particle number) has compelled practitioners to consider lipid-lowering therapy in a greater number of their patients, as one in two individuals over age 50 has the metabolic syndrome (He et al. 2012a, b). Individuals with the metabolic syndrome typically have normal low-density lipoprotein cholesterol levels, and current lipid-lowering guidelines may underestimate their cardiovascular risk. Two subgroups of patients with the metabolic syndrome are at particularly high risk for premature CAD. One, individuals with type 2 diabetes, accounts for 20–30 % of early cardiovascular disease. The second, familial combined hyperlipidemia, accounts for an additional 10–20 % of premature CAD. Familial-combined hyperlipidemia is characterized by the metabolic syndrome in addition to a disproportionate elevation of apolipoprotein B levels. The measurement of fasting glucose and apolipoprotein B, in addition to the fasting lipid profile, can help to estimate CAD risk in patients with the metabolic syndrome.

In recent years, amino acids, especially functional amino acids have attracted increased attention among the public and the medical community because of accumulated evidence from a large body of literature suggesting that functional amino acids may benefit human health (Newsholme et al. 2011; Kim et al. 2007a, b, c; Tan et al. 2009, 2011; Wu et al. 2007; Yao et al. 2008; Yin and Tan 2010; Li et al. 2011; Yin et al. 2010). Amino acid metabolism can be split into those 20 amino acids used for protein biosynthesis and play important role related with a wide variety of cells and tissues and relate these changes to alterations in endocrine, physiologic, and immune function (Newsholme et al. 2011). Substantial data from epidemiologic surveys and nutritional intervention studies in humans and animals suggest that dietary amino acids have protective effects against a variety of disorders, including cardiovascular disease, hyperlipidemia, cancer, osteoporosis, and various forms of chronic renal disease (Eizirik et al. 1988; Sulochana et al. 1998).

Amino acid metabolism is complex because of the large number of metabolites involved. Evidence is also emerging that consumption or supplementation of amino acids may have a beneficial effect on diabetes mellitus and obesity in animals and humans. Results showed that plasma spectrum of 20 kinds of amino acid metabolism and blood glucose values have a strong correlation between high and low, of which 7 amino acids (Arg, Cit, Asp, Asn, Thr, Leu, Trp) on the blood glucose level of grouping the greatest contribution. Several studies in humans and animals have shown that amino acids reduce plasma total cholesterol and LDL cholesterol (Hagemeister et al. 1990).

This chapter examines the evidence for a possible role of dietary amino acids in diabetes mellitus and obesity and discusses various mechanisms by which this class of amino acids may affect glucose and lipid metabolism and improve the control of body weight and glucose homeostasis as well as dyslipidemia.

14.2 Amino Acids and Obesity

The last 25 years have seen a great increase of the incidence of obesity in the world. Obesity is increasing, probably as a consequence of easily available, fat-rich food and an increasingly sedentary lifestyle. Obesity has reached epidemic proportions and is strongly linked to a number of chronic diseases, including diabetes, dyslipidemia, cardiovascular disease, and so on (Hedley et al. 2004).

Branched-chain amino acids (BCAA) cannot be made by the body so they are labeled "essential amino acids" and include leucine, isoleucine, and valine. It has been found that increasing intake of branched-chain amino acids may help in maintaining a healthy weight (Qin et al. 2011). Possible proposed mechanisms of branched-chain amino acids on obesity include the effect of leucine supplementation on the anti-hunger hormone leptin (Lynch et al. 2006), as well as the ability of branched-chain amino acids to participate in the control of blood sugar and in the maintenance of a healthy weight (Mobbs et al. 2005).

Leucine supplementation markedly reduced diet-induced obesity when fed with high-fat feed. Supplementation with leucine leads to a 32 % reduction of weight gain and a 25 % decrease in adiposity in mice fed high-fat diet. Leucine also decreases food intake and BW in normal rats and *ob/ob* mice. It is similar to that achieved by high protein diet, also indicating that leucine is a major component of the effects of a high protein diet. Moreover, leucine and high protein diet reduce AMPK but promote mTOR activity in the hypothalamus, resulting in inhibition of neuropeptide Y and stimulation of pro-opiomelanocortin expression. It suggests that a cross-regulation between AMPK and mTOR exists to regulate feeding in a leucine-dependent manner.

Furthermore, the relationship between obesity and amino acid metabolism displays important physiological and clinical implications. Patterson et al. (2002) studied the effect of obesity on regional skeletal muscle and adipose amino acid metabolism using a combination of stable isotope tracer and arterio-venous balance methods (Patterson et al. 2002). The rate of leucine release from forearm and adipose tissues in obese women is lower than in lean subjects, and obesity is associated with a lower fractional contribution from skeletal muscle to systemic leucine rate of appearance. It helps to illustrate the reasons why obese persons are more effective than lean ones in preserving body protein during fasting.

For amino acids, changes in blood concentrations of select essential amino acids and their derivatives, in particular BCAA, sulfur amino acids, tyrosine, and phenylalanine, are apparent with obesity (Adams 2011). Metabolomic profiling of obese versus lean humans reveals a BCAA-related metabolite signature that is suggestive of increased catabolism of BCAA and correlated with insulin resistance. This comprehensive metabolic profiling was performed on74 obese (median BMI of 36.6 kg/m^2) and 67 lean (median BMI of 23.2 kg/m^2) subjects (Table 14.1) (Newgard et al. 2009). The data showed that levels of the BCAAs valine and leucine/isoleucine were 20 % and 14 % higher, respectively, in obese compared to lean subjects ($P < 0.0001$). It is hypothesized that these changes may reflect an overload of BCAA catabolism in obese subjects. In growing pigs, by comparing genetically obese and lean strains, it was found that threonine, tyrosine, and creatine are higher in obese than in lean pigs, while urea is lower in obese pigs suggesting that obese pigs have different nitrogenous metabolism when compared with lean animals (He et al. 2012a, b).

Table 14.1 Amino acids concentration in the serum detected by tandem mass spectrometry

Amino acids (mM)	Obese ($n = 74$)	Lean ($n = 67$)	*P* value
Valine	281.4 (249.2, 332.9)	235.3 (204.1, 257.0)	<0.0001
Leucine/Isoleucine	170.0 (150.2, 200.8)	149.0 (132.5, 176.6)	<0.0001
Glutamate/Glutamine	118.4 (91.4, 143.7)	81.2 (66.7, 95.2)	<0.0001
Glycine	282.6 (245.6, 319.6)	328.4 (265.6, 403.0)	0.0007
Alanine	433.4 (394.5, 492.3)	367.3 (297.1, 420.0)	<0.0001
Phenylalanine	72.6 (66.3, 78.9)	61.6 (55.1, 68.8)	<0.0001
Tyrosine	79.5 (68.5, 90.0)	67.1 (56.7, 73.5)	<0.0001
Aspartate/Asparagine	20.1 (17.3, 23.8)	16.5 (13.5, 19.7)	<0.0001
Arginine	135.2 (116.5, 148.5)	115.3 (101.6, 137.0)	0.0007
Citrulline	32.0 (27.9, 40.3)	36.3 (30.5, 40.7)	0.04
Histidine	81.6 (73.5,88.9)	81.9 (71.9, 91.6)	0.57
Methionine	27.5 (25.1, 30.7)	27.6 (24.2, 30.9)	0.68
Omithine	69.6 (55.3, 80.8)	64.8 (53.3, 71.8)	0.18
Proline	176.4 (155.4, 231.5)	158.1 (138.4, 201.7)	0.02
Serine	115.6 (100.7, 131.9)	116.7 (102.2, 138.6)	0.69

Note: The obese subjects were comprised of 70 % women and 41 % African Americans, whereas the lean subjects were 57 % women and 45 % African Americans (Newgard et al. 2009)

Tryptophan and phenylalanine may help in the fight against obesity by burning fat more efficiently and suppressing appetite. Many researchers suggest that certain supplements, amino acids in particular, may be effective in weight loss programs. Tryptophan supplements tend to diminish the desire for carbohydrates. And since most overweight people seem to favor sweet, processed carbohydrate foods, it may be helpful in weight loss. Phenylalanine suppresses the appetite by increasing the production of the neurotransmitter norepinephrine.

In another hand, the prevalence of pre-pregnancy obesity has increased and is estimated to be almost 30 % among reproductive aged women (Artal et al. 2010; Kim et al. 2007a, b, c; Ogden et al. 2006).

Maternal obesity is associated with increased placental amino acid transport and hyperleptinemia. The study of placental amino acid transport and the effect of leptin on placental amino acid transport in vitro in the setting of maternal obesity was conducted. The obese group had decreased placental sodium-dependent neutral amino acid transporter (SNAT) activity (Fig. 14.1), maternal hyperleptinemia, and decreased syncytiotrophoblast expression of leptin receptor and SNAT-4. Placental amino acid uptake was significantly stimulated by leptin in the lean group as compared to the obese group. Maternal weight gain and offspring birth weights were not different between groups. Maternal obesity was accompanied by decreased placental SNAT activity associated with maternal hyperleptinemia and placental leptin resistance in spite of appropriate maternal weight gain and normally grown neonates (Farley et al. 2010). These findings suggest altered placental function that may have clinical implications in obese pregnant women.

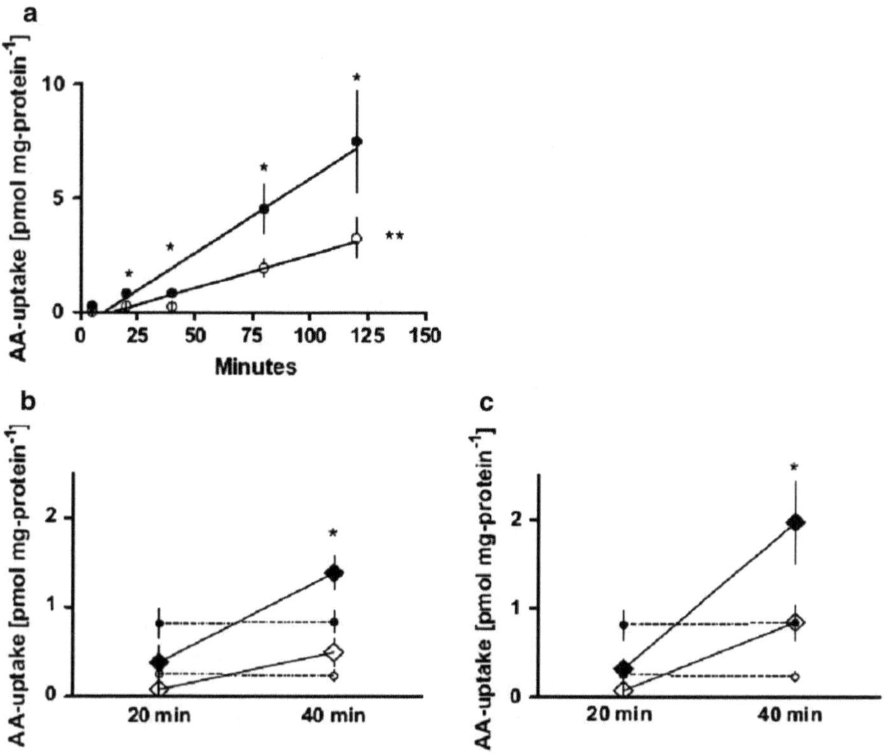

Fig. 14.1 (**a**) Placental SNAT activity (pmol mg/protein/min), term placental villous fragments. Lean (**c**, $n = 1{,}017$), obese (**b**, $n = 7$). *$P < 0.05$ (20, 40, 80, 120 min time points e lean vs. obese), **$P < 0.005$ (SNAT activity lean vs. obese), amino acid (AA). (**b**, **c**) Effect of leptin stimulation on placental SNAT activity (pmol mg/protein/min), at concentrations of 100 ng/mL (**b**, lean vs. obese $P < 0.07$), 500 ng/mL (**c**, lean vs. obese $P < 0.10$) in term placental villous fragments (Farley et al. 2010)

14.3 Amino Acids and Diabetes

Diabetes is a typical metabolic disease, with a variety of complications, an important factor in the quality of life of patients, early diabetes. Thus ultra-early detection and prevention are particularly important.

Amino acids are the building blocks of proteins and are also used in the synthesis of DNA, glucose, and ATP. Amino acids as important metabolic network metabolites are associated with diabetes mellitus (Chen et al. 2011; Newsholme et al. 2011). Amino acids may play a direct or indirect (via generation of putative messengers of mitochondrial origin) role in insulin secretion (Newsholme et al. 2007). Amino acids have antidiabetic effects (Sulochana et al. 1998). Specific amino acids are known to acutely and chronically regulate insulin secretion from pancreatic cells in vivo and in vitro (Smith et al. 1997). Mitochondrial metabolism is crucial for the coupling of amino acid and glucose recognition to exocytosis of insulin granules. Mitochondria generate ATP, which is the main coupling messenger in insulin secretion, and other coupling factors, which serve as sensors for the control of the exocytotic process.

Numerous studies have sought to identify the factors that mediate the key amplifying pathway over the Ca^{2+} signal in nutrient-stimulated insulin secretion. Predominantly, these factors are nucleotides (ATP, GTP, cAMP, and NADPH), although metabolites have also been proposed, such as long-chain acyl-CoA derivatives and glutamate (Newsholme et al. 2005). This scenario further highlights the importance of the key enzymes or transporters, e.g., glutamate dehydrogenase, the aspartate and alanine aminotransferases, and the malate–aspartate shuttle in the control of insulin secretion. In addition, after chronic exposure, amino acids may influence gene expression in the pancreatic beta cells, which subsequently alters levels of insulin secretion. Only a relatively small number of amino acids promote or synergistically enhance insulin release from pancreatic beta cells (Fajans et al. 1967). Four amino acids were found to be particularly important for stimulating beta cell electrical activity, essential for insulin secretion (leucine, isoleucine, alanine, and arginine) (Bolea et al. 1997).

The mechanisms by which amino acids enhance insulin secretion are varied. The cationically charged amino acid, L-arginine, does so by accumulating in beta cells provoking direct depolarization of the plasma membrane at neutral pH but only in the presence of glucose (Blachier et al. 1989), whereas other amino acids, which are co-transported with Na^{+}, can also depolarize the cell membrane as a consequence of Na^{+} transport and thus induce insulin secretion by activating voltage-dependent calcium channels. Metabolism, resulting in partial oxidation, e.g., L-alanine (Brennan et al. 2002), may initially increase the cellular content of ATP, leading to closure of the ATP-sensitive K^{+} (KATP) channel, depolarization of the plasma membrane, activation of the voltage-activated Ca^{2+} channel, Ca^{2+} influx, and insulin exocytosis. Additional mitochondrial signals may be generated that affect insulin secretion (Maechler 2002). A summary of potential regulatory mechanisms with respect to amino acid-stimulated insulin secretion is illustrated in Fig. 14.2.

In vitro studies have shown that amino acids generate signals transduced by the serine/threonine kinase mammalian target of rapamycin (mTOR) to regulate protein synthesis and cell proliferation as mediated by S6 kinase 1 (S6K1) and eukaryotic initiation factor 4EeC binding protein 1 (4E-BP1) (Reimann et al. 2004). Thus, mTOR appears to sense the availability of nutrients, most notably amino acids. It is interesting that activation of the mTOR pathway by prolonged insulin stimulation or increased amino acid availability induces insulin resistance in muscle cells and adipocytes in vitro (10eC12). Activation of mTOR and/or its downstream target, S6K1, leads to serine/threonine phosphorylation and consequently proteosomal degradation of insulin receptor substrate (IRS)-1 (10eC12). This in turn results in inhibition of phosphoinositide (PI) 3-kinase activity (Fajans et al. 1967), an essential step in insulin-mediated glucose metabolism. We further demonstrated that amino acid-induced mTOR/S6K1 activation speeds up insulin-dependent PI3-kinase deactivation in cultured myocytes even before detectable IRS-1 degradation, suggesting that uncoupling of IRS-1/PI3-kinase signaling is

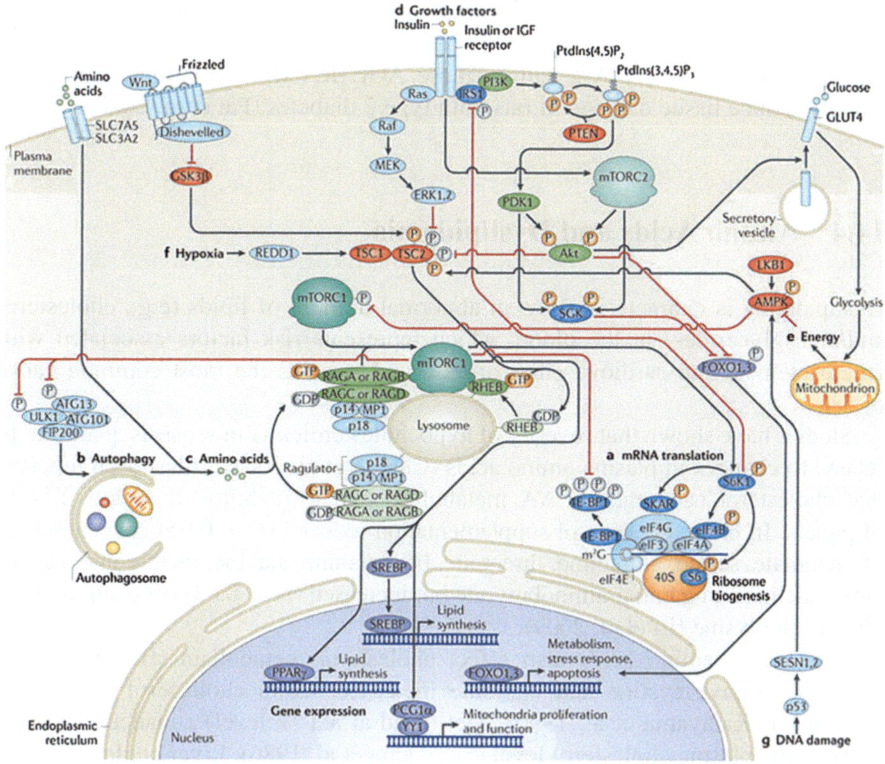

Nature Reviews | Molecular Cell Biology

Fig. 14.2 ThemTORsignalling pathway. Mammalian target of rapamycin complex 1 (mTORC1) promotes mRNA translation (**a**) and inhibits autophagy (**b**) by integrating nutrient signals that are generated by amino acids (**c**), growth factors such as insulin and insulin-like growth factors (IGFs) (**d**), energy signals that act through AMP-activated kinase (AMPK) (**e**), and various stressors including hypoxia (**f**) and DNA damage (**g**). Signal integration occurs at the level of the tuberous sclerosis 1 (TSC1; also known as hamartin)-TSC2 (also known as tuberin) complex. Akt and extracellular regulated kinase 1 (ERK1) and ERK2 phosphorylate TSC2, thus inhibiting the GTPase activating protein (GAP) activity of TSC1–TSC2 towards Ras homologue enriched in brain (RHEB). By contrast, phosphorylation of TSC2 by AMPK and glycogen synthase kinase 3β (GSK3β) results in the activation of TSC1–TSC2. The hypoxic factor protein regulated in development and DNA damage response 1 (REDD1; also known as DDIT4) promotes the assembly and activation of TSC1–TSC2. A second level of integration occurs at the lysosome: the Rag GTPases (which are held in place by the Ragulator, which consists of p18, p14, and MAPK scaffold protein 1 (MP1)) recruit mTORC1 to the lysosomal surface in response to amino acids (**c**); in turn, lysosomal recruitment enables mTORC1 to interact with GTP-bound RHEB, the end point of growth factor (**d**), energy (**e**), and stress (**f**, **g**) inputs. Growth factor receptors activate mTORC2 near the plasma membrane (**d**), where mTORC2 may be recruited through binding of mammalian stress-activated map kinase-interacting protein 1 (mSIN1; also known as MAPKAP1) to phospholipids. Because of its role in phosphorylating and activating Akt, mTORC2 forms a core component of the phosphoinositide 3-kinase (PI3K) pathway. Activating and inhibitory phosphates are *orange* and *gray*, respectively. *4E-BP1* eIF4E-binding protein 1, *ATG* autophagy-related, *CBP80* 80 kDa nuclear cap-binding protein, *eEF2K* eukaryotic elongation factor 2 kinase, *eIF* eukaryotic translation initiation factor, *FIP200* 200 kDa FAK family kinase-interacting protein, *FOXO* forkhead

an early event in the interaction between amino acids and insulin action (Fajans et al. 1967).

Supplemental dietary Arg can decrease AGE–RAGE interactions and consequently reduce tissue damage in rats with type 2 diabetes (Pai et al. 2010).

14.4 Amino Acids and Dyslipidemia

Dyslipidemia is characterized by an abnormal amount of lipids (e.g., cholesterol and/or triglycerides) in the blood, which represents risk factors associated with atherosclerosis and cardiovascular disease and remains the most common cause of death.

Studies have shown that severity of hypocholesterolemia in sepsis is quantifiably related to changes in plasma amino acids (Chiarla et al. 2004). It has been reported that cholesterol can regulate AA metabolism and availability in various tissues of piglets. In brain, cholesterol supplementation reduced ($P < 0.05$) concentrations of glutamate, serine, glutamine, threonine, beta-alanine, alanine, methionine, isoleucine, leucine, and gamma-aminobutyrate but increased ($P < 0.05$) concentrations of glycine and lysine (Li et al. 2009).

Also, some amino acids also affect cholesterol metabolism (Blachier et al. 2010). Excessive cystine and histidine increased serum cholesterol and alpha-tocopherol (Katayama et al. 1990). Met (used at a 1 % level) enhanced and Cys reduced the plasma cholesterol level (Sugiyama et al. 1986). Liver cholesterol was also markedly decreased by the addition of Gly to casein, but not methionine nor cystine (Tanaka and Sugano 1989). Adding methionine or cystine to a 10 % soybean protein isolate diet increases cholesterol 7alpha-hydroxylase activity (Taniguchi et al. 2008).

14.5 Summary and Perspectives

As the fundamental building blocks of protein and as precursors of numerous metabolites, amino acids metabolism play important roles related with a wide variety of cell and tissue functions and relate these changes to alterations in endocrine, physiologic, and immune functions. Further research works are needed

---

Fig. 14.2 (continued) box protein O, *IRS1*, insulin receptor substrate 1, *MEK* MAPK/ERK kinase, *PDK1* 3-phosphoinositide-dependent protein kinase 1, *PGC1α* PPARγcoactivator 1α, *PKC* protein kinase C, *PPARγ* peroxisome proliferator-activated receptor-γ, *PTEN* phosphatase and tensin homologue, *S6K1* S6 kinase 1, *SESN* sestrin, *SGK* serum- and glucocorticoid-regulated kinase, *SREBP* sterol regulatory element-binding protein, *SKAR* S6K1 Aly/REF-like target (also known as POLDIP3)

which would aim at deciphering effects of amino acids in the variety of disorders, including cardiovascular diseases, hyperlipidemia, osteoporosis, and various forms of chronic renal disease.

References

Adams SH (2011) Emerging perspectives on essential amino acid metabolism in obesity and the insulin-resistant state. Adv Nutr 2:445–456

Artal R, Lockwood CJ, Brown HL (2010) Weight gain recommendations in pregnancy and the obesity epidemic. Obstet Gynecol 115:152–155

Blachier F, Mourtada A, Sener A, Malaisse WJ (1989) Stimulus-secretion coupling of arginine-induced insulin release. Uptake of metabolized and nonmetabolized cationic amino acids by pancreatic islets. Endocrinology 124:134–141

Blachier F, Lancha AH Jr, Boutry C, Tome D (2010) Alimentary proteins, amino acids and cholesterolemia. Amino Acids 38:15–22

Bolea S, Pertusa JA, Martin F, Sanchez-Andres JV, Soria B (1997) Regulation of pancreatic beta-cell electrical activity and insulin release by physiological amino acid concentrations. Pflugers Arch 433:699–704

Brennan L, Shine A, Hewage C, Malthouse JP, Brindle KM, McClenaghan N, Flatt PR, Newsholme P (2002) A nuclear magnetic resonance-based demonstration of substantial oxidative L-alanine metabolism and L-alanine-enhanced glucose metabolism in a clonal pancreatic beta-cell line: metabolism of L-alanine is important to the regulation of insulin secretion. Diabetes 51:1714–1721

Chen T, Zhang X, Long Y, Yu H, Ran X, Gao Y, Lu H, Xie X, Chen X, Ren Y, Shi J, Tian H (2011) The association of plasma free amino acids with liver enzymes in type 2 diabetes patients. J Endocrinol Invest 35(8):772–5

Chiarla C, Giovannini I, Siegel JH (2004) The relationship between plasma cholesterol, amino acids and acute phase proteins in sepsis. Amino Acids 27:97–100

Eizirik DL, Germano CM, Migliorini RH (1988) Dietetic supplementation with branched chain amino acids attenuates the severity of streptozotocin-induced diabetes in rats. Acta Diabetol Lat 25:117–126

Fajans SS, Floyd JC Jr, Knopf RF, Conn FW (1967) Effect of amino acids and proteins on insulin secretion in man. Recent Prog Horm Res 23:617–662

Farley DM, Choi J, Dudley DJ, Li C, Jenkins SL, Myatt L, Nathanielsz PW (2010) Placental amino acid transport and placental leptin resistance in pregnancies complicated by maternal obesity. Placenta 31:718–724

Franssen R, Monajemi H, Stroes ES, Kastelein JJ (2011) Obesity and dyslipidemia. Med Clin North Am 95:893–902

Hagemeister H, Scholz-Ahrens KE, Schulte-Coerne H, Barth CA (1990) Plasma amino acids and cholesterol following consumption of dietary casein or soy protein in minipigs. J Nutr 120:1305–1311

He Q, Ren P, Kong X, Wu Y, Wu G, Li P, Hao F, Tang H, Blachier F, Yin Y (2012a) Comparison of serum metabolite compositions between obese and lean growing pigs using an NMR-based metabonomic approach. J Nutr Biochem 23:133–139

He Q, Ren P, Kong X, Wu Y, Wu G, Li P, Hao F, Tang H, Yin Y (2012b) Comparison of serum metabolite compositions between obese and lean growing pigs. J Nutr Biochem 23:133–139

Hedley AA, Ogden CL, Johnson CL, Carroll MD, Curtin LR, Flegal KM (2004) Prevalence of overweight and obesity among US children, adolescents, and adults, 1999–2002. J Am Med Assoc 291:2847–2850

Katayama T, Hayashi J, Kishida M, Kato N (1990) Effects of dietary excess amino acids on the concentrations of cholesterol, alpha-tocopherol, ascorbic acid, and copper in serum and tissues of rats. J Nutr Sci Vitaminol (Tokyo) 36:485–495

Kim SW, Mateo RD, Yin YL, Wu GY (2007a) Functional amino acids and fatty acids for enhancing production performance of sows and piglets. Asian Australas J Anim Sci 20:295–306

Kim SW, Mateo RD, Yin YL, Wu GY (2007b) Functional amino acids and fatty acids for enhancing production performance of sows and piglets. Asian Australa J Anim Sci 20:295–306

Kim SY, Dietz PM, England L, Morrow B, Callaghan WM (2007c) Trends in pre-pregnancy obesity in nine states, 1993–2003. Obesity (Silver Spring) 15:986–993

Li P, Kim SW, Li X, Datta S, Pond WG, Wu G (2009) Dietary supplementation with cholesterol and docosahexaenoic acid affects concentrations of amino acids in tissues of young pigs. Amino Acids 37:709–716

Li F, Yin Y, Tan B, Kong X, Wu G (2011) Leucine nutrition in animals and humans: mTOR signaling 3 and beyond. Amino Acids 41:1185–1193

Lynch CJ, Gern B, Lloyd C, Hutson SM, Eicher R, Vary TC (2006) Leucine in food mediates some of the postprandial rise in plasma leptin concentrations. Am J Physiol Endocrinol Metab 291:E621–630

Maechler P (2002) Mitochondria as the conductor of metabolic signals for insulin exocytosis in pancreatic beta-cells. Cell Mol Life Sci 59:1803–1818

Mobbs CV, Isoda F, Makimura H, Mastaitis J, Mizuno T, Shu IW, Yen K, Yang XJ (2005) Impaired glucose signaling as a cause of obesity and the metabolic syndrome: the glucoadipostatic hypothesis. Physiol Behav 85:3–23

Newgard CB, An J, Bain JR, Muehlbauer MJ, Stevens RD, Lien LF, Haqq AM, Shah SH, Arlotto M, Slentz CA, Rochon J, Gallup D, Ilkayeva O, Wenner BR, Yancy WS Jr, Eisenson H, Musante G, Surwit RS, Millington DS, Butler MD, Svetkey LP (2009) A branched-chain amino acid-related metabolic signature that differentiates obese and lean humans and contributes to insulin resistance. Cell Metab 9:311–326

Newsholme P, Brennan L, Rubi B, Maechler P (2005) New insights into amino acid metabolism, beta-cell function and diabetes. Clin Sci (Lond) 108:185–194

Newsholme P, Bender K, Kiely A, Brennan L (2007) Amino acid metabolism, insulin secretion and diabetes. Biochem Soc Trans 35:1180–1186

Newsholme P, Abdulkader F, Rebelato E, Romanatto T, Pinheiro CH, Vitzel KF, Silva EP, Bazotte RB, Procopio J, Curi R, Gorjao R, Pithon-Curi TC (2011) Amino acids and diabetes: implications for endocrine, metabolic and immune function. Front Biosci 16:315–339

Nguyen NT, Magno CP, Lane KT, Hinojosa MW, Lane JS (2008) Association of hypertension, diabetes, dyslipidemia, and metabolic syndrome with obesity: findings from the National Health and Nutrition Examination Survey, 1999 to 2004. J Am Coll Surg 207:928–934

Ogden CL, Carroll MD, Curtin LR, McDowell MA, Tabak CJ, Flegal KM (2006) Prevalence of overweight and obesity in the United States, 1999–2004. J Am Med Assoc 295:1549–1555

Pai MH, Huang KH, Wu CH, Yeh SL (2010) Effects of dietary arginine on inflammatory mediator and receptor of advanced glycation endproducts (RAGE) expression in rats with streptozotocin-induced type 2 diabetes. Br J Nutr 104:686–692

Patterson BW, Horowitz JF, Wu G, Watford M, Coppack SW, Klein S (2002) Regional muscle and adipose tissue amino acid metabolism in lean and obese women. Am J Physiol Endocrinol Metab 282:E931–936

Qin LQ, Xun P, Bujnowski D, Daviglus ML, Van Horn L, Stamler J, He K (2011) Higher branched-chain amino acid intake is associated with a lower prevalence of being overweight or obese in middle-aged East Asian and Western adults. J Nutr 141:249–254

Reimann F, Williams L, da Silva XG, Rutter GA, Gribble FM (2004) Glutamine potently stimulates glucagon-like peptide-1 secretion from GLUTag cells. Diabetologia 47:1592–1601

Repas T (2011) Obesity and dyslipidemia. S D Med 64:241–243, 245, 247 passim

Smith PA, Sakura H, Coles B, Gummerson N, Proks P, Ashcroft FM (1997) Electrogenic arginine transport mediates stimulus-secretion coupling in mouse pancreatic beta-cells. J Physiol 499(Pt 3):625–635

Sugiyama K, Mizuno M, Muramatsu K (1986) Effect of individual amino acids on plasma cholesterol level in rats fed a high cholesterol diet. J Nutr Sci Vitaminol (Tokyo) 32:623–633

Sulochana KN, Punitham R, Ramakrishnan S (1998) Beneficial effect of lysine and amino acids on cataractogenesis in experimental diabetes through possible antiglycation of lens proteins. Exp Eye Res 67:597–601

Tanaka K, Sugano M (1989) Effects of addition of sulfur-containing amino acids and glycine to soybean protein and casein on serum cholesterol levels of rats. J Nutr Sci Vitaminol (Tokyo) 35:323–332

Tan B, Yin YL, Liu ZQ, Li XG, Xu HJ, Kong XF, Huang RL, Tang WJ, Shinzato I, Smith SB, Wu GY (2009) Dietary L-arginine supplementation increases muscle gain and reduces body fat mass in growing-finishing pigs. Amino Acids 37:169–175

Tan BE, Yin YL, Liu ZQ, Tang WJ, Xu HJ, Kong XF, Li XG, Yao K, Gu WT, Smith SB, Wu GY (2011) Dietary L-arginine supplementation differentially regulates expression of fat-metabolic genes in porcine adipose tissue and skeletal muscle. J Nutr Biochem 22:441–445

Taniguchi M, Nagao K, Inoue K, Imaizumi K (2008) Cholesterol lowering effect of sulfur-containing amino acids added to a soybean protein diet in rats. J Nutr Sci Vitaminol (Tokyo) 54:448–453

Wu GY, Bazer FW, Davis TA, Johnson GA, Kim SW, Knabe DA, Spencer TE, Yin YL (2007) Important roles for arginine-family amino acids in swine nutrition and production. Livest Sci 122:8–22

Yao K, Yin YL, Chu WY, Liu ZQ, Dun D, Li TJ, Huang RL, Zhang JS, Tan B, Wang WC, Wu GY (2008) Dietary arginine supplementation increases mTOR signaling activity in skeletal muscle of neonatal pigs. J Nutr 138:867–872

Yin YL, Tan BE (2010) Manipulation of dietary nitrogen, amino acids and phosphorus to reduce environmental impact of swine production and enhance animal health. J Food Agr Environ 8:447–462

Yin YL, Yao K, Liu ZJ, Gong M, Ruan Z, Deng D, Tan BE, Liu ZQ, Wu GY (2010) Supplementing L-leucine to a low-protein diet increases tissue protein synthesis in weanling pigs. Amino Acids 39:1477–1486

Part IV
Methodology for Research on Amino Acid

Chapter 15
Methods of Amino Acid Analysis

Xia Xiong, Yulong Yin, Yiqiang Huang, Yongfei Wang, Qingqi Wen, Yuyun Mu, Xugang Shu, Zhichun Zhan, Ying Zhou, and Guixiong Qiu

Contents

X. Xiong • Y. Yin (✉)
Research Center for Healthy Breeding of Livestock and Poultry, Hunan Engineering and Research Center of Animal and Poultry Science, Key Laboratory of Agro-ecological Processes in Subtropical Region, Institute of Subtropical Agriculture, Scientific Observing and Experimental Station of Animal Nutrition and Feed Science in South-Central, Ministry of Agriculture, Chinese Academy of Sciences, Changsha, Hunan 410125, People's Republic of China
e-mail: yinyulong@isa.ac.cn

Y. Huang
Xingjia Bio-engineering Co., Ltd, Changsha, Hunan 410011, China

Y. Wang
Research and Development Center, Twins Group Co., Ltd, Nanchang, Jiangxi 330096, China

Q. Wen
Research and Development Center, Twins Group Co., Ltd, Nanchang, Jiangxi 330096, China

Fujian Aonong Biotechnology Corporation, Xiamen, Fujian 361007, China

Y. Mu
Shangshai Xinnong Feed Co. Ltd, Shanghai 201613, China

X. Shu • G. Qiu
Guang Zhou Tanke International Ltd, GuangZhou, Guangdong 510627, China

Z. Zhan • Y. Zhou
Wuhan Sunhy Biology Co., Ltd, Wuhan 430070, Hubei

F. Blachier et al. (eds.), *Nutritional and Physiological Functions of Amino Acids in Pigs*,
DOI 10.1007/978-3-7091-1328-8_15, © Springer-Verlag Wien 2013

Abbreviations

AA	Amino acid
AAA	Amino acid analysis
HPLC	High-performance liquid chromatography
NMR	Nuclear magnetic resonance
GC-MS	Gas chromatography-mass spectrometry
GE-MS	Capillary electrophoresis mass spectrometry
LC-MS/MS	Liquid chromatography-tandem mass spectrometry
CZE	Capillary zone electrophoresis
AQC	6-Aminoquinolyl-*N*-hydroxysuccinimidyl carbamate
OPA	*o*-Phthalaldehyde reagent
PITC	Phenyl isothiocyanate
FMOC-Cl	9-Fluorenylmethyloxycarbonyl chloride
Dansyl-Cl	Dansyl chloride

15.1 Introduction

There is growing recognition that besides amino acids (AA) as building blocks of proteins and polypeptides, AA act also as cell signaling molecules and regulators of gene expression (Wu 2009; Yin and Tan 2010; Zeng et al. 2012; Kim et al. 2007). Additionally, they are also a main source of energy and serve as precursors for the biosynthesis of numerous compounds including neurotransmitters, porphyrins, polyamines, nitric oxide, etc. (Kaspar et al. 2009a, b; Zhang et al. 2012). AA are important compounds for metabolic profiling, and their quantitative analysis is important for some disease diagnostics as well as for elucidating nutritional influences on physiology. Amino acid analysis (AAA) is used to determine the AA content of amino acid-, peptide-, and protein-containing samples.

Since Moore and Stein (1948) released AA from proteins using 6 M HCl acid at 110 °C for 22 h and separated the AA by ion-exchange chromatography with postcolumn derivatization in the early 1950s, different methods for analysis of amino acids have been developed, including amino acid analyzer, gas chromatography, reversed-phase (RP) high-performance liquid chromatography (HPLC) with precolumn derivatization, capillary electrophoresis, and so on. For several technical and practical advantages, amino acid separations have been carried out mainly by ion-exchange chromatography with postcolumn derivatization and RP-HPLC with precolumn derivatization. Many reviews concerning the analytical methods of AA have been published (Kaspar et al. 2009a, b; Davidson and O'Connor 2008; Rutherfurd and Gilani 2009). The analytical methods of choice for the analysis of AA include gas chromatography-mass spectrometry (GC-MS) , nuclear magnetic resonance (NMR) spectroscopy, capillary electrophoresis mass spectrometry (CE-MS), and liquid chromatography-tandem mass spectrometry(LC-MS/MS). The goal of the present chapter is to briefly compare the established methods and to present new approaches to amino acid analysis.

Quantitative analysis of amino acids is required in several fields, including clinical diagnostics of inborn errors of AA metabolism, biomedical research, bioengineering, and food science (Kaspar et al. 2009a, b). The amino acid analysis in animal nutrition is mainly used not only to determine the composition of AA in feed ingredients for animal diets but also for AA determination in physiological fluids. The AA analysis includes sample preparation and analysis.

15.2 Sample Preparation

15.2.1 Hydrolysis of Proteins and Peptides

Digestion of protein to amino acids requires hydrolysis. Accurate results from AAA require purified protein and peptide samples. Buffer components (e.g., detergents, urea, salts) can interfere with the amino acid analysis and must be removed from the sample before analysis. There are many useful methods including acidic hydrolysis, basic hydrolysis, enzymatic hydrolysis, microwave radiation-induced hydrolysis, and so on. Comparative studies have shown that errors in the performance of the hydrolysis are primarily responsible for variations in the determined compositions (Yuksel et al. 1995; Yin et al. 1991).

15.2.1.1 Acid Hydrolysis

Acid hydrolysis is the most common method for hydrolyzing protein in a biological sample before amino acid analysis. Protein in samples can be hydrolyzed in 6 M HCl at 110 °C for 24 h under N_2 (Wu 1993, 1997; Yin et al. 1994; Fan et al. 1996; Fan and Sauer 2002; Li et al. 2011). However only aspartic acid, glutamic acid, proline, glycine, alanine, leucine, phenylalanine, histidine, arginine, and hydroxyproline can be determined quantitatively during acid hydrolysis (Rutherfurd and Gilani 2009). Under the condition of acidic hydrolysis, asparagine and glutamine are hydrolyzed quantitatively to aspartic acid and glutamic acid. Darragh et al applied a series of hydrolysis times to study the amino acid losses during acid hydrolysis. Most of the residues underwent some degree of loss and the greatest loss was observed with cysteic acid and serine (Darragh et al. 1996). Tryptophan is also largely destroyed, and cysteine cannot be directly determined from the acid-hydrolyzed samples. Alkaline hydrolysis is most commonly used to quantitatively hydrolyze tryptophan (Wu et al. 1999; Li et al. 2011). Methionine and cysteine contents can be determined as methionine, sulfone, and cysteic acid after oxidation with performic acid. The oxidized samples can be then hydrolyzed and analyzed in the same manner as the samples which were not oxidized (Fan et al. 1996; Fan and Sauer 2002). Losses of threonine and serine also occur, and represents usually around 5 and 10 %, respectively (Ozols 1990). To reduce AA losses, various reagents, such as phenol, mercaptoethanol, thioglycolic acid, and indole (Ravindran and Bryden 2005) can be added to the HCl to stabilize the side chains of specific amino acids.

15.2.1.2 Alkaline Hydrolysis

For alkaline hydrolysis, the samples are hydrolyzed at 110 °C for 20 h in 4.2 M NaOH plus 25 % thiodiglycol. The disadvantage of basic hydrolysis is that the serine, threonine, arginine, and cysteine are destroyed. Alkaline hydrolysis is applied if the protein sample contains a large percentage of carbohydrates and is also used to analysis the phospho-amino acid (Ekman and Jager 1993; Fountoulakis and Lahm 1998). In the recent years, only few articles on alkaline hydrolysis of proteins for amino acid analysis were published.

15.2.1.3 Enzymatic Hydrolysis

Enzymatic hydrolysis has the advantage that it allows the quantification of asparagine and glutamine and of other sensitive residues, which are otherwise destroyed during chemical digestion of residues carrying side-chain modifications and this kind of hydrolysis does not induce racemization during digestion. Enzymatic hydrolysis of proteins for amino acid analysis is seldom applied.

15.2.1.4 Microwave Hydrolysis

Acid hydrolysis of protein samples has traditionally been the rate limiting step in amino acid analysis. Several reports over the past 20 years have advocated the use of microwave irradiation as a tool to speed up the hydrolysis of proteins and peptides under acidic conditions (Weiss et al. 1998; Zhong et al. 2005; Damm et al. 2011).

Some samples contain complex mixture of compounds, including protein, carbohydrate, fat, vitamins, and minerals. If the feeds contain >10 % fat, it may be necessary to defat the sample prior to hydrolysis, particularly for tryptophan analysis (Rutherfurd and Gilani 2009). For example, the milk needs to be defatted by centrifugation in studies of free and protein-bound amino acids in sow's colostrum and milk (Wu and Knabe 1994). Then analysis of free amino acids in biological fluids and tissue often requires deproteinization prior to analysis by compounds such as sulfosalicylic acid, trichloroacetic acid, or $HClO_4$, followed by ultrafiltration.

15.2.2 Derivatization of Free Amino Acids

Since amino acids are nonvolatile compounds and most of them have little UV absorbance, amino acids have been commonly analyzed by ion-exchange chromatography with postcolumn derivatization or high-performance liquid chromatography (HPLC) methods with precolumn derivatization using UV chromophore or fluorophore reagents (Yin et al. 1993a, b).

15.3 Chromatographic Separation and Detection

Many techniques for amino acid analysis exist, and the choice of one given technique often depends on the sensitivity required for the assay.

15.3.1 Ion-Exchange Chromatography with Postcolumn Derivatization

For over 30 years, most AA analysis has been performed by ion-exchange chromatography with postcolumn derivatization with ninhydrin or *o*-phthalaldehyde. Ion-exchange chromatography with postcolumn ninhydrin detection is one of the most common methods employed for quantitative amino acid analysis (Yin et al. 2010, 2011). As a rule, a Li-base cation-exchange system is employed for the analysis of the more complex samples, and the faster Na-based cation-exchange system is used for the more simple amino acid mixtures obtained with protein hydrolysates (typically containing 17 amino acid components). Separation of the amino acids on an ion-exchange column is accomplished through a combination of changes in pH and cation strength.

The reaction of ninhydrin with primary amino groups to form the purple dye called Ruhemann's purple (RP) was discovered by Siegfried Ruhemann in 1910. Ninhydrin reacts with primary amines at elevated temperatures (100–120 °C) and approximately pH 5. The absorption peak is found at 570 nm, and the absorption intensity is proportional to the concentration of amino acids. In addition, ninhydrin also reacts with secondary amines in compounds such as proline, arginine, asparagine, tryptophan, and cysteine. It is worth noting that ninhydrin reagent is unstable, limiting the use of the ninhydrin/acetate buffer mixture to ~2 weeks. The classic method has been considerably improved with the availability of modern dedicated AA analyzers (Osowska et al. 2004; Miyaji et al. 2010; Boutry et al. 2011, 2012). Despite considerable improvements (Cynober et al. 1985), the run time is still lengthy and when ninhydrin-based detection is used, the method is poorly sensitive for AA present at low concentrations, such as aspartic acid, citrulline, methionine, and tryptophan (Cynober et al. 1987). Detection limit is considered to be 10 pmol for most of the amino acid derivatives, but 50 pmol for proline.

In addition, OPA is also often used as precolumn derivatization reagent. OPA-amino acid derivative is a fluorogenic compound. OPA is 5–10 times more sensitive than either fluorescamine or ninhydrin (Bensen and Hare 1975). However, OPA and its analogs fail to react with secondary amino acids, and the derivatives are often unstable. Postcolumn derivatization reaction occurs after the separation of the amino acids, avoiding the interference with other substances. This technique is suitable for the analysis of amino acids in complex samples and can be used with sample that contain small amounts of buffer components, such as salts and urea, and generally require between 5 and 10 μg of protein sample per analysis. This method is often cited for its accuracy and reproducibility, but RP-HPLC methods with precolumn derivatization are often used because they are faster, more sensitive, and less costly for the analysis of amino acids.

15.3.2 Reversed-Phase Liquid Chromatography with Precolumn Derivatization

In comparison with ion-exchange chromatographic RP-HPLC with precolumn derivatization has the advantages of reducing analysis times, enhancing sensitivity and flexibility, and lowering cost of instrumentation and maintenance (Ziegler et al. 1992; Jacques et al. 1997). However the precolumn derivatization methods have some disadvantages. This includes derivative instability, derivatization by-product interference with chromatographic separation, and lack of reaction with secondary amino acids. Also, precolumn derivatization techniques may be influenced by buffer salts in the samples. RP chromatography methods were also applied to the separation of amino acid enantiomers (Waldhier et al. 2009). Factors such as mobile phase composition, flow rate, temperature, pH, gradient program, and column type and length affect the performance of RP-HPLC with different degrees of sensitivities. To optimize precolumn derivatization with phenylisothiocyanate, the methodological design has been used to optimize amino acid separation by RP-HPLC (Gheshlaghi et al. 2008).

Different reagents have been used for precolumn derivatization reagents such as 6-aminoquinolyl-*N*-hydroxysuccinimidyl carbamate (AQC) (Bosch et al. 2006; Cohen 1994), *o*-phthalaldehyde reagent (OPA) (Wu and Knabe 1994; Molnár-Perl 2001, 2003), phenyl isothiocyanate (PITC) (Battaqlia et al. 1999; Komarova et al. 2004), 9-fluorenylmethyloxycarbonyl chloride (FMOC) (Jambor and Molnar-Perl 2009) and Dansyl chloride (Dansyl-Cl) (Marquez et al. 1986; Furst et al. 1990). The standard to select derivatization reagents is to consider that derived products have different degrees of stability and do not produce interferences. Moreover, the reagent itself should not be fluorescent and consequently should not produce interfering peaks (Liu 2000). Depending on the number of compounds to be measured, chromatographic run time ranges for instance from 13 min for 23 compounds to 95 min for 38 compounds (Kaspar et al. 2009a, b). Some derivatization reagents are summarized in Table 15.1. OPA and FMOC-Cl are the most frequently applied derivatization reagents in the analysis of AA because of their selectivity and sensitivity (Molnar-Perl 2011). OPA method for routine analyses of primary free amino acids (except cystine) in biological fluids. Combination of the OPA and FMOC-Cl techniques are used to determined the secondary amino acids. The use of the dansyl-Cl technique is suggested when the determination of free cystine or cystine-containing short-chain peptides is required (Furst et al. 1990).

Table 15.1 Comparison of precolumn derivativation reagents for analysis of amino acids

Derivative reagent	Detection	Sensitivity	Reproducibility	Stable derivative	Removal of redundant reagent by drying	Removal of redundant reagent by solvent extraction	Interfering side-products	Secondary amino acids detected
OPA	Fluor	fmol	Good	No	No	No	No	No
PITC	UV	pmol	Good	Yes	Yes	No	No	Yes
FMOC-Cl	Fluor	fmol	Good	Yes	No	Yes	Yes	Yes
Dansyl-CL	UV/Fluor	pmol	Good	Yes	No	No	Yes	Yes
FDNB	UV	pmol	Good	Yes	Yes	No	Yes	Yes
FDNPAA	UV	pmol	Good	Yes	Yes	No	Yes	Yes
FDNDEA	UV	pmol	Good	Yes	Yes	No	Yes	Yes
DABS-Cl	UV/Fluor	pmol	Good	Yes				Yes
AQC	UV/Fluor	fmol	Good	Yes	No	Yes	No	Yes
PITC	UV	pmol	Good	Yes	Yes	No	No	Yes

15.3.3 Capillary Electrophoresis

Capillary electrophoresis (CE) has been developed in the 1980s as a high-performance liquid technology. It is based on the use of an electric field to separate charged compounds such as amino acids. It has proven to be a powerful technique for the rapid and highly efficient separation of minute amounts of protein and amino acid sample (Issaq 2001). The analysis of AA by CE has been mainly performed with precolumn, postcolumn, and on-column derivatization techniques with UV chromophore or fluorophore reagents to provide better resolution and sensitivity. Komarova et al. (2004) used pre-capillary derivatization of amino acids with PITC and separation of PTC-derivatives by capillary zone electrophoresis (CZE) and detection at 254 nm. This method allows to widen a list of detectable components up to 19 (without tryptophan) and significantly improve detection limits. Thorough discussion of this method for AAA has been published (Poinsot et al. 2012). Derivatization of amino acids prior to chromatography with chiral reagents has gained in popularity, because the diastereomers formed can be resolved on conventional RP-HPL. Compared with conventional electrophoresis and HPLC, this method has the advantages of high sensitivity and high degree of automation. CE does not require derivatization, but sensitivity of CE-UV analysis can be increased by introduction of a UV-active label. Most of derivatization reagents used for HPLC method can be used for CE method.

The literature of amino acid analysis using CE has been extensively reviewed (Smith 1997, 1999; Poinsot et al. 2006). The new developments of CE, especially those concerning the novelties in detection methods and chirality and the different biological applications have been recently described (Poinsot et al. 2012).

15.3.4 Mass Spectrometry

Ion-exchange chromatography is limited by long run time and by the fact that co-eluting compounds may prevent accurate quantification in some cases. Some amino acids commonly have overlapping retention times by ion-exchange chromatography with postcolumn ninhydrin detection. For example, this is the case for methionine, phenylalanine, and histidine. Mass spectrometry methods, for instance GC-MS (Revelsky et al. 2003; Fiamegos and Stalikas 2006), LC-MS (Molnár-Perl 2003), and GE-MS have the potential to avoid these limitations.

15.3.4.1 GC-MS

Derivatization of amino acids needs to be performed before analysis to produce volatile compounds which are analyzed by gas chromatographic method. The main derivatization methods include silylation, alkylation (Marquez et al. 1994), and acylation. The silylation is the main derivatization method. Silylation is carried out using BSTFA and MTBSTFA under anhydrous conditions and heating or esterification/acylation in two successive steps (Fiamegos and Stalikas 2006). GC is easily combined with mass spectrometry and a vast number of methods for the preparation of volatile derivatives suitable for GC and GE-MS analysis of amino acids have been reported. Kataoka et al have summarized these approaches (Kataoka et al. 2000). GC/MS methods are known to have good precision and accuracy. Duncan et al have employed isotope dilution gas chromatography/electron capture negative ionization/mass spectrometry (GC/ECNI/MS) to provide accurate and reliable data on <100 fmol of material (Duncan and Poljak 1998; Fan et al. 2006).

15.3.4.2 LC-MS/MS

LC-MS/MS is comparable to traditional LC-ninhydrin detection method. Mass spectral detection shortens analysis times and reduces interference in analysis involved in the detection of inborn metabolic errors (Dietzen et al. 2008). Both GC-MS and iTRAQ-LC-MS/MS are suited for high-throughput amino acid analysis, with the former offering at present higher reproducibility in a completely automated sample pretreatment method, while the latter covers more amino acids and related amines (Kaspar et al. 2009a, b).

15.3.4.3 CE-MS

CE-MS is based on CE as a front-end fractionation coupled to a mass spectrometer. This method provides fast separation and high resolution and is compatible with most buffers and compounds to be analyzed. The method for the determination of underivatized amino acids based on capillary electrophoresis coupled to electrospray ionization mass spectrometry (CE-ESI-MS) has been described (Soga 2000; Soga et al. 2004). GE-MS is a valid alternative to GC-MS for targeted profiling of metabolites, such as amino acids, and possesses some significant advantages over GC-MS (Williams et al. 2007). A review related to the use of CE-MS in the field of metabolomics has been published (Ramautar et al. 2009).

15.3.4.4 NMR-Spectroscopy

Metabonomics provides a useful systems approach to understand global changes in metabolites in animals in response to alterations in genetics, nutrition, environments, and gut microbiota. He et al. have, for instance, compared the metabonome of serum between the genetically obese and lean pigs using a NMR-based metabolomic method (He et al. 2009, 2012) and compared serum and jejunum metabonome in the situation of intrauterine growth restriction of piglets (He et al. 2011a, b, c)

References

Battaqlia A, Bertoluzza A, Calbucci F, Eusebi V, Giorqianni P, Ricci R, Tosi R, Tuqnoli V (1999) High-performance liquid chromatographic analysis of physiological amino acids in human brain tumors by pre-column derivatization with phenylisothiocyanate. J Chromatogr B 730:81–93

Bensen JR, Hare PE (1975) O-phthaladehyde: fluorogenic detection of primary amines in the picomole range. Comparison with fluorescamine and ninhydrin. Proc Natl Acad Sci USA 72:619–622

Bosch L, Alegria A, Farre R (2006) Application of the 6-aminoquinolyl-N-hydroxysccinimidyl carbamate(AQC) reagent to the RP-HPLC determination fo amino acids in infant foods. J Chromatogr B 831:176–183

Boutry C, Matsumoto H, Airinei G, Benamouzig R, Tome D, Blachier F, Bos C (2011) Monosodium glutamate raises antral distension and plasma amino acid after a standard meal in humans. Am J Physiol Gastrointest Liver Physiol 300:137–145

Boutry C, Matsumoto H, Bos C, Moinard C, Cynober L, Yin YL, Tome D, Blachier F (2012) Decreased glutamate, glutamine and citrulline concentrations in plasma and muscle in endotoxemia cannot be reversed by glutamate or glutamine supplementation: a primary intestinal defect? Amino Acids 43:1485–1498

Cohen SA (1994) Application for amino acid derivatization with 6-aminoquinolyl-N-hydroxysuccinimidyl carbamate: analysis of feed grains, intravenous solutions and glycoproteins. J Chromatogr A 661:25–34

Cynober L, Coudray-lucas C, Ziegler F et al (1985) High performance ion-exchange chromatography of amino-acids in biological fluids using chromakon 500-performance of the apparatus. J Automat Chem 7:201–203

Cynober L, Ziegler F, Coudray-lucas C, Chauffert M et al (1987) Comparison of methods of assay of amino acids by gas chromatography and ion exchange chromatography. Ann Biol Clin 45:537–540

Damm M, Michael H, Radspieler G, Marsche G, Kappe O (2011) Microwave-assisted high-throughput acid hydrolysis in silicon carbide microtiter platforms-A rapid and low volume sample preparation technique for total amino acid analysis in proteins and peptides. J Chromatogr A 1217:7826–7832

Darragh AJ, Garrick DJ, Moughan PJ et al (1996) Correction for amino acid loss during acid hydrolysis of a purified protein. Anal Biochem 236:199–207

Davidson I, O'Connor P (2008) Amino acid analysis. In: J. M. Walker and R. Rapley (eds) Molecular biomethods handbook (2nd edn), Humana Press, Totowa, NJ, pp 793–808

Dietzen DJ, Weindel AL, Carayannopoulos MO et al (2008) Rapid comprehensive amino acid analysis by liquid chromatography/tandem mass spectrometry:comparison to cation exchange with post-column ninhydrin detection. Rapid Commun Mass Spectrom 22:3481–3488

Duncan MW, Poljak A (1998) Amino acid analysis of peptide and proteins on the femtomole scale by gas chromatography/mass spectrometry. Anal Chem 70:890–896

Ekman P, Jager O (1993) Quantification of subnanomolar amounts of phosphate bound to seryl and threonyl residues in phosphoproteins using alkaline hydrolysis and malachite green. Anal Biochem 214:138–141

Fan MZ, Sauer WC (2002) Determination of true ileal amino acid digestibility and the endogenous amino acid outputs associated with barley samples for growing-finishing pigs by the regression analysis technique. J Anim Sci 80:1593–1605

Fan MZ, Sauer WC, Gabert VM (1996) Variability of apparent ileal digestibility of amino acids in different samples of canola meal for growing-finishing pigs. Can J Anim Sci 76:563–569

Fan MZ, Chiba LI, Matzat PD, Yin YL (2006) Measuring nitrogen-containing polymer synthesis rates by using stable isotope tracers. J Anim Sci 84(E. Suppl):E79–E93

Fiamegos YC, Stalikas CD (2006) Gas chromatographic determination of amino acids via one-step phase-transfer catalytic pentafluorobenzylation-preconcentration. J Chromatogr A 1110:66–72

Fountoulakis M, Lahm HW (1998) Hydrolysis and amino acid composition analysis of proteins. J Chromatogr A 826:109–134
Furst P, Pollack L, Graser TA, Godel H, Stehle P (1990) Appraisal of four pre-column derivatization methods for the high-performance liquid chromatographic determination of free amino acids in biological materials. J Chromatogr 499:557–569
Gheshlaghi R, Scharer JM, Young MM, Douglas PL (2008) Application of statistical design for the optimization of amino acid separation by reverse-phase HPLC. Anal Biochem 383:93–102
He QH, Kong XF, Wu GY, Ren PP, Tang HR, Hao FH, Huang RL, Li TJ, Tan BE, Li P, Tang ZR, Yin YL, Wu YN (2009) Metabolomic analysis of the response of growing pigs to dietary L-arginine supplementation. Amino Acids 37:199–208
He QH, Ren PP, Zhao F, Kong XF, Wu GY, Yin YL (2011a) Metabonomics and its role in amino acid nutrition research. Front Biosci 16:2451–2460
He QH, Ren PP, Kong XF, Xu W, Tang HR, Yin YL, Wang Y (2011b) Intrauterine growth restriction alters the metabonome of the serum and jejunum in piglets. Mol Biosyst 7:2147–2155
He Q, Tang H, Ren P, Kong X, Wu G, Yin Y, Wang Y (2011c) Dietary supplementation with L-arginine partially counteracts serum metabonome induced by weaning stress in piglets. J Proteome Res 10(11):5214–5221
He QH, Ren PP, Kong XF, Wu YN, Wu GY, Li P, Hao FH, Tang HR, Blachier F, Yin YL (2012) Comparison of serum metabolite compositions between obese and lean growing pigs using an NMR-based metabonomic approach. J Nutr Biochem 23:133–139
Issaq HJ (2001) The role of separation science in proteomics research. Electrophoresis 22:3629–3638
Jacques LB, Christelle C, Colette CL, Giboudeau J, Cynober L (1997) Amino acid determination in biological fluids by automated ion-exchange chromatography:performance of Hitachi L-8500A. Clin Chem 43:1421–1428
Jambor A, Molnar-Perl I (2009) Quantitation of amino acids in plasma by high performance liquid chromatography: simultaneous deproteinization and derivatization with 9-fluorenylmethyloxycarbonyl chloride. J Chromatogr A 1216:6218–6223
Kaspar H, Dettmer K, Gronwald W, Oefner PJ (2009a) Advances in amino acid analysis. Anal Bioanal Chem 393:445–452
Kaspar H, Dettmer K, Chan Q, Daniels S, Nimkar S, Daviglus ML, Stamler J, Elliott P, Oefner PJ (2009b) Urinary amino acid analysis: a comparison of iTRAQ-LC-MS/MS, GC-MS, and amino acid analyzer. J Chromatogr B 877:1838–1846
Kataoka H, Matsumura S, Yamamoto S (2000) Amino acid analysis protocols. Methods Mol Biol 159:101–122
Kim SW, Mateo RD, Yin YL, Wu GY (2007) Functional amino acids and fatty acids for enhancing production performance of sows and piglets. Asian Australas J Anim Sci 20:295–306
Komarova NV, Kamentsev JS, Solomonova AP, Anufrieva RM (2004) Determination of amino acids in fodders and raw materials using capillary zone electrophoresis. J Chromatogr B 800:135–143
Li XL, Rezaei R, Li P, Wu GY (2011) Composition of amino acids in feed ingredients for animal diets. Amino Acids 40:1159–1168
Liu H (2000) Measurement of blood plasma amino acids in ultrafiltrates by high-performance liquid chromatography with automatic precolumn o-phthaldialdehyde derivatization. Methods Mol Biol 159:123–140
Marquez FJ, Quesada AR, Sanchez-Jimenez F, Nunez de Castro I (1986) Determination of 27 dansyl amino acid derivatives in biological fluids by reversed-phase high-performance liquid chromatography. J Chromatogr 380:275–283
Marquez CD, Weintraub ST, Smith PC (1994) Femtomole detection of amino acids and dipeptides by gas chromatography-negative-ion chemical ionization mass spectrometry following alkalation with pentafluorobenzyl bromide. J Chromatogr B 658:213–221

Miyaji K, Nagao K, Bannai M, Asakawa H, Kohyama K, Ohtsu D, Terasawa F, Ito S, Iwao H, Ohtani N, Ohta M (2010) Characteristic metabolism of free amino acids in cetacean plasma: cluster analysis and comparison with mice. Plos ONE 5:E13808

Molnár-Perl I (2001) Derivatization and chromatographic behavior of the o-phthaldialdehyde amino acid derivatives obtained with various SH-group-containing additives. J Chromatogr A 913:283–302

Molnár-Perl I (2003) Quantitation of amino acids and amines in the same matrix by high-performance liquid chromatography either simultaneously or separately. J Chromatogr A 987:291–309

Molnar-Perl I (2011) Advancement in the derivatizations of the amino groups with the o-phthaldehyde-thiol and with the 9-flluorenylmethyloxycarbonyl chloride reagents. J Chromatogr B 879:1241–1269

Moore S, Stein WH (1948) Photometric ninhydrin method for use in the chromatography of amino acids. J Biol Chem 176:367–388

Osowska S, Moinard C, Loi C, Neveux N, Cynober L (2004) Citrulline increases arginine pools and restores nitrogen balance after massive intestinal resection. Gut 53:1781–1786

Ozols J (1990) Amino acid analysis. Methods Enzymol 182:587–601

Poinsot V, Lacroix M, Maury D, Chataigne G, Feurer B, Couderc F (2006) Recent advances in amino acid analysis by capillary electrophoresis. Electrophoresis 27:176–94

Poinsot V, Carpene MA, Bouajila J, Gavard P, Feurer B, Couderc F (2012) Recent advances in amino acid analysis by capillary electrophoresis. Electrophoresis 33:14–35

Ramautar R, Somsen GW, De Jong GJ (2009) GE-MS in metabolomics. Electrophoresis 30:276–291

Ravindran G, Bryden WL (2005) Tryptophan determination in proteins and feedstuffs by ion exchange chromatography. Food Chem 89:309–314

Revelsky IA, Yashin YS, Sobolevsky TG, Revelsky AI (2003) Electron ionization and atmospheric pressure photochemical ionization in gas chromatography-mass spectrometry analysis of amino acids. Eur J Mass Spectrom 9:497–507

Rutherfurd SM, Gilani GS (2009) Amino acid analysis. Curr Protoc Protein Sci. doi:10.1002/0471140864.ps1109s58

Smith JT (1997) Developments in amino acid analysis using capillary electrophoresis. Electrophoresis 18:2377–2392

Smith JT (1999) Recent advancements in amino acid analysis using capillary electrophoresis. Electrophoresis 20:3078–3083

Soga T (2000) Amino acid analysis by capillary electrophoresis electrospray ionization mass spectrometry. Anal Chem 72:1236–1241

Soga T, Kakazu Y, Robert M, Tomita M, Nishioka T (2004) Qualitative and quantitative analysis of amino acids by capillary electrophoresis-electrospray ionization-tandem mass spectrometry. Electrophoresis 25:64–1972

Waldhier MC, Gruber MA, Dettmer K, Oefner PJ (2009) Capillary electrophoresis and column chromatography in biomedical chiral amino acid analysis. Anal Bioanal Chem 394:695–706

Weiss M, Manneberg M, Juranville JF, Lahm HW, Fountoullakis M (1998) Effect of the hydrolysis method on the determination of the amino acid composition of proteins. J Chromogr A 795:263–275

Williams BJ, Cameron CJ, Workman R, Broeckling CD, Sumner LW, Smith JT (2007) Amino acid profiling in plant cell cultures: an inter-laboratory comparison of GE-MS and GC-MS. Electrophoresis 28:1371–1379

Wu GY (1993) Determination of proline by reverse-phase high-performance liquid chromatography with automated pre-column o-phthaldialdehyde derivatization. J Chromatogr A 641:168–175

Wu GY (1997) Synthesis of citrulline and arginine from proline in enterocytes of postnatal pigs. Am J Physiol Gastrointest Liver Physiol 272:1382–1390

Wu GY (2009) Amino acids: metabolism, functions, and nutrition. Amino Acids 37:1–7

Wu GY, Knabe DA (1994) Free and protein-bound amino acids in sow's colostrum and mild. J Nutr 124:415–424

Wu GY, Ott TL, Knabe DA, Bazer FW (1999) Amino acid composition of the fetal pig. J Nutr 129:1031–1038

Yin YL, Tan BE (2010) Manipulation of dietary nitrogen, amino acids and phosphorus to reduce environmental impact of swine production and enhance animal health. J Food Agric Environ 8:447–462

Yin YL, Huang RL, Zhong HY, Chen CM, Dai H (1991) Influence of different cannulation techniques on the pre-cecal digestibility of protein, amino acids and cell wall constituents from diets, containing different protein meal, in pigs. Anim Feed Sci Technol 35:271–281

Yin YL, Zhong HY, Huang RL, Chen CM, Li TJ, Pai YF (1993a) Nutritive value of feedstuffs and diets for pigs. I. Chemical composition, apparent ileal and fecal digestibility. Anim Feed Sci Technol 44:1–27

Yin YL, Huang RL, Zhong HY (1993b) Effect of autoclaving on ileal digestibility of crude protein, amino acid, urease activity of beans. Zeitschrift für Tierphysiologie, Tierernährung und Futtermittelkunde 71:65–74

Yin YL, Chen CM, Zhong HY, Huang RL (1994) Digestibility of energy, cell wall constituents, crude protein and amino acids of the Chinese oil seed meals for pigs. Anim Feed Sci Technol 45:283–298

Yin FG, Zhang ZZ, Huang J, Yin YL (2010) Digestion rate of dietary starch affects systemic circulation of amino acids in weaned pigs. Br J Nutr 103:1404–1412

Yin F, Yin Y, Li T, Ren W, Blachier F, Huang R (2011) Developmental changes of serum amino acids in suckling piglets. J Food Agric Environ 9:322–327

Yuksel KU, Andersen TT, Apostol I et al (1995) Techniques in protein chemistry VI. Academic, San Diego, CA, p 185

Zeng LiMing, Tan BiE, Xiao Hao, Yin YuLong, LU XiangYang, Fang Jun (2012) Amino acid sensing signaling induced by amino acid transporters. Scientia Sinica Vitae 42: 699–708

Zhang Y-G, Yin Y-L, Fang J, Wang Q (2012) Pig production in subtropical agriculture. J Sci Food Agric 92:1016–1024

Zhong H, Marcus SL, Li L (2005) Microwave-assisted acid hydrolysis of proteins combined with liquid chromatography MALDI MS/MS for protein identification. J Am Soc Mass Spectrom 16:471–481

Ziegler F, Le Boucher J, Coudray-Lucas C, Cynober L (1992) Plasma amino-acid determinations by reversed-phase HPLC: improvement of the orthophthalaldehyde method and comparison with ion exchange chromatography. J Automat Chem 4:145–149

Chapter 16
Surgical Techniques Used in Amino Acid Nutrition Research

Yonggang Zhang, Yulong Yin, Bo Deng, Ruilin Huang, Tiejun Li, Xugang Shu, Guixiong Qiu, and Martin Nyachoti

Contents

Y. Zhang • Y. Yin (✉) • R. Huang • T. Li
Chinese Academy of Sciences, Institute of Subtropical Agriculture, Research Center for Healthy Breeding of Livestock and Poultry, Hunan Engineering and Research Center of Animal and Poultry Science, Key Laboratory for Agri-ecological Processes in Subtropical Region, Scientific Observing and Experimental Station of Animal Nutrition and Feed Science in South-Central, Ministry of Agriculture, Furong Road #644, Changsha 410125, Hunan, People's Republic of China
e-mail: yinyulong@isa.ac.cn

B. Deng
Institute of Animal Husbandry and Veterinary Science, Zhejiang Academy of Agricultural Sciences, Hang Zhou, Zhe Jiang Province 310021, Peoples Republic of China

X. Shu • G. Qiu
Guangzhou Tanke Bio-tech Co., LTD, Room 2801, East Tower of Huihao Building, No. 519 Machang Road, Pearl River New City, Guangzhou, People's Republic of China

M. Nyachoti
Department of Animal Science, University of Manitoba, Winnipeg, MB, Canada R3T 2N2

F. Blachier et al. (eds.), *Nutritional and Physiological Functions of Amino Acids in Pigs*, DOI 10.1007/978-3-7091-1328-8_16, © Springer-Verlag Wien 2013

16.1 Introduction

The measurement of the ileal digestibility of amino acids is a convenient method for measuring bioavailability of amino acids in feedstuffs for pigs (Yin 2008). Because of the presence and metabolic activity of microbes, as well as the absence of amino acid absorption by the large intestine (except in the neonatal period), amino acid digestibility determined at the distal ileum is more accurate than fecal digestibility (Zebrowska 1973; Yin and McCracken 1996). That is why many scientists developed numerous surgical techniques to collect ileal digesta, including the slaughter and cannulation techniques. These latter cannulation techniques include:

1. Blood catheter
2. T-cannulation

   Simple T-cannulation (STC)
   Post-cecum valve T-cannulation (PVTC)
   Steered ileo-cecal valve cannulation (SICV)

3. Re-entrant cannulation

   Fore-valve re-entrant cannulation (FVRC)

   Ileo-ileal re-entrant cannulation (IIRC)
   Ileo-cecal re-entrant cannulation (ICRC)
   Ileo-colic re-entrant cannulation (IORC)
   Ileo-rectal re-entrant cannulation (IRRC)

   Post-ileo-colic valve cannulation (PICVC)

4. Ileo-rectal anastomosis

   Simple end-to-side without ileo-cecal valve (ES-IRA)
   End-to-end without ileo-cecal valve (EE-IRA)
   End-to-side including ileo-cecal valve (ESV-IRA)
   End-to-end including ileo-cecal valve (EEV-IRA)

16.2 Blood Catheter Methods

The method of using catheters for blood collection was described firstly by Rerat et al. (1980) and was further developed and improved by Manolas et al. (1983) and Yen and Killefer (1987). Further improvements were then introduced by Huang et al. (2003) and Yin et al. (2010).

The hepatic portal vein carries blood coming from the gastrointestinal tract, pancreas, and spleen; thus it is the major pathway for transporting the nutrients absorbed from the gastrointestinal tract (Huang et al. 2003). Amino acids absorption into the portal vein per unit of time can be measured using the portal blood flow rate (PVBF) and the differences in nutrient concentrations between the portal vein and systemic arterial blood (Rerat et al. 1980; Yen and Killefer 1987).

The whole surgical procedure including preoperative preparation of the animal, implantation of catheters into portal and ileal mesenterium veins and transonic probe, implantation of catheters into the carotid artery, final exteriorization and anchoring of these catheters, postoperative management, and measurement of portal vein blood flow rate, can be done according to the procedure described by Huang et al. (2003).

Furthermore, the portal vein net amino acid or related metabolite net absorption can be calculated according to the following two equations:

$$\mathrm{PVBF} = C_\mathrm{i} \times \mathrm{IR} \times (\mathrm{PAH_{pv}}(100 - \mathrm{PCVpv}) - \mathrm{PAH_a}(100 - \mathrm{PCV_a}))^{-1\times 100} \qquad (16.1)$$

$$q = (C_\mathrm{p} - C_\mathrm{a}) \times \mathrm{PVBF} \times \mathrm{d}t \qquad (16.2)$$

where PVBF is portal vein blood flow (ml/min); C_i is concentration of PAH infusion solution (mg/ml); IR is portal plasma PAH concentration (ml/min); $\mathrm{PAH_{pv}}$ is portal plasma PAH concentration (mg/ml); $\mathrm{PAH_a}$ is carotid arterial plasma PAH concentration (mg/ml); PCVpv is portal venous blood packed volume, $\mathrm{PCV_a}$ is arterial blood packed volume; q is net absorption; C_p is concentration of the nutrient in portal plasma; and C_a is concentration of the nutrient in carotid arterial plasma (Huang et al. 2003).

16.2.1 Advantages of the Blood Catheter Method

The blood catheter method allows us to measure the net absorption of amino acids and their metabolites from the gastrointestinal tract.

16.2.2 Disadvantages of the Blood Catheter Method

The whole procedure is much complicated. It needs sophisticated methods and great technical expertise to do the surgery. Also, the expense of the whole surgery is much higher than other methods.

16.3 Slaughter Method

The slaughter method was first developed by Cori (1925). This procedure includes euthanasia of the pig through an approved method followed by removal of the ileum and subsequent collection of digesta (Yin et al. 2001; Tan et al. 2009).

The length of the small intestine removed may affect protein digestibility estimates. So the longer the section of intestine removed, the greater the chances that complete absorption has not occurred. After eating, feed takes about 6 h to reach the distal ileum in pigs (Donkoh et al. 1994). So the time point for digesta collecting is very important.

16.3.1 Advantages of the Slaughter Method

1. Although the slaughter method does have the advantage of causing minimal interference with the animal's digestive tract prior to the time of sampling, it also allows us to take digesta from several parts of the digestive tract (Donkoh et al. 1994).
2. Any kind of diet can be used to determine amino acid digestibilities.
3. Experimental period is shorter and the procedure is relatively simple.

16.3.2 Disadvantage of the Slaughter Method

1. It is difficult to obtain representative digesta samples, because the optimal sampling time is easily influenced by diet type and other factors (Buraczewski et al. 1971; Rerat 1972). In addition, only one sample of ileal digesta can be obtained per animal. So this method needs more animals than other methods to obtain reliable results.
2. Furthermore, there may be shedding of mucosal cells into the gut lumen at death, which interferes with digesta nitrogen (N) content (Badawy et al. 1957, 1958; Fell 1961). Especially electrical current may cause cell sloughing in the gastro-intestinal tract leading to contamination of the amino acid content of the digesta (Sauer and de Lange 1993; Albin et al. 1999).

16.4 Cannulation Technique

Due to the disadvantages of slaughter method, cannulation techniques were developed. Cannulation techniques consist of inserting a device (cannula) into the ileum to allow access to digesta in the gastrointestinal tract from the outside of the body directly and it has become a popular method in the determination of nutrient digestibility in pigs (Yin et al. 1993a, 2002).

16.4.1 Simple T-Cannulation

The Simple T-cannulation (STC) method was invented by Cunninghem et al. (1963) and later improved by Dierick et al. (1983). This procedure mainly includes insertion of a T-cannula into the terminal ileum. The T-cannula could be a full- or a half-round

T-cannula. The full round T-cannula forces digesta to travel through the lumen of the cannula at all times, which can lead to increased blockage problems, especially with high fiber diets. In order to reduce the incidences of blockage of full-round T-cannula, the half-round T-cannula was developed (Sauer and de Lange 1993; Albin et al. 1999). The flange of the cannula is inserted into the ileal lumen and the barrel is exteriorized through the body wall, usually on the right side in a region bordered by the last rib, kidney, and the point of the hip (Sauer and de Lange 1993).

16.4.1.1 Advantage of the STC Technique

The STC method is widely used and has many beneficial characteristics, including the following:

1. It is a relatively quick and simple surgical procedure.
2. It little influences on the process of digestion when sampling.
3. It maintains a normal physiological state, which is advantageous for research involving the gastrointestinal tract (Sauer and de Lange 1992).
4. Digesta blockage and leakage risks remain low when compared with the FVRC technique, especially when half-round T-cannula is used.
5. It does not transect the small intestine wall, a practice that leaves the myoelectric complex intact (Sauer and de Lange 1992).

16.4.1.2 Disadvantages of the STC Technique

1. It remains difficult to quantitatively sample ileal digesta because of the unequal distribution of the liquid and solid phase of digesta and the possibility of blocking digesta when a full-round T-cannula is used (Graham and Aman 1986; Zebrowska 1978; Sauer and Ozimek 1986; Schroeder et al. 1989; Leterme et al. 1990). In addition, because of the position of the cannula, the digesta collected has not passed through the entire length of the small intestine.
2. In order to measure digestibility, an indigestible marker has to be used.
3. When high levels of fibrous dietary ingredients are used, blockage and leakage occur frequently so that the accuracy of results may not be optimal.

16.4.1.3 Cannula Design

Some points must be considered in order to design the cannula so as to improve its practical use.

1. In order to make room for more threads making the ring more stable, a thicker inner edge incorporated into the ring has to be designed.
2. Taking into account the animal care and protection, the ring must have a smooth outer edge in order to protect the skin.

3. Two holes can be placed in the ring to facilitate the use of a spanner wrench for ring adjustments, thus eliminating the need for notches on the outer edge of the ring.
4. Generally, a cannula barrel plug is made from nylon which can fit the cannula barrel very well. So the plug prevents digesta from entering and blocking the cannula barrel between collection periods.
5. The length and width of the cannula flange must be designed according to the pig age and body size to make sure it is suitable for placement in the distal ileum (Wubben et al. 2001; Walker et al. 1986).

16.4.2 Postvalvular T-Cecum Cannulation

The Postvalvular T-cecum Cannulation (PVTC) was developed by van Leeuwen et al. on the basis of the simple T-cannula method (van Leeuwen et al. 1988a, b). Later on, de Lange et al. (1989), Jørgensen et al. (1992), and Yin et al. (2000a) used these techniques to measure the digestibility of different nutrients and experienced some problems such as.... According to the limitations and problems, Jennifer et al. (2001) added some modifications to the PVTC technique with success to determine nutrient digestibility in pigs.

This technique involves placement of a full-round T-cannula at the ileo-cecal junction (Köhler et al. 1992a, b; Sauer and de Lange 1993; Albin et al. 1999). After removal of the cecum, the T-cannula is placed between the ileum and the large intestine, allowing digesta to flow from the ileum to the large intestine or out of the pig into a collection bag.

16.4.2.1 Advantage of the PVTC Technique

This technique has similar advantages to those given above for the STC technique. However, additional advantages include the following:

1. It can allow almost quantitative samples collection (Yin et al. 2000b).
2. In the PVTC technique, a larger diameter T-cannula can be used (Sauer and de Lange 1993; Albin et al. 1999).
3. This method reduces variation among animals and the number of animals required to complete an experiment (Wubben et al. 2001)

16.4.2.2 Disadvantage of the PVTC Technique

1. It includes a more complex surgical procedure and removal of the cecum (Radcliffe 1999) which represents an unphysiological situation.
2. As well as in the situation of ileo-rectal anastomosis (IRA), it remains unclear if the remaining intestine has similar function when a segment of the gastro-intestinal tract is removed (Radcliffe 1999).

16.4.2.3 Surgical Procedures for the PVTC Technique

Surgery Preparation

Surgery is performed on healthy pigs that have been fasted for 12 h prior to surgery. Pigs are premedicated by intramuscular injection of 4 mg Azaperone (Stresnil®) and 0.05 mg Atropine sulfate per kg bodyweight (Van Leeuwen et al. 1991) or 1.5 ml of a telazol–ketamine–xylazine solution (Wubben et al. 2001). The surgical table is prepared with a heating pad covered by a clean towel to keep the animal warm and dry during surgery (Wubben et al. 2001; Walker et al. 1986). The pigs are placed in the left lateral recumbency and the right side of the animals is washed with a detergent solution to remove any external manure or dirt. The surgical area (a 30 × 30-cm area surrounding the last two ribs) is shaved by using an electric razor (Wubben et al. 2001, Walker et al. 1986).

After anesthesia, the pigs are left with inhalation of O_2/NO in a ratio of 3:2, Halothane (Fluothane) (Van Leeuwen et al. 1991; Wubben et al. 2001). Wubben et al. (2001) has suggested to maintain the initial 5 % (v/v) halothane concentration for 20 min. The halothane concentration can then be reduced to 1.5–2 % to maintain general anesthesia for the remainder of the surgery. Wubben et al. (2001) suggested that to provide a snug fit, the mask should include a flexible rubber portion next to the pig's snout. Of course, it is difficult to be completely sure that there is no anesthesia gases escaping into the surgery room, so it is necessary that safety procedures are put in place to protect surgeons from the anesthesia gases. Such procedures may include an exhaust fan with an attached exhaust tube placed near the pig's head to vent the gas outside (Wubben et al. 2001). Pig's breathing must be monitored during the entire surgery.

Obviously, all the surgical instruments, surgical drapes, and gowns must be sterilized by autoclaving to avoid infection during and after surgery. Cannulas are sterilized by immersion into a 2 % (v/v) solution of glutaraldehyde for at least 60 min.

Before cannulas are inserted into the ileum, they must be rinsed with sterile saline. The skin around the surgical area on the animal must be washed with 70 % ethanol and iodine. Sterile drapes are placed over the surgical area and the front and back legs (Wubben et al. 2001).

Surgical Procedures

Lidocaine with noradrenaline are administered in the area of the incision subcutaneously and intramuscularly for local anesthesia (Van Leeuwen et al. 1991). The detail surgical procedure can be done according to the procedures described by Wubben et al. (2001).

16.4.3 Steered Ileo-Cecal Valve Cannulation

A more recent technique for total collection of ileal digesta is the ileo-colic post-valve procedure which was developed by Darcy et al. (1980). This method maintains the integrity of the ileum and preserves the functional role of the ileo-cecal sphincter. Later Mroz et al. (1996) implemented a new cannulation technique called steered ileo-cecal valve cannulation (SICV). This method allows for a quantitative collection of ileal digesta.

The SICV cannula comprises of five parts: an inner cannula barrel, an outer cannula barrel, an internal ring attached to a nylon cord, an external ring, and a cylindrical stopper (Radcliffe et al. 2005). The cecum is not cut off when using the SICV cannulation method. The SICV cannula can be moved by using two rings; the inner ring is for the ileum and its distance to the ileo-cecal valve must be about 10 cm, depending on the age of the pigs. To let inner ring stay at the end of ileum, the external ring is made much smaller than inner ring. The external ring is at the end of ileum and attaches to the wall of cecum tightly. The T-cannula is fixed outside of the cut which is about 2–3 cm in cecum. At the time of sampling, the ileo-cecal valve will get into the T-cannula by pulling the nylon cord, meanwhile the external ring will block the space between the colon and the cecum. Mroz et al. (1991) compared the effect of the SICV and PVTC methods on growth performance and digestibility and found that the growth performance of SICV group is better, but the apparent digestibility of DM, organic matter and chloride is lower. The reason is likely that PVTC cannot get total digesta or that flow rate of digesta is different. Mroz et al. (1991) propose to use the SICV method only for 6–8 weeks after surgery because fibrous tissue between two rings provokes hyperplasia 13–14 weeks after surgery. Later Zhang et al. (2004) compared the method of SICV-cannula and T-cannula and found that ileal digesta samples from SICV-cannula are more homogenous than those from the T-cannula.

16.4.3.1 Advantages of the SICV Method

1. Sample collection is easy.
2. This method can maintain the integrity of the ileum and preserves the functional role of the ileo-cecal sphincter.
3. This method allows for a quantitative collection of ileal digesta.

16.4.3.2 Disadvantages of the SICV Method

1. SICV needs complicated surgery procedure and more labor and time cost than other methods.
2. SICV could only be used for 6–7 weeks after surgery, because of the fibrous tissue that develops between two rings due to hyperplasia 13–14 weeks after surgery (Mroz et al. 1991).

16.4.4 The Re-entrant Cannulation Technique

The re-entrant cannulation technique is also one of those that allow for total ileal digesta sample collection. This method consists of fitting two T-cannulas at two different sites of intestine and connecting the two cannulas using a bridge cannula. When getting samples, the bridge cannula is removed. Then, total digesta can be recovered from the proximal cannula. The digesta after analysis is warmed and put back into intestine through the distal cannula which allows for a situation that is close to the normal physiology and metabolism in pigs (Yin et al. 1991).

Depending on to the position of the cannulas, the re-entrant cannulation technique can be divided into two fore-valve re-entrant cannulation (FVRC) and post-ileo-colic valve cannulation (PICV).

16.4.4.1 Fore-Valve Re-entrant Cannulation Technique

The FVRC is when the proximal cannulation is made before the ileo-cecal valve. The FVRC can be divided into IIRC, ICRC, IORC, and IRRC. The detailed operation procedure can be found in previous studies (Cunninghem et al. 1962; Easter and Tanksley 1973; Laplace et al. 1985). van Leeuwen et al. (1988a, b) who designed the ileo-cecal re-entrant cannula for piglets have described in detail the technique used.

Advantages of the Re-entrant Cannulation Technique

1. Re-entrant cannulation is easy to operate.
2. The accuracy of data obtained by using re-entrant cannulation is very good.
3. When using the re-entrant cannulation technique, there is no need for a digestibility marker, thus avoiding the challenges associated with maker analysis and recovery.

Disadvantages of the FVRC Technique

1. Because the small intestine is completely cut off, the ileo-cecal valve does not work, which affects the normal excitation–contraction coupling which control the normal digesta flow in the small intestine.
2. The bridge cannula can easily drop off, with resultant problems of blockage and leakage, especially with diets containing high levels of non-starch polysaccharides.

16.4.4.2 Post-ileo-colic Valve Cannulation

Darcy et al. (1980) developed the ileo-colic post-valve cannulation technique. PICVC is when the proximal cannula is placed after the ileo-cecal valve. The technique involves the transsection of the intestine and the collection of digesta after the ileo-cecal valve. Although two cannulas are used in this technique, the re-entrant flow of digesta into the colon is not spontaneous. Also the surgery involved in this technique is rather difficult (Darcy et al. 1980). PICVC keeps the integrity of whole ileum, including the ileo-colic valve. However, the cecum is almost cut off totally and the proximal cannula is put into the residential cecum which is behind ileo-colic valve.

Advantages of PICVC

1. PICVC keeps the integrity of the whole ileum.
2. The result accuracy is higher than FVRC.

Disadvantages of PICVC

1. Cecum is cut off totally, so the digestibility of amino acids in proteins is impacted.
2. Blockage can happen.
3. The digesta can easily be contaminated by microbes.
4. Surgery procedure is much complicated, hard to operate, and is time consuming.

16.4.5 Ileo-Rectal Anastomosis

As a novel means of avoiding problems with cannulation, the IRA technique has been reported by Fuller and Livingstone (1982), Picard et al. (1984), Darcy-Vrillon and Laplace (1985), Souffrant et al. (1985), Hennig et al. (1986), and Yin et al. (1993a).

The procedure for IRA involves transecting the small intestine just proximal to the ileo-cecal valve and reattaching it to the distal colon (Sauer and de Lange 1993). Picard et al. (1984) also reported another modified version of this technique in which a T-cannula is placed in the colon to evacuate fermentation gases. Pigs with IRA are relatively easy to maintain and can be fed with diets of any texture (Sauer and de Lange 1992).

The function of the remaining intact intestine is altered over time due to the removal of the large intestine and cecum. The reason of such alteration may be related to the fact that the functional role of the small intestine is changed to compensate for the missing colon. This may alter the amino acid composition of the digesta and observed digestibilities compared with that obtained when using other methods (Fuller and Cadenhead 1991). Köhler et al. (1992a, b) found that N retention of ileo-rectal anastomosed pigs decreased compared to normal pigs and pigs fitted with post-valve T-cecal cannulas, meanwhile Sauer and de Lange (1993) found that nutrients absorption was changed.

According to the position of anastomosis and to the maintenance of the ileo-cecal valve, IRA could be divided into ES-IRA, EE-IRA, ESV-IRA, and EEV-IRA (Laplace et al. 1994). The characteristics of IRA is summarized in Table 16.1.

16.4.5.1 Advantages of the IRA Technique

1. The most important advantage of IRA is that digesta can be collected quantitatively via the anus.

Table 16.1 The kind and features of IRA

Kind of IRA	ES-IRA	ESV-IRA	EE-IRA	EEV-IRA
Style of anastomosis	End-to-side	End-to-side	End-to-end	End-to-end
Position of anastomosis	Ileum end–rectum side	Ileum end–rectum side	Ileum end–rectum end	Ileum end–rectum end
Descending colon is opened or not to rectum	Opened	Opened	No	No
With or without ileo-caecal valves post-surgery	Without	With	Without	With
Degree of backflow of digesta	Significant	medium	Little significant	Not significant
Side effects[a]	Significant	medium	Little significant	Not significant
Data resource	Fuller and Livingstone (1982)	Souffrant et al. (1985)	Picard et al. (1984) and Green et al. (1987)	Green et al. (1988)

[a]Side effects are related to the level of the changes in the microflora activity in digesta and to the digestibility of amino acids with the extension of time post-surgery. Redlich et al. (1997) have compared the morphometry of the small intestine in pigs with four different IRA (i.e., EEV, EE, ESV, and ES) with intact pigs. They concluded that there was no major disturbance after IRA, and the different procedure used are not deleterious to the structural integrity of the small intestine

2. Using this technique, problems such as blockages and leakages which have been described for re-entrant cannulation (Sauer 1976; Just et al. 1980) do not occur.
3. The surgery is simpler compared with other techniques, and pigs recover rapidly.
4. All kinds of diets can be used for tests

16.4.5.2 Disadvantages of the IRA Technique

The normal physiology of swine is altered after surgery, because large intestine is dissociated. As a result, the energy metabolism as well as nitrogen metabolism and water and electrolyte movements may change.

Green et al. (1987) used IRA with the technique of complete isolation of the large intestine to study the digestibility of amino acids in maize, wheat, or barley meals. Endogenous secretion was different when fed different diets. The choice of factor for the correction of apparent digestibility to true digestibility was based upon CF intake of cereal diet.

Redlich et al. (1997) measured villus length, crypt depth, and mucosal thickness and found that despite several minor changes in the morphological characteristics of the small intestine after IRA, there was no major disturbance undermining the validity of the IRA method. Among the four different surgical procedures tested, the data obtained suggested that the end to end procedure (EE) is the most beneficial for the maintenance of the structural integrity of the small intestine.

16.5 Comparison of Different Techniques

16.5.1 Intact Pig and STC in Pigs

Some studies (Laplace and Borgida 1976; Sauer et al. 1979) found some adverse effects of cannulation on nutrient digestibility. Livingstone and Mcwilliam (1985) showed an effect of STC on pig growth which may have been due to decreased nutrient digestibility or utilization. Moughan and Smith (1987) compared the apparent ileal amino acid digestibility between normal pigs and pigs with T-cannulation. They found that there was good agreement between the ileal digestibilities of the amino acids as determined with intact and cannulated pigs with no significant effect of animal type on digestibility.

16.5.2 Slaughter Method and the STC Technique

Donkoh et al. (1994) studied the effect of slaughter method and STC on terminal ileum amino acid digestibility. They found that there was no significant effect of digesta collection procedure on the apparent ileal nitrogen or amino acid digestibility coefficients.

16.5.3 Comparison Among IRA Methods

Green (1988) measured and compared amino acids digestibility in pigs with EE-IRA and in pigs with EEV-IRA. There was no difference in amino acid digestibility when using these two anastomosis. Later Hennig et al. (1992) compared amino acid absorption of swine with ESV-IRA and EEV-IRA, and found about 5 % difference in amino acids absorption and finally proposed to use the EEV-IRA. Laplace et al. (1994), when comparing four IRA (i.e., EEV, ESV, EE, and ES), found that the ES models do not provide correct digestibility values due to backflow and bacterial metabolism. The EE models seem to be the most appropriate to study ileal digestibility of proteins, even in the long term. The EE models can also be used to measure the ileal digestibility of fiber, but only in short term experiments and, in this case, the preservation of the valve appears of interest.

16.5.4 IRA and STC Methods

Leterme et al. (1990) compared the effect of IRA and T-cannulas on the digestibility of amino acids. These authors found that among amino acids, the differences were only significant for the apparent as well as the true digestibilities of methionine, proline, and tryptophan. Apparently, the observed differences originated from the method used for the collection of the samples from the cannulated pigs. In order to understand why there are differences in amino acid digestibility between IRA and T-cannulas, Leterme et al. (1991) compared the emptying rates with both methods and found as a result that the emptying rate of IRA is significantly faster than T-cannulas. However it is worth to note that Fuller et al. (1994) found that there was no significant difference in ileal amino acid digestibility between IRA and T-cannulas.

16.5.5 PVTC and IRA Methods

Köhler et al. (1991a, b, c, 1992a, b) have compared the effect of PVTC and IRA on the digestibility of nutrients in pigs. They found that there were significant differences in the digestibility of dry matter, nitrogen, and amino acids between PVTC-pigs and IRA-pigs. Meanwhile, they indicated that PVTC is much better for determining the digestibility of nutrients than IRA.

16.5.6 IRA and ICPV Methods

Darcy-Vrillon and Laplace (1985, 1990) have studied the effects of IRA and ICPV on the digestibility of nutrients. They found that the values which were determined by using ICPV were higher than the values which were determined by using ICPV. In addition, the differences were found to change with different diets. Later on, Fuller et al. (1994) reported similar results in their study.

16.5.7 IRA and FVRC Methods

Picard et al. (1984) conducted 53 experiments to compare the difference of IRA and IIRC in the digestibility of amino acids. They found differences only in the digestibility of cystine. Later Hennig et al. (1991) have compared the difference of ES-IRA and ICRC by using ten diets. In all the 20 amino acids of the 10 diets, the differences were higher than 5 %. Especially, the digestibility of methionine which was determined by using ES-IRA was 15 % lower than the value which was determined by using the ICRC technique. Hennig et al. (1993) when studying the effect of ES-IRA and ICRC on the digestibility of amino acids got similar results, but nonetheless proposed to use EE-IRA to replace ES-IRA, because the former may decrease the effect of intestine microbiota on amino acid digestibility.

Yin et al. (1993a, b) also compared ES-IRA and ICRC methods for the measurement of amino acid digestibility and found that there was no significant difference between ES-IRA and ICRC techniques.

16.5.8 STC and FVRC Methods

Zebrowska (1978) compared the difference between STC and FVRC on amino acids digestibility and found some differences, although the reasons explaining these differences remain unclear. Michael et al. (1983) compared the STC and FVRC techniques using different diets and found that the values of amino acid digestibility determined with the STC technique were higher than the values determined with the FVRC technique.

den Hartog et al. (1988) reported that there was no significant difference in the apparent digestibility of dry matter and nitrogen when using PVTC, STC, or FVRC. Later, Köhler et al. (1990) compared the effects of STC, PVTC, and ICRC techniques on nutrients digestibility. Some differences among different techniques were noted, and for some items there were indeed significant differences. The authors suggested that the differences observed originate from the recovery of indicator. In fact, in practice, the recovery of the indicator (i.e., digestibility marker) was lower than 100 %, but when the recovery of indicator was corrected up to 100 %, the difference was found to decrease.

Yin et al. (1991) found that the ICRC technique could not replace STC when abundant cell wall is present in experimental diets. Fuller et al (1994) reported that the values of protein digestibility which were determined by using PVTC are higher than the value which were determined by using T-cannula.

16.6 Conclusion

Above all, it is much likely that the measurement of amino acid digestibility gives different results according to the different techniques used. As shown above, each technique has its advantages and disadvantages. Extensive comparisons between different methods do not allow to define precisely what is the best method to be used when measuring the digestibility of amino acids using different methods.

Only optimal physiological status of animals after surgery allows the measurement of "physiological" digestibility. In other words, it is mandatory to maintain the function and structure of the gastrointestinal tract as normal as possible. In the process of quantitative and representative sample collection, T-cannulation and slaughter technique only get a part of digest. Furthermore, these two methods require the utilization a digestibility marker. In contrast, IRA and re-entrant cannulation allow to collect the whole digesta. In terms of operability, IRA is more simple compared with other techniques, and pigs can recover quickly. In terms of the type of diets tested, only IRA and slaughter technique can be used with any kind of diets. So IRA appears one good choice in determining amino acids digestibility, but the effect of IRA on the animal physiology should be further studied.

References

Albin DM, Wubben JE, Gabert VM (1999) Approaches to collecting ileal digesta from swine. University of Illinois Swine, Research Reports, p 21

Badawy AM, Campbell RM, Cuthbertson DP, Fell BF (1957) Changes in the intestinal mucosa of the sheep following death by humane killer. Nature 180:756–757

Badawy AM, Campbell RM, Cuthbertson DP, Fell BF, Mackie WS (1958) Further studies on changing composition of the digesta along the alimentary tract of the sheep. Total and non-protein nitrogen. Br J Nutri 12:367–383

Buraczewski S, Porter JW, Rolls BA, Zebrowska T (1971) The course of digestion of different food proteins in the rat. 2. The effect of feeding carbohydrates with proteins. Br J Nutr 25:299–306

Cori F (1925) The fate of sugar in the animal body. J Biol Chem 66:691–715

Cunninghem HM, Friend W, Nicholson JWG (1962) Note on a re-entrant fiscual for digestion studies with pigs. Can J Anim Sci 42:112–113

Cunninghem HM, Friend DW, Nicholson JWG (1963) Observation of digestion in the pig using a re-entrant intestinal fistula. Can J Anim Sci 43:215–225

Darcy B, Laplace JP, Villiers PA (1980) Collection of digesta flowing into the large intestine after post-ileocolic valve fistulation: preliminary results. Reprod Nutr Dev 20(4B):1197–1202

Darcy-Vrillon B, Laplace JP (1985) Ileal amino acid digestibility measurement in pigs fed high-fibre diet: ileo-rectal anastomosis versus ileo-colic post-valve fistulation. In: Just A, Jsrgensen H, Fernandez LA (eds) Digestive physiology in the pig. Report no. 580. National Institute of Animal Science, Copenhagen, Denmark, pp 184–187

Darcy-Vrilton B, Laplace JP (1990) Digests collection procedure may affect ileal digestibility in pigs led diets based on wheat bran or beet pulp. Anim Feed Sci Technol 27:307–316

De Lange CFM, Sauer WC, Mosenthin R, Souffrant WB (1989) The effect of feeding different protein-free diets on the recovery and amino acid composition of endogenous protein collected from the distal ileum and feces in pigs. J Anim Sci 67:746–754

den Hartog LA, van Leeuwen P, Huisman J, Zandstra T, van Heugten E, van Ummeren HJ, van Kleef DJ (1988) Comparison of ileal digestibility data obtained from pigs provided with a different type of cannula. In: Buraczewska L, Byraczewski S, Pastuszewska B, Zebrowka T (eds) Digestive physiology in the pig. Institute of Animal Physiology and Nutrition, Jablonna, Poland, pp 275–282

Dierick N, Vervaeke L, Decuypere J, Hemderickx HK (1983) Invloed van de aard en het niveau van ruwvezel op de schijnbare ileale en faecale droge stof, eiwit en aminozuurverteerbaarheid en op de N-retentie bij varkens. Landbouwtijdschrift 36:1661–1682

Donkoh A, Moughan PJ, Smith WC (1994) Comparison of the slaughter method and simple T-piece cannulation of the terminal ileam for determining ileal amino acid digestibility in meat and bone meal for the growing pig. Anim Feed Sci Technol 49:43–56

Easter RA, Tanksley TD (1973) A technique for re-entrant ileo-caecal cannulation of swine. J Anim Sci 36:1099–1103

Fell BF (1961) Cell shedding in the epithelium of the intestinal mucosa: fact and artifact. J Pathol Bacteriol 81:251–254

Fuller MF (1988) Methods of protein evaluation for non-ruminants. In: Orskov ER (ed) World animal science B4. Feed Science. Elsevier, Amsterdam, pp 81–101

Fuller MF, Cadenhead A (1991) Effect of the amount and composition of the diet on galactosamine flow from the small intestine. In: Verstegen MWA, Huisman J, den Hartog LA (eds) Proceedings of the 5th international symposium on digestive physiology in pigs. 24–26 April 1991. Pudoc, Wageningen, The Netherlands, pp 330–333

Fuller MF, Livingstone RM (1982) Annual report of studies Page 39 in Animal nutrition and allied sciences. Rowett Research Institute, Scotland, 45

Fuller MF, Darcy-Vrillon B, Laplace JP, Picard M, Cadenhead A, Jung J, Brown D, Franklin MF (1994) The measurement of dietary amino acid digestibility in pigs, rats and chickens: a comparison of methodologies. J Anim Feed Sci Technol 48(3–4):305–324

Graham H, Aman P (1986) Circadian variation in composition duodenal and ileal digesta from pigs fitted with T-cannula. Anim Prod 43:133–140

Green S (1988) A note on amino acids digestibility measured in pigs with pro- or post valve ileo-rectal anastomosis, fed soya-bean pea and meat meals. Anim Prod 47:317–320

Green S, Bertrand SL, Duron MJC, Maillard RA (1987) Digestibility of amino acids in maise wheat and barley meal, measured in pig with ileo-rectal anastomosis and isolation of the large intestine. J Sci Food Agric 41:29–43

Hennig U, Noel R, Herrmann U, Wiinsche J, Mehnert E (1986) Ernahrungsphysiologische Untersuchungen an Schweinen mit Ileo-Rektal- Anastomoscn. 1. Mitteilung (Nutritional physiological studies in pigs with ileo-rectal anastomoses. I. Communication). Archiv Fiir Tiererniihrung 36: 585–596

Hennig U, Wünsche J, Souffrant WB, Kreienbring F (1991) Precaecal nutrient digestibility and amino acid absorption in pigs with ileo-rectal anastomosis and i1eo-caecal re-entrant cannulas. In: Verstegen MWA, Huisman I, den Hartog LA (eds.) Digestive physiology in the pig. EAAP Publication No. 54. PUDOC Wageningen, The Netherlands, pp 304–309

Hennig U, Wünsche J, Souffrant WB, Borgmann E, Kreienbring F (1992) Comparison of methods for determining the prececal nutrient digestibility and amino acid absorption of swine with ileorectal anastomoses and ileocecal bridge cannulas. Arch Tierernahr 42(3–4):197–211

Hennig U, Wuensche J, Souffrant WB, Kreienbring F (1993) Comparison of the precaecal nutrient digestibility and amino acid absorption between pigs with ileo-rectal anastomosis and ileo-caecal re-entrant cannulae. Arch Aaim Nutr 42:197–211

Huang RL, Yin YL, Li TJ, Ti XW, Gao JP, Tao LH (2003) Techniques for implanting chronic hepatic portal vein transonic flow meter and hepatic portal vein, ileal mesenterium vein and carotid artery catheters in pigs. Acta Zoonutrimenta Sinica 15(s1):10–20

Huang RL, Yin YL, Wu GY, Zhang YG, Li TJ, Li LL, Li MX, Tang ZR, Zhang J, Wang B, He JH, Nie XZ (2005) Effect of dietary oligochitosan supplementation on ileal digestibility of nutrients and performance in broilers. Poult Sci 84:1383–1388

Jennifer E, Wubben MS, Micheller R, Smiricky MS, David M, Albin MS, Vince M, Gabert (2001) Improved procedure and cannula design for simple-T cannulation at the distal ileum in growing rigs. J Anim Sci 40(6):27–31

Jørgensen H, Jakobsen K, Eggum BO (1992) The influence of different protein, fat and mineral levels on the digestibility of fat and fatty acids measured at the terminal ileum and in feces of growing pigs. Acta Agric Scand Sect A Anim Sci 42:177–184

Just A, Sauer WC, Bech-Andersen S, Jorgensen HH, Eggum BO (1980) The influence of the hind gut microflora on the digestibility of protein and amino acids in growing pigs elucidated by addition of antibiotics to different fractions of barley. Z Tierphysiol Tierernahr U Futtermittelkd 43:83–91

Köhler T (1992) Evaluation of techniques to detect ileal digesta in pigs. PhD thesis of Animal Nutrition Department, Wageningen Agriculture University

Köhler T, Huisman J, den Hartog LA, Mosenthin R (1990) Comparison of different digesta collection methods to determine the apparent digestibility of the nutrients at the terminal ileum in pigs. J Sci Food Agric 53:465–475

Köhler T, Verstegen MWA, Huisman J, van Leeuwen P, Mosenthin R (1991a) Comparison of various techniques for measuring ileal digestibility in pigs. In: Verstegen MWA, Huisman J, den Hartog LA (eds) Digestive physiology in the pig. EAAP Publication No. 54, PUDOC, Wageningen, The Netherlands, pp 296–303

Köhler T, den Hartog A, Huisman J, Mosenthin R, Verstegen MWA (1991b) Digestibility measurements in pig by using post-valve T-caecum cannulation or end-to-side ileo-rectal anastomosis. J Anim Physiol Anim Nutr 6B:278–286

Köhler T, Denhartog LA, Huiman J, Mosenthin R, Verstegen MWA (1991c) Adaptative effect of ileo-rectal anastomosis on digestibility. In: Verstegen MWA, Huisman J, den Hartog LA (eds) Digestive physiology in the pig. EAAP Publication 563 No. 54. PUDOC, Wageningen, The Netherlands, pp 346–365

Köhler T, Mosenthin R, Verstegen MWA (1992a) Effect of ileo-rectal anastomosis and post-valve T-caecum cannulation on growing pigs. 1. Growth performance, N-balance and intestinal adaptation. Brit. J Nutr 68(1):293–303

Köhler T, Mosenthin R, Verstegen MWA (1992b) Effect of ileo-rectal anastomosis and post-valve T-caecum cannulation on growing pigs. 2. Blood variables and minerals balances. Brit J Nutr 68(1):305–315

Laplace JP, Borgida LP (1976) Physiological problems involved with re-entrant cannulas in the pig ileum: review and experimental studies. Ann Zootech 25:361–371

Laplace JP, Darcy-Vrlllon B, Picard M (1985) Evaluation de la disponibilite des acides amines: choix raisonne dune methode. J Rech Porc Fr 17:353–370

Laplace JP, Souffrant WB, Hennig U, Chabeauti E, Fevrier C (1994) Measurement of precaccal dietary protein and plant cell wall digestion in pigs; comparison of four surgical procedures for ileorectal anastomosis. Livest Prod Sci 40:313–328

Leterme P, Thewis A, Beckersa Y, Baudartb E (1990) Apparent and true ileal digestibility of amino acids and nitrogen balance measured in pigs with ileo-rectal anastomosis or T-cannulas, given a diet containing peas. J Sci Food Agric 52:485–497

Leterme P, Pirara L, Thewis A, Francois E (1991) Comparison of the rate of passage of digesta in pigs modified by ileo-rectal anastomosis or fitted with an ileal T-cannula. Digestive physiology in pigs. Proceedings of the 5th international symposium on digestive physiology in pigs, Wageningen (Doorwerth), Netherlands, vol 52, pp 361–364

Livingstone RM, Mcwilliam R (1985) The effect of terminal ileum cannulation on the performance of growing pigs. Br Vet J 141:186–191

Manolas KJ, Farmer HM, Cussen M, Welbourn RB (1983) An experimental model for simultaneous chronic sampling of portal and systemic blood and gastrointestinal lymph via cannulae in conscious swine. Cornell Vet 73:333

Moughan PJ, Smith WC (1987) A note on the effect of cannulation of the terminal ileum of the growing pig on the apparent ileal digestibility of amino acids in ground barley. Anim Prod 44:319–321

Mroz Z, Bakker GCM, Jongbloed AW, Dekker RA, Jongbloed R, van Beers A (1996) Apparent digestibility of nutrients in diets with different energy density, as estimated by direct and marker methods for pigs with or without ileo-cecal cannulas. J Anim Sci 74:403–412

Mroz Z, Jongbloed AW, Kemme PA, Everts H, Vuuren AM, van Hoste R (1991) Preliminary evaluation of a new cannulation technique (steered ileo-caecal valve) for quantitative collection of digesta from the small intestine of pigs. In: Verstegen MWA, Huisman J, den Hartog LA (eds) Digestive phyalology in the pig, EAAP Publication No. 54, PUDOC. Wageningen, The Netherlands, pp 334–339

Picard M, Bertrand S, Genin F, Maillard R (1984) Digestibility of amino acids: interest of the ileo-rectal shunt technique in pig. Journk de la Recherche Portin, en France 16:355–360

Radcliffe JS (1999) The use of steered ileo-cecal valve cannulated pigs to evaluate the effects of adding phytase or b-mannanase to the diet on amino acid, mineral and energy utilization. Doctor thesis. Virginia Polytechnic Institute and State University

Radcliffe JS, Rice JP, Pleasant RS, Apgar GA (2005) Technical note: improved technique for fitting pigs with steered ileocecal valve cannulas. J Anim Sci 83:1563–1567

Redlich J, Souffrant WB, Laplace JP, Hennig U, Berg R, Mouwen JM (1997) Morphometry of the Small Intestine in Pigs with Ileo-Rectal Anastomosis. Can J Vet Res 61:21–27

Rerat A (1972) Protein nutrition and metabolism in the growing pig. Nutr Abst Rev 42:13–39

Rerat A, Vaugelade P, Villiers P (1980) A new method for measuring the absorption of nutrients in pig: critical examination. In: Low AG, Partridge IG (eds) Current concepts of digestion and

absorption in pigs. National Institute for Research in Dairying. Reading, England, Hannah Research Institute, Ayr, Scotland, Technical Bulletin 3
Rerat AA, Vaissade P, Vaugelade P (1984) Absorption kinetics of some carbohydrates in conscious pigs. 1. Qualitative aspects. Br J Nutr 51(3):505–515
Sauer WC (1976) Factors influencing amino acid digestibility for cereal grains and their components for growing monogastric animals. Ph.D. Thesis, University of Manitoba, Winnipeg, Manitoba
Sauer WC, de Lange K (1992) Novel methods for determining protein and amino acid digestibilities in feedstuffs. In: Nissen S (ed) Modern methods in protein nutrition and metabolism. Academic, San Diego, CA, pp 87–120
Sauer WC, de Lange K (1993) Novel methods for determining protein and amino acid digestibilities in feedstuffs. In: Nissen S (ed) Modern methods in protein metabolism. Academic, New York, p 87
Sauer WC, Ozimek L (1986) Digestibility of amino acids in swine: results and their practical application. A review. Livest Prod Sci 15:378–388
Sauer WC, Aherne FX, Thacker PA (1979) Comparison of amino acid digestibilities in normal and cannulated pigs. The University of Alberta 58th Annual Feeder's Day Report, The University of Alberta Faculty of Extension, Edmonton, pp 28–29
Schroeder H, Schulz E, Oslage HJ (1989) Influence of different cannulae-techniques—simple cannula vs re-entrant cannula—on the precaecally measured digestibility of N-compounds. J Anim Physiol Anim Nutr 61(4):169–178
Souffrant WB, Schumann B, Matkowitz R, Gebhardt G (1985) Untersuchungen zur Stickstoff- und Aminos urenresorption im Diinndarm von wachsenden Schwcinen (Studies on the absorption of nitrogen and amino acids in the small intestine of growing pigs). Archiv, fir Tieremu/irung 35:781–789
Tan B, Yin YL, Liu ZQ, Li XG, Xu HJ, Kong XF, Huang RL, Tang WJ, Shinzato I, Smith SB, Wu GY (2009) Dietary L-arginine supplementation increases muscle gain and reduces body fat mass in growing-finishing pigs. Amino Acids 37:169–175
Taverner MR, Curic DM, Rayner CJ (1983) A comparison of the extent and site of energy and protein digestion of wheat, lupin and meat and bone meal by pigs. J Sci Food Agric 34:122–128
van Leeuwen P, Huisman J, Verstegen MWA (1988a) A new technique for the collection of ileal chyme. In: Buraczewska L, Buraczewska S, Pastuszewska B, Zebrowska T (eds) Proceedings of the 4th international seminar on digestive physiology in the pig. Polish Academy of Sciences, Jablonna, Poland
Van Leeuwen P, Sauer WC, Huisman J, Van Kleef D, Den Hartog LA (1988b) Ileo-cecal reentrant cannulation in baby pigs. J Anim Physiol Anim Nutr 59:59–63
Van Leeuwen P, van Kleef DJ, van Kempen GJM, Huisman J, Verstegen MWA (1991) The post valve T-caecum cannulation technique in pigs applicated to determine the digestibility of amino acids in maize, groundnut and sunflower meal. J Anim Physiol Anim Nutr 65:183–193
Walker WR, Morgan GL, Maxwell CV (1986) Ileal cannulation in baby pigs with a simple T-cannula. J Anim Sci 62:407–411
Wubben JE, Smiricky MS, Albin DM, Gabert VM (2001) Improved procedure and cannula design for simple-T cannulation at the distal ileum in growing pigs. Contemp Top Lab Anim Sci 40 (6):27–31
Yen JT, Killefer J (1987) A method for chronically quantifying net absorption of nutrients and gut metabolites into hepatic portal vein in conscious swine. J Anim Sci 64(3):923–934
Yin Y (2008) Amino acids nutrition and metabolism in pigs. Scientia Sinica, Beijing, China
Yin YL, McCracken KJ (1996) Methodological aspects of the in vivo measurement of ileal amino acid digestibility in pigs—a review. Asian-Australas J Anim Sci 9:495–502
Yin YL, Huang RL, Zhong HY, Chen CM, Dai H (1991) Influence of different cannulation techniques on the pre-cecal digestibility of protein, amino acids and cell wall constituents from diets, containing different protein meal, in pigs. Anim Feed Sci Technol 35:271–281

Yin YL, Huang RL, Zhong HY (1993a) Comparison of ileo-ctel anastomosis and the conventional method for the measurement of ileal digestibility of protein sources and mixed diets in growing pigs. Anim Feed Sci Technol 42:297–308

Yin YL, Zhong HY, Huang RL, Chen CM, Li TJ, Pai YF (1993b) Nutritive value of feedstuffs and diets for pigs. I. Chemical composition, apparent ileal and fecal digestibility. Anim Feed Sci Technol 44:1–27

Yin YL, McEvoy J, Souffrant WB, Schulze H, McCracken KJ (2000a) Apparent digestibility (ileal and overall) of nutrients as evaluated with PVTC-cannulated or ileo-rectal anastomised pigs fed diets containing two indigestible markers. Livest Prod Sci 62:133–141

Yin YL, McEvoy J, Schulze H, McCracken KJ (2000b) Studies on cannulation method and alternative indigestible markers and the effects of food enzyme supplementation in barley-based diets on ileal and overall apparent digestibility in growing pigs. Anim Sci 70:63–72

Yin YL, Baidoo SK, Schulze H, Simmins PH (2001) Effect of supplementing diets containing hulless barley varieties having different levels of non-starch polysaccharides with β-glucanase and xylanase on the physiological status of gastrointestinal tract and nutrient digestibility of weaned pigs. Livest Prod Sci 71:97–107

Yin YL, Gurung Nar K, Jeaurond EA, Sharpe PH, de Lange CFM (2002) Nutrient digestibility of Canadian-developed sorghum and pearl millet grains fed to growing pigs compared to traditional cereal grains. Can J Anim Sci 82:385–391

Yin YL, Huang RL, Li TJ, Ruan Z, Xie MY, Deng ZY, Hou YQ, Wu GY (2010) Amino acid metabolism in the portal-drained viscera of young pigs: effects of dietary supplementation with chitosan and pea hull. Amino Acids 39:1581–1587

Yin FG, Zhang ZZ, Huang J, Yin YL (2010a) Digestion rate of dietary starch affects systemic circulation of amino acids in weaned pigs. Br J Nutr 103:1404–1412

Zebrowska T (1973) Digestion and absorption of nitrogenous compounds in the large intestine of pigs. Roczn Nauk Roln 95B:85–90

Zebrowska T (1978) Effect of diet and method of collection on amino acid composition of ileal digesta and digestibility of nitrogen and amino acids in pigs. Roczn Nauk Roln 99B:75–83

Zhang YC, Jrgensen H, Fernandez JA, Bach Knudsen KE (2004) Digestibility of carbohydrates in growing pigs: a comparison between the t-cannula and the steered ileo-caecal valve cannula. Arch Anim Nutr 58(3):219–231

Chapter 17
Determination of Protein Digestibility in the Growing Pig

Paul J. Moughan and Warren Miner-Williams

Contents

17.1 Introduction

The cost of feed is usually the most significant variable cost in modern intensive pork production, and protein is the most expensive component of the diet. There is thus a considerable financial incentive to maximise the efficiency of utilisation of the dietary amino acids, and especially the dietary first-limiting amino acid. The dietary first-limiting amino acid for the weaned and growing pig is often lysine. It is particularly important that the supply of dietary amino acids to the body tissues (i.e. absorbed amino acids) closely matches the amino acid requirements of the tissues for protein synthesis, to avoid unnecessary wastage.

P.J. Moughan (✉) • W. Miner-Williams
Riddet Institute, Massey University, Private Bag 11-222, Palmerston North, New Zealand
e-mail: P.J.Moughan@massey.ac.nz

F. Blachier et al. (eds.), *Nutritional and Physiological Functions of Amino Acids in Pigs*,
DOI 10.1007/978-3-7091-1328-8_17,

For the first-half of the twentieth century, animal nutritionists focussed on dietary crude protein content and its optimal supply. With the advent of the automated analysis of amino acids, following the groundbreaking work of Moore and Stein in the USA in the 1950s and 1960s, emphasis in the latter half of the twentieth century shifted to dietary amino acids. Initially, faecal amino acid digestibility was determined with animals to allow the prediction of amino acid uptake from the digestive tract, but this was superseded by analysis of amino acid uptake based on amino acids remaining undigested at the end of the small intestine (ileal digestibility). This significant technical advance was based on work from a number of groups undertaken mainly in the 1980s and 1990s.

The aim of this contribution is to review aspects of the ileal measure of amino acid digestibility, and discussion will relate especially to methods that have been developed for the collection of ileal digesta from pigs.

17.2 Ileal Amino Acid Digestibility

From a nutritional standpoint, the digestion of dietary protein is largely complete by the end of the ileum. Protein, peptides and amino acids entering the hindgut (colon and caecum) of the growing pig are subjected to extensive microbial metabolism. Much material is catabolised with end-products such as ammonia and amines. Also, nitrogen (both from the breakdown of undigested dietary and endogenous protein and from urea entering the digestive tract from the blood) is used by the hindgut microbes to synthesise bacterial protein.

Any uptake of amino acids (as such) from the hind gut of the pig is believed to be nutritionally inconsequential. Moreover, faecal protein is predominantly microbial protein, the amino acid composition of which bears no direct relationship to the amino acid composition of the undigested dietary protein that enters the colon. Thus, the determination of amino acid digestibility based on the total collection of faecal output (faecal digestibility) is considered to be flawed (Hodgkinson and Moughan 2000a; Moughan 2003; Fuller and Tomé 2005; Hendriks et al. 2012). Determination of the amounts of dietary amino acids remaining unabsorbed at the terminal ileum, leading to coefficients of ileal digestibility is the preferred approach. There is comprehensive published evidence demonstrating that ileal amino acid digestibility coefficients are more accurate than their faecal counterparts for predicting amino acid uptake into the pig's body (Just et al. 1985; Moughan and Smith 1985; Moughan et al. 1991; Rutherfurd et al. 1997a, b; Columbus and de Lange 2012).

It is now common practice in pig nutrition for protein and amino acid digestibility to be determined based on the collection of ileal digesta rather than faeces (Moughan 2003). Faecal amino acid digestibility values may be quite misleading, either overestimating (which is usually the case) or underestimating (sometimes seen especially for methionine) digestibility. Whenever digestibility is determined based on digesta contents collected from the terminal ileum, however, the digesta amino acids need to be corrected for amino acids of endogenous origin present in the digesta along with undigested dietary amino acids (Skilton et al. 1988).

If total amino acids are deducted from the dietary amino acids ingested, without any correction for the endogenous amino acids, then "apparent" estimates of digestibility are obtained. Apparent estimates of digestibility are influenced by the protein content of the test diet (Donkoh and Moughan 1994) and can be misleading especially for low dietary protein or amino acid contents. "True" estimates of amino acid digestibility (corrected for the endogenous amino acids), however, are independent of the dietary protein concentration and are a fundamental property of the food itself. True ileal digestibility is the preferred approach for determining amino acid digestibility in the growing pig (Boisen and Moughan 1996; Hodgkinson and Moughan 2000b; Moughan 2003).

Where there is an agreed approach to determining ileal endogenous amino acids, and the ileal endogenous component is corrected for, the term "standardised" digestibility is used instead of "true" (Stein et al. 2007a, b).

Currently in pig production and where the "standardised" amino acid digestibility method is adopted, the traditional protein-free diet method is used to derive estimates of ileal endogenous amino acid flow (Stein et al. 2007a). This is justified based on the work of Jansman et al (2002), where a comparison was made of the overall mean ileal endogenous amino acid losses for pigs given a protein-free diet versus comparable flows for pigs given protein-containing diets, determined over several studies. The flows for the protein-containing diets were higher, but not greatly so compared to when a protein-free diet had been given. There is a need for caution, however, as when one examines data within studies, for protein-free versus protein-containing diets, in most cases dietary protein causes a statistically significantly higher endogenous amino acid flow at the terminal ileum, and furthermore such differences are practically meaningful.

Quite in addition to an increased amount of endogenous protein it may well be that the pattern of amino acids in the ileal endogenous protein also differs when the animal consumes a protein-containing rather than a protein-free diet, thus affecting relative differences among amino acid digestibility coefficients. This aspect is worthy of further investigation. Not only is endogenous ileal amino acid flow higher when protein is present in the diet, but it has also been shown in both the growing rat and pig that the dietary concentration of protein influences the endogenous amino acid loss with greater losses at higher dietary protein or peptide concentrations (Hodgkinson et al. 2000; Hodgkinson and Moughan 2007). The various methods for determining ileal endogenous amino acid flows have been reviewed (Hodgkinson and Moughan 2000a; Moughan et al. 1998).

In the case of processed feedstuffs or feedstuffs that have been stored for long time periods, proteins may have undergone structural changes, which have implications for the determination of amino acid digestibility. This is particularly important for the amino acid lysine. Here, true ileal lysine digestibility or standardised ileal lysine digestibility are still appropriate measures, but the traditional chemical analysis of lysine shouldn't be used. Rather the protein should be reacted with the reagent *o*-methylisourea before acid hydrolysis of the protein, to allow detection of "reactive lysine", and the ileal digestibility of reactive lysine

Table 17.1 Mean true ileal digestible lysine contents[a] compared with true ileal digestible reactive lysine[b] in some processed feedstuffs

	True ileal digestible lysine (g/kg)	
	Traditional analysis	Reactive lysine
Wheat	3.2	2.9***
Maize	2.6	1.9***
Skim milk powder	19.8	16.6***
Cottonseed meal	12.9	10.3***
Evaporated milk	23.4	20.5***
Whole milk powder	26.2	24.0***
Lactose hydrolyzed milk powder	27.2	25.1***
Popped rice cereal product	0.7	0.3***
Grain-based cereal product	1.2	0.5***
Whole grain bread	2.4	2.0**

*$P < 0.05$; **$P < 0.01$; ***$P < 0.001$

[a]Based on traditional amino acid analysis

[b]Based on reactive lysine determined after reacting the protein with *o*-methylisourea (Moughan and Rutherfurd 1996)

should be determined (Rutherfurd and Moughan 1990, 1998; Moughan and Rutherfurd 1996, 2008; Rutherfurd et al. 1997a, b; Moughan 2005). Damaged proteins are commonly encountered in the feed proteins used in preparing compound pig feeds, and the differences between traditionally determined ileal lysine digestibility and the true ileal digestibility of reactive lysine (available lysine) can be considerable (Table 17.1).

The determination of standardised ileal amino acid digestibility coefficients for pig feeds is one of the major technical advances in pig science over the last few decades. There are economically important relative differences in determined amino acid contents between gross and digestible amino acids. However, to be able to determine ileal amino acid digestibility, digesta from the terminal small intestine (final 10–20 cm of ileum) need to be collected either in total or by sampling along with the use of a marker compound in the feed. The collection of digesta from the pig has spawned much research over the years, and there is a considerable body of knowledge on the subject. Digesta can be obtained either by dissecting the ileum from the body of the pig while under deep anaesthesia (sometimes called the "slaughter method") or by collecting digesta from the ileo-rectal anastomised animal or from a cannulated animal. The various collection methods used in the pig have been the subject of review (Fuller 1991; Yin and McCracken 1996)

17.3 Collection of Digesta: Slaughter Method

The most straightforward way of collecting digesta from the end of the ileum is to dissect the ileum from a deeply anaesthetised animal or immediately upon euthanasia. This method is commonly used in smaller animals such as rats and chickens where the animal is killed (Payne et al. 1968; Moughan et al. 1984), and in early work with the growing pig using this approach, digesta were removed immediately after death of the animal. It is for this reason that the method is often referred to as the "slaughter method". Because only a sample of digesta is collected with this technique, a marker compound (such as chromic oxide or titanium dioxide) needs to be included in the test diet. With the growing pig an adequate amount of digesta is usually found in the final 10–20 cm of ileum, unless a very highly digestible diet has been fed to the animal, whereupon only small quantities of digesta may be recovered. The technique has successfully been used with milk-fed piglets (Moughan et al. 1990).

There is a need, with this technique, for careful handling of the ileal tissue and attention to how the animal is killed or anaesthetised. Mucosal cells are rapidly shed into the gut lumen following death, thus increasing the protein content of the digesta. The potential problem of cell shedding can be avoided if barbiturates are used to kill the animal and digesta are collected immediately (Badawy et al. 1957; Thorpe and Tomlinson 1967), or if the ileum is removed from the anaesthetised animal.

When using the slaughter method, a frequent (usually hourly) feeding regimen may be employed, or digesta may be collected at a set time after ingestion of a single meal, to coincide with the peak digesta flow. Work has been conducted, often with the rat, studying aspects of digesta sampling (Donkoh et al. 1994a; Van Wijk et al. 1998; James et al. 2002a, b; Butts et al. 2002; Hodgkinson et al. 2002).

The advantage of this approach is its simplicity. There is minimal interference with the digestive tract prior to sampling, there are no limitations on the type of diet fed to the animal (confer cannula blockages), and it is perhaps more acceptable ethically in comparison with cannulation and anastomosis. Digesta can be sampled from several parts of the digestive tract, which is sometimes useful, but dietary markers must be used and only one terminal ileal digesta sample can be obtained per animal, so the animal cannot be used as its own control. However, the major advantage remains, that the anatomical, histological, microbial and physiological changes that can occur over time with cannulation or anastomosis are avoided with this technique.

Controlled studies with the growing pig comparing dietary protein digestibility values obtained either with the "slaughter method" or by using simple T cannulated animals have shown similar mean values and similar variances around the mean (Moughan and Smith 1987; Donkoh et al. 1994b).

17.4 Collection of Digesta: Ileo-Rectal Anastomosis and Cannulation Techniques

17.4.1 Ileo-Rectal Anastomosis

To avoid interference from colonic microflora and to allow repeated sampling of digesta over time, ileo-rectal anastomoses (IRA), have in the past been used to collect pre-caecal digesta from the terminal ileum. Numerous IRA configurations have been tried since they were first proposed in the early 1980s (Fuller and Livingstone 1982; Laplace et al. 1985; Picard et al. 1984). Digestibility estimates obtained using these procedures vary depending on the configuration used (Darcy-Vrillon and Laplace 1990). In all of the ileo-rectal configurations the normal functions associated with the large intestine are interrupted, and will therefore affect the absorption of water and minerals (Köhler et al. 1991). Since the first use by Fuller and Livingstone (1982) little research has been conducted to determine the effects of IRA on intestinal morphology. Salgado et al. (2002) found that anastomosed pigs had higher spleen and small intestine weights and lower large intestine weights than intact pigs. They also discovered that IRA influenced intestinal villus and crypt architecture although it had no significant effect on the activities of intestinal enzymes. Marinho et al. (2007) also found that IRA modified the intestinal villus and crypt architecture in the duodenum and ileum.

Although there are advantages of IRA there are also many disadvantages. The method has been deemed unethical by some researchers and is not permitted in some jurisdictions. Despite this the method still finds application (Branner et al. 2004; Bohmer et al. 2005; Fontes et al. 2007; Hennig et al. 2006, 2008; Hackl et al. 2010).

17.4.2 T-Cannulae

Classically, methods to obtain ileal digesta samples have involved the use of cannulae, where the lumen of the distal ileum has been exteriorised.

Cannulation techniques can be broadly divided into two categories, those where the experimental animal is fitted with a single cannula and those fitted with two, the so called re-entrant variants. Compared with single cannulation techniques, re-entrant cannulation has been thought to give more reliable nutrient digestibility values since this technique allows the collection of total ileal digesta samples within the experimental time period and this obviates the need for an indigestible marker (Yin and McCracken 1996). However, both approaches have been criticised on the grounds that such procedures have a modifying effect on the processes of digestion and absorption (Sauer and Ozimek 1986) and that differences in the determined digestibility may occur depending on the technique used (Darcy-Vrillon and Laplace 1990). Double re-entrant cannulae are rarely used today because of the complex and expensive surgical procedures.

Studies comparing the different techniques have been undertaken by a number of researchers who have all reported some disparity in the calculated digestibility coefficients (Donkoh et al. 1994b; Köhler et al. 1990; Yin and McCracken 1996; Yin et al. 2000a, b), although some of these differences may be attributable to other secondary factors including the composition of the diet, the amount of dietary fibre or the nature of the indigestible marker (Laplace et al. 1994; Sauer and Ozimek 1986; Yin et al. 2000a, b). Further discussion on the merits of these techniques is therefore required.

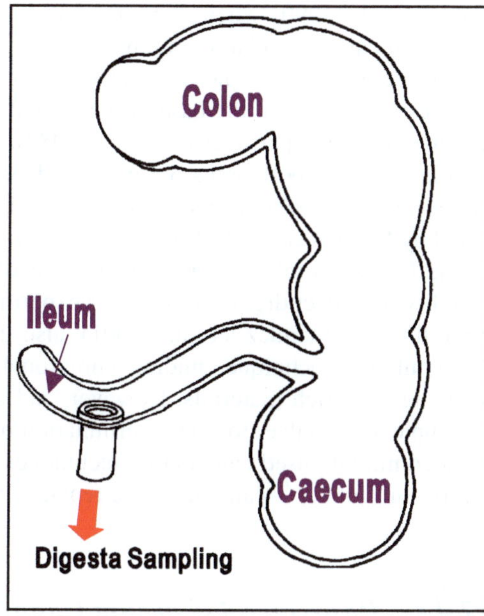

Fig. 17.1 Simple "T" cannula

17.4.3 Simple "T" Cannula

First proposed by Livingstone et al. (1977a) this technique utilised a "T"-shaped cannula made of annealed Pyrex glass that had an internal diameter of just 16–20 mm (Fig. 17.1). The cannula was secured from becoming internalised by one or more Perspex washers and a self-locking nylon strap. The shank of the cannula was threaded to take an impact-resistant plastic cap. The cannula was inserted 15 cm proximal to the ileo-caecal valve. Livingstone reported that the cannulae remained intact and functional when fitted to pigs, housed in smooth-walled pens, growing to a live weight of 175 kg. Trauma from the surgery of fitting the cannulae was reported to be minimal and the intestinal activity, appetite, and growth rate of the animals was unaffected by the cannulation. They concluded that pigs fitted with such a cannula were physiologically similar to intact pigs (Livingstone et al. 1977a, b). As with all single cannula techniques, not all of the digesta flows through the cannula and the incorporation of an indigestible dietary marker to determine the total flow of digesta is essential. It is assumed that the composition of the digesta collected via the T-cannula is representative of the total quantity of digesta flowing through the distal part of the ileum, and that no fractionation of its contents, particularly between the solid and liquid phases occurs. However, this has been questioned by a number of researchers (Fuller et al. 1994; Sauer et al. 1989; Yin and McCracken 1996). Differences in the recovery of indigestible markers have indicated that phase separation may indeed occur and that this may be one reason why a greater variability in results was observed by

Köhler et al. (1990) when they compared a simple T-cannula with a post-valve T-caecum cannula and a re-entrant cannula. They suggested that a reduction in the homogeneity of the samples recovered through the T-cannula may be attributable to changes in pressure at the base of the open cannula, forcing a separation of coarse and fine insoluble particles (Schröder 1988), and this might explain differences in the digestibility coefficients observed in diets rich in fibre. Although no incidence of the cannula blocking was reported by Livingstone et al. (1977a, b) this has been noted by Potkins et al. (1991), who had to change their experimental design owing to the frequency of blockages at the site of the cannula and the consequential inability to collect digesta when their experimental animals were fed a diet rich in bran fibre. Fernández et al. (2001) also reported problems with the simple T-cannulation technique, finding unreasonably large deviations in results when studying fibre rich materials. However, and despite its few disadvantages the use of simple T cannulae, for the determination of nutrient digestibilities, is one of the most commonly used cannulation techniques (Goerke et al. 2012). Today, flexible plastic and rubber T cannulae have replaced the early glass models.

17.4.4 Post-valve T-Caecum Cannula

Developed by van Leeuwen et al. (1988) in the late 1980s at the Institute for Animal Nutrition and Physiology in the Netherlands, the Post-valve T-caecum cannula (PVTC) is made of medical grade silicon tubing, with an internal diameter of 25 mm and is secured by an exterior ring and self-locking nylon straps (Fig. 17.2). The technique involves the removal of part of the caecum and installation of the cannula opposite the ileo-caecal valve. One hour before the collection of digesta starts, the cannula is opened to allow the ileo-caecal valve to protrude into the lumen of the cannula. Although this ensures that more of the digesta flowing from the ileum is collected, this technique cannot guarantee a complete quantitative collection of ileal digesta and so the inclusion of an indigestible dietary marker is advised (van Leeuwen et al. 1991). Having studied more than 100 animals fitted with PVTC cannulae, with live weights ranging from 8 to 100 kg, van Leeuwen's group found that post-operative recovery was swift and that the animals displayed no signs of discomfort or a reduction in appetite, even over long periods of time. Following the partial caecectomy and insertion of the cannula into what remains of the caecum, the physiology of both the ileum and the colon appear to be normal. As the gut is not transected, there appears to be no interference with the myoelectric innervation of the intestines that may reduce intestinal motility, points that have been corroborated by a number of other researchers (Köhler et al. 1990, 1992a, 1992b; Yin and McCracken 1996; Yin et al. 2000b). Although few digestibility studies have compared PVTC cannulae with other methods of collection, those that have concluded that results, using the PVTC technique, are comparable with those obtained using simple T-cannula (Köhler et al. 1990; Yin and McCracken 1996; Yin et al. 2000b).

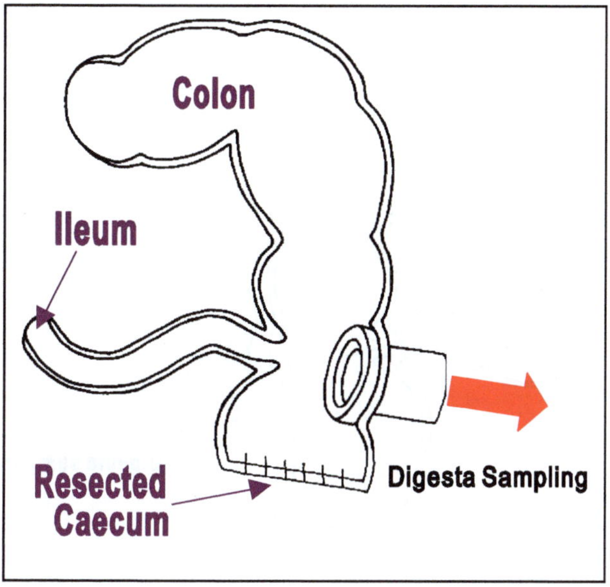

Fig. 17.2 Post-valve T-caecum cannula

In one long-term study by Kohler et al. (1992a, b) using 23 pigs (9 with IRAs, 7 with PVTCs and 6 intact pigs), no differences were found in growth performance, body nitrogen retention, blood variables and mineral balances, between intact animals and those fitted with PVTCs.

Greater precision can be achieved using the PVTC method for most diets and this is thought to be attributable to greater consistency in the recovery of the indigestible marker, specifically titanium dioxide. However, with high fibre diets some variability in the amount of digesta recovered has been recorded, which confirms the necessity of using an indigestible marker (Köhler et al. 1990, 1991; van Leeuwen et al. 1991; Yin et al. 2000a) Although the PVTC cannulation technique has been found to have a minor effect on total tract nutrient digestibility (Lindberg 1997), it is at present considered to give the most satisfactory results for ileal digestibility determination (Hodgkinson and Moughan 2000a; Presto et al. 2011).

17.4.5 Steered Ileo-Caecal Valve

Mroz et al. (1994, 1996) developed a variant of the PVTC technique which they termed the steered ileo-caecal valve (SICV), cannulation (Fig. 17.3). This technique permits the collection of digesta, with both low and high fibre diets, and maintains the normal physiological function of the gut as it does not require the removal or transection of any part of the GIT. Although this technique was thought to allow the

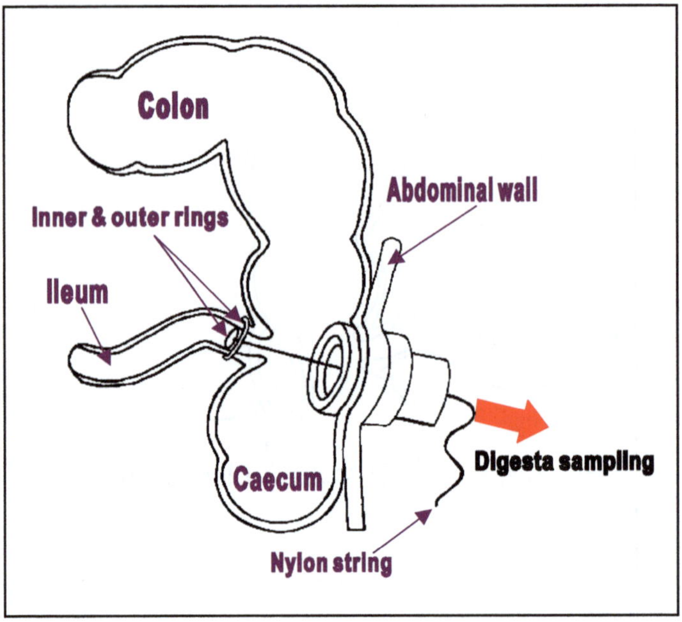

Fig. 17.3 Steered ileo-caecal valve

quantitative collection of digesta, the recovery of a chromium marker from the faeces of cannulated pigs was approximately 20 % less than in intact pigs. Therefore under the conditions of their study, the use of an inert marker was deemed advisable for determination of the total tract digestibility and ileal digestibility of nutrients. Changes proposed by Radcliffe et al. (2005) resulted in a decrease in post-surgical complications, notably fewer and less severe adhesions, together with less leakage of digesta around the cannulae. The surgery for this technique is more complex than the PVTC method, but it appears to be less prone to blockages.

In comparison to the simple T cannula, Zhang et al. (2004) concluded that ileal digesta samples collected using the SICV-cannula were more homogenous.

17.4.6 Simple Ileo-Ileal Re-entrant Cannula

Although re-entrant fistulae have been commonly used to study the digestive process in ruminants (Phillipson 1952), Cunningham et al. (1962, 1963) used a similar re-entrant cannula to compare the digestibility and rate of passage of a variety of purified and natural feed materials in the gut of the pig (Fig. 17.4). Blockages would often occur in the cannulae, particularly, if coarsely ground feeds were administered, but the cannulae were well tolerated by the pigs. Kowalik et al. (2004) studying the digestion of starch and crude fibre in segments of the digestive tract of sheep reported no problems with blockages.

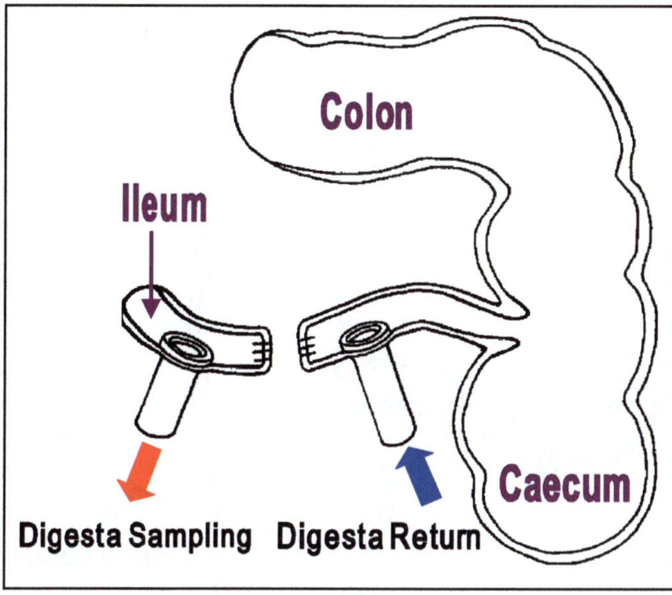

Fig. 17.4 Simple ileo-ileal re-entrant cannula

17.4.7 Ileo-Caecal Cannula

Both Cunningham et al. (1963) and later Cho et al. (1971) found that the flow of digesta through the externally joined ileo-ileal re-entrant cannulae was not uniform. Easter et al. (1972) and Easter and Tanksley (1973) believed that the surgically disrupted small intestine was not capable of forcing digesta through the externalised cannulae and the ileo-caecal valve. As the ileo-caecal valve was not essential to the digestion process (Davenport 1966), changing the configuration of the re-entrant cannula from ileo-ileal to ileo-caecal offered a solution to this resistance (Fig. 17.5). Although the use of single caecal cannulae in the digestion studies of pigs had been demonstrated by earlier researchers (Henderickx et al. 1964; Redman et al. 1964), this was the first time that a double, ileo-caecal re-entrant had been used. Although not widely used, ileo-caecal re-entrant cannulae are still employed particularly in studies involving dietary fibre (Bartelt et al. 1999; Rapp et al. 2001).

17.4.8 Ileo-Colic Post-valve Fistulation

First described by Darcy et al. (1980), Ileo-colic post-valve fistulation (ICPV), which they considered to be a reference method (Fig. 17.6), is surgically difficult and too time consuming for routine ileal digestibility measurement. Early

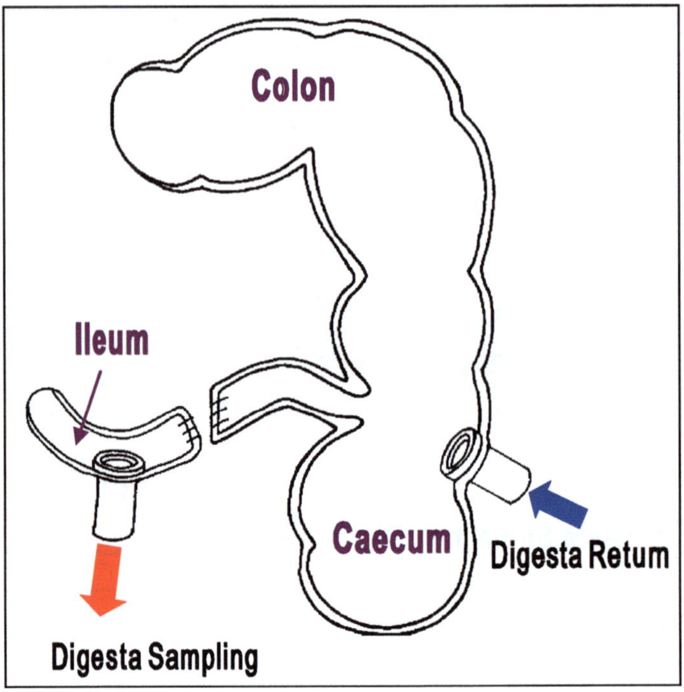

Fig. 17.5 Ileo-caecal cannula

suggestions that this method could be used in diets rich in fibre (Sauer and Ozimek 1986) were recanted by Fuller et al. (1994) who reported evidence to the contrary for diets rich in barley.

17.4.9 Comparative Summary of Ileal Digesta Collection Techniques

There is little agreement between researchers as to which digesta collection technique is the best, although lately the PVTC method has been popular. Variations in determined ileal amino acid digestibility values have been noted due to differences in:

- Collection techniques (Fuller et al. 1994).
- Basal diets (Viljoen et al. 2000; Yin and McCracken 1996).
- Dietary fibre (Darcy-Vrillon and Laplace 1985; Laplace et al. 1985).
- Laboratories (Radcliffe et al. 2005; Yin and McCracken 1996).

Indeed Fuller et al. (1994), state that each method has its strengths and limitations such that no single method is suited for all purposes, a point echoed

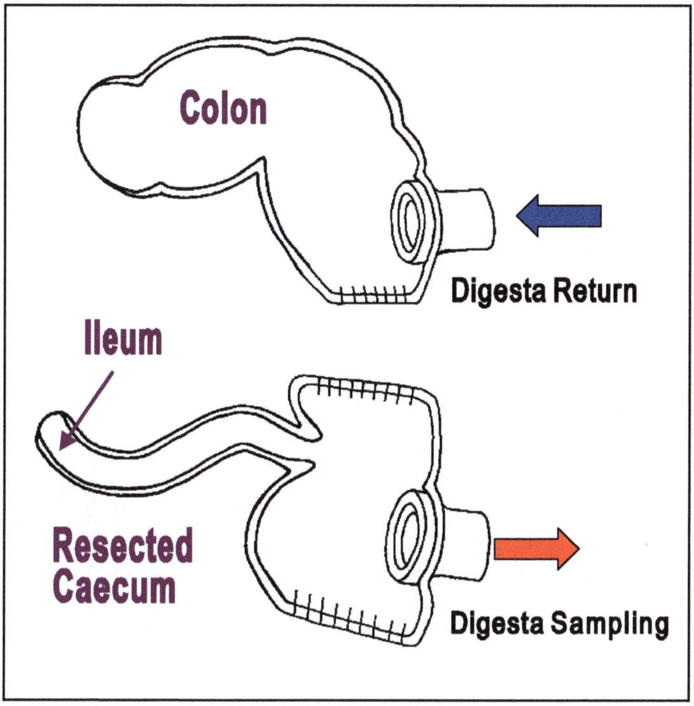

Fig. 17.6 Ileo-colic post-valve fistulation

more recently by Knudsen et al. (2006) when reviewing in vivo methods of studying the digestion of starch in pigs and poultry. Such variation is also compounded by the differing performance of indigestible markers necessary to quantify the total flow of digesta (Yin et al. 2000b). A comparative summary of the advantages and disadvantages of the various collection techniques is given in Tables 17.2 and 17.3.

Some digesta collection techniques are not suitable when the dietary fibre content is high. To avoid potentially fatal intestinal blockages some researchers have overcome this problem by regrinding the feed to a smaller mesh size (Fuller et al. 1994), but this of course may alter the digestibility of the feed.

Köhler et al. (1992a) question whether experimental animals with re-entrant cannulae or ileo-rectal anastomoses (IRA), compared to those with intact gastrointestinal tracts have similar metabolic rates. MacRae et al. (1982) found that sheep fitted with re-entrant cannulae had a higher metabolic rate than those of intact animals. With the IRA technique, Köhler et al. (1992a) argue that an increased metabolic rate would result in a lower live weight gain for the same metabolizable energy intake. Such differences in live weight gain were reported by Köhler et al. (1992a), with IRA-animals gaining only 47 % of the body mass gained by intact animals. In contrast, animals fitted with PVTC cannulae gained 93 % of the body weight gain achieved by intact animals.

Table 17.2 Advantages and disadvantages of different cannulation techniques

Ileal digesta collection method	Advantages	Disadvantages
Simple "T" cannula	Minimal disruption of normal gut activity Minimal surgical trauma Appetite and rate of growth unaffected Ileo-caecal valve remains intact	Indigestible marker required Use of indigestible marker is likely to induce errors with heterogeneous digesta Possibility of unrepresentative sampling and marker recovery Blockages with high fibre diets
Post-valve T-caecum cannula	Semi-quantitative sampling Greater precision and less labour intensive than simple T-cannulation Greater recovery of indigestible marker No interference with gut wall, ileo-caecal valve or colon	Involves the removal of gut tissue. High fibre diets cause blockages and interfere with the normal rate of digesta passing through the gut Variable recovery of digesta confirms the need for an indigestible marker
Steered ileo-caecal valve	Semi-quantitative sampling Normal flow through the gut when not sampling Minimal blockages	Invasive, costly and complex surgery Possibility of adhesions causing fatal gut obstructions
Simple ileo-ileal re-entrant cannula	Quantitative collection of digesta samples	Assumption of 100 % marker recovery incorrect Transection of the gut interrupts migrating myoelectric complex necessary for the normal passage of digesta through the gut Complex surgery and associated trauma
Ileo-caecal cannula	Quantitative collection of digesta samples Placement of the distal cannula in the caecum reduces the number of blockages	Assumption of 100 % marker recovery incorrect Inhibition of the myoelectric complex Complex surgery and associated trauma
Ileo-colic post-valve fistulation	Fewer blockages Claims to be a reference method	Surgical technique "too difficult and time consuming for routine measurements" Inhibition of the myoelectric complex Complex surgery and associated trauma

Table 17.3 Advantages and disadvantages of different ileal digesta collection techniques

Ileal digesta collection method	Advantages	Disadvantages
Cannulation	Simple T-cannulae cause minimal disturbance to the digestion of protein throughout the digestive tract	Use of an indigestible marker likely to introduce errors with heterogeneous digesta
	Surgery straightforward and less invasive for simple T-cannulae	Re-entrant cannulation reduces gut motility which can cause blockages
	Less adverse affects upon gut physiology for simple T-cannulae	Cannot be used for diets rich in fibre unless particle size is reduced
		Marker recovery less than 100 %
Ileorectal anastomosis	Quantitative digesta collection via the anus	Ethically less acceptable, prohibited in the Netherlands
	Use of an indigestible marker not required	Complex surgery with longer animal recovery time
	Minimal blockage and leakage	Colon has no digestive/absorption role, therefore need for electrolyte replacement
	No dietary restrictions	Colonic bypass may affect animal physiology which leads to some compensatory adaptation
		Increased microbial colonisation in gut
		Possibility of digesta reflux into colon and microbial fermentation
Slaughter technique	Technique straightforward	Possible sloughing of epithelial tissue, although this can be avoided with the correct technique
	Time required for study reduced	
	Ethically more acceptable	
	Particularly suitable for small lab animals and birds	Small volume of digesta recovered
	No dietary restrictions	Difficulty in determining the optimal time for slaughter of test animals
		One sample per animal

Using the diaminopimelic acid (DAPA), and volatile fatty acid (VFA) levels, as a measure of microbial activity in the gut, Köhler et al. (1992a) found the digesta concentration of VFA in IRA-animals was 3–8 times higher than in PVTC-animals and DAPA 1.3–2.3 times higher. Higher microbial digestive activity, he argued, might therefore affect the composition of digesta collected.

Although both re-entrant cannulae and ileo-rectal anastomoses are claimed to be quantitative methods of digesta collection, where the inclusion of an indigestible marker is unnecessary, both Köhler et al. (1990) and Fuller et al. (1994) reported that when a marker is included less than 100 % is recovered for both techniques. This may be a property of the marker itself, [e.g., the recovery of chromic III oxide (Cr_2O_3), appears to be less than that of titanium IV oxide (TiO_2)] or consequent to the digesta collection method used or the diet being studied. The percentage recovery of Cr_2O_3 is lower for T-cannulae than for PVTC cannulae and the recovery of Cr_2O_3 is lower for diets high in fibre than for those with less fibre (Yin et al. 2000b).

For many researchers the chosen method of digesta collection is the PVTC cannula as it allows the semi-quantitative sampling of digesta (Köhler et al. 1992a) with greater precision and is less labour intensive than simple T-cannulation (Donkoh et al. 1994b). There is proportionally a greater recovery of indigestible marker (Sauer and Ozimek 1986) and there is no interference with the gut wall, ileo-caecal valve or colon (Köhler et al. 1992a). Finally, the PVTC cannula appears to interfere less with the determination of microbial activity in the GIT and the composition of the digest collected (Köhler et al. 1992a). It seems that the advantages of using the PVTC cannula far outweigh any disadvantages.

17.5 Conclusion

The digestibility of amino acids and proteins varies greatly among different pig food ingredients and between batches of a given food ingredient. Recognising and managing this variability, during the formulation of pig rations, has important economic implications. The development and practical application of true or standardised ileal amino acid digestibility (and for lysine, availability) bioassays has been one of the major technical successes in animal science of the twentieth century. All feed manufacturers should use standardised ileal amino acid digestible amino acid contents, when formulating pig feeds. Similar gains in efficiency can also be made by using an ileal as opposed to faecal (TAAA) amino acid digestibility system with poultry.

References

Badawy AM, Campbell RM, Cuthbertson DP, Fell BF (1957) Changes in the intestinal mucosa of the sheep following death by humane killer. Nature 189:756–757

Bartelt J, Drochner W, Gotz KP, Szakacz J, Ceresnakova Z, Sommer A (1999) Determination of endogenous nitrogen associated with bacteria in ileal digesta of pigs receiving cereal-based diets with or without fish meal and various fibre supplements by using a simple N-15-dilution technique. J Anim Feed Sci 8(3):425–440

Bohmer BM, Branner GR, Roth-Maier DA (2005) Precaecal and faecal digestibility of inulin (DP 10-12) or an inulin/Enterococcus faecium mix and effects on nutrient digestibility and microbial gut flora. J Anim Physiol Anim Nutr 89(11–12):388–396

Boisen S, Moughan PJ (1996) Different expressions of dietary protein and amino acid digestibility and their application in protein evaluation: A theoretical approach. Acta Agric Scand 46: 165–172

Branner GR, Bohmer BM, Erhardt W, Henke J, Roth-Maier DA (2004) Investigation on the precaecal and faecal digestibility of lactulose and inulin and their influence on nutrient digestibility and microbial characteristics. Arch Anim Nutr 58(5):353–366

Butts CA, James KAC, Koolaard JP, Booth CL, Donaldson HE, Scott MF, Moughan PJ (2002) The effect of digesta sampling time and dietary protein source on ileal nitrogen digestibility for the growing rat. J Sci Food Agric 82:343–350

Cho CY, Horney FD, Bayley HS (1971) Examination of amino acids in ileal digesta and feces of cannulated pigs. Fed Proc 30(2):A235

Columbus D, de Lange CFM (2012) Evidence for validity of ileal digestibility coefficients in monogastrics. Br J Nutr 108:S264–272

Cunningham HM, Friend DW, Nicholson JWG (1962) Note on a re-entrant fistula for digestion studies with pigs. Can J Anim Sci 42:112–113

Cunningham HM, Friend DW, Nicholson JWG (1963) Observations on digestion in the pig using a re-entrant intestinal fistula. Can J Anim Sci 43:215–225

Darcy B, Laplace JP, Villiers PA (1980) Digestion in the Small-Intestine of the Pig.3. Kinetics of Passage of Starch and Crude Protein in the Ileum According to the Mode of Fistulation, the Type of Starch and Protein Level of the Diet. Annales De Zootechnie 29(3):277–298

Darcy-Vrillon B, Laplace JP (1985) Ileal amino acid digestibility measurement in pigs fed high-fibre diet: ileo-rectal anastomosis versus ileo-colic post-valve fistulation. In: Just A, Jsrgensen H, Fernandez LA (eds) Digestive physiology in the pig, Vol. Report no. 580. National Institute of Animal Science, Copenhagen, Denmark, pp 184–187

Darcy-Vrillon B, Laplace JP (1990) Digesta collection procedure may affect ileal digestibility in pigs fed diets based on wheat bran or beet pulp. Anim Feed Sci Technol 27:307–316

Davenport HW (1966) Physiology of the digestive tract year book, 2nd edn. Medical Publishers Inc., Chicago, IL

Donkoh A, Moughan PJ (1994) The effect of dietary crude protein content on apparent and true ileal nitrogen and amino acid digestibilities. Br J Nutr 72:59–68

Donkoh A, Moughan PJ, Smith WC (1994a) True ileal digestibility of amino acids in meat and bone meal for the growing pig—application of a routine rat digestibility assay. Anim Feed Sci Technol 49:73–86

Donkoh A, Moughan PJ, Smith WC (1994b) Comparison of the slaughter method and simple T-piece cannulation of the terminal ileum for determining the ileal digestibility of amino acids in meat and bone meal for the growing pig. Anim Feed Sci Technol 49:43–56

Easter RA, Tanksley TD (1973) Technique for reentrant ileocecal cannulation of swine. J Anim Sci 36(6):1099–1103

Easter RA, Driedger A, Tanksley TD (1972) Technique for reentrant ileocecal cannulation of swine. J Anim Sci 35(1):215–216

Fernández JA, Jørgensen H, Mroz S, Jongbloed AW (2001) Ileal digestibility determination of different diets by two fistulation techniques (T-cannula vs. steered ileocaecal valve cannula) and two markers (chromic oxide vs. titanium oxide). Paper presented at the 8th symposium of digestive physiology of pigs, Uppsala, Sweden

Fontes DO, Donzele JL, Mascarenhas AG, Apolonio LR, Silva FCO, Fontes FAPV, Correa GSS (2007) Amino acids composition and ileal digestibility of amino acids in energetics feedstuffs measured in pigs with ileo-rectal anastomosis. Arq Bras Med Vet Zootec 59(1):196–202

Fuller MF (1991) Methodologies for the measurement of digestion. In: Verstegen MWA, Huisman J, de Hartog LA (eds) Digestive physiology in pigs. Proceedings of the 5th international symposium. Wageningen Academic Publishers, Wageningen, pp 273–288

Fuller MF, Livingstone RM (1982) Effect of ileorectal anastomosis on intestinal propulsion. Annual report of studies in animal nutrition and allied sciences. Rowett Research Institute, pp 38–39

Fuller MF, Tomé D (2005) In vivo determination of amino acid bioavailability in humans and model animals. J AOAC Int 88(3):923–934

Fuller MF, Darcy-Vrillon B, Laplace JP, Picard M, Cadenhead A, Jung J, Brown D, Franklin MF (1994) The measurement of dietary amino-acid digestibility in pigs, rats and chickens—a comparison of methodologies. Anim Feed Sci Technol 48(3–4):305–324

Goerke M, Eklund M, Sauer N, Rademacher M, Piepho HP, Borner C, Mosenthin R (2012) Effect of feed intake level on ileal digestibilities of crude protein and amino acids in diets for piglets. J Sci Food Agric 92(6):1261–1266
Hackl W, Pieper B, Pieper R, Korn U, Zeyner A (2010) Effects of ensiling cereal grains (barley, wheat, triticale and rye) on total and pre-caecal digestibility of proximate nutrients and amino acids in pigs. J Anim Physiol Anim Nutr 94(6):729–735
Henderickx HK, Teague HS, Redman DR, Grifo AP (1964) Absorption of vitamin B12 from colon of pig. J Anim Sci 23(4):1036–1038
Hendriks WH, van Baal J, Bosch G (2012) Ileal and faecal protein digestibility measurement in humans and other non-ruminants—a comparative species view. Br J Nutr 108:S247–257
Hennig U, Kuhla S, Souffrant WB, Tuchscherer A, Metges CC (2006) Effect of partial dehulling of two- and six-row barley varieties on precaecal digestibility of amino acids in pigs. Arch Anim Nutr 60(3):205–217
Hennig U, Hackl W, Priepke A, Tuchscherer A, Souffrant WB, Metges CC (2008) Comparison of ileal apparent, standardized and true digestibilities of amino acids in pigs fed wheat and lupine seeds. Livest Sci 118(1–2):61–71
Hodgkinson SM, Moughan PJ (2000a) The collection of ileal digesta and characterisation of the endogenous component. In: Moughan PJ, Verstegen MWA, Visser-Reyneveld MI (eds) Feed evaluation—principles and practice. Wageningen Pers, Wageningen, pp 105–118
Hodgkinson SM, Moughan PJ (2000b) Amino acids—digestibility, availability and metabolism. In: Moughan PJ, Verstegen MWA, Visser-Reyneveld MI (eds) Feed evaluation—principles and practice. Wageningen Pers, Wageningen, pp 125–132
Hodgkinson SM, Moughan PJ (2007) An effect of dietary protein content on endogenous ileal lysine flow in the growing rat. J Sci Food Agric 87:233–238
Hodgkinson SM, Moughan PJ, Reynolds GW, James KAC (2000) The effect of dietary peptide concentration on endogenous ileal amino acid loss in the growing pig. Br J Nutr 83:421–430
Hodgkinson SM, Moughan PJ, Morel PCH, Reynolds GW (2002) The diurnal pattern of ileal dry matter and endogenous ileal nitrogen flows for the growing pig. J Sci Food Agric 82: 1860–1866
James KAC, Butts CA, Koolaard JP, Donaldson HE, Scott MF, Moughan PJ (2002a) The effect of feeding regimen on apparent and true ileal nitrogen digestibility for rats fed diets containing different sources of protein. J Sci Food Agric 82:1050–1060
James KAC, Butts CA, Koolard JP, Donaldson HE, Scott MF, Moughan PJ (2002b) The effect of food dry matter intake on the flow of amino acids at the terminal ileum for rats fed an enzyme hydrolysed casein-based diet. J Sci Food Agric 82:1128–1135
Jansman AJM, Smink W, van Leeuwen P, Rademacher R (2002) Evaluation through literature data of the amount and amino acid composition of basal endogenous crude protein at the terminal ileum of pigs. Anim Feed Sci Technol 98:49–50
Just A, Jorgensen H, Fernandez JA (1985) Correlations of protein deposited in growing female pigs to ileal and faecal digestible crude protein and amino acids. Livest Prod Sci 12:145–159
Knudsen KEB, Laerke HN, Steenfeldt S, Hedemann MS, Jorgensen H (2006) In vivo methods to study the digestion of starch in pigs and poultry. Anim Feed Sci Technol 130(1–2):114–135
Köhler T, Huisman J, Denhartog LA, Mosenthin R (1990) Comparison of different digesta collection methods to determine the apparent digestibilities of the nutrients at the terminal ileum in pigs. J Sci Food Agric 53(4):465–475
Köhler T, Denhartog LA, Huisman J, Mosenthin R, Verstegen MWA (1991) Digestibility measurements in pigs by using post-valve T-cecum cannulation or end-to-side ileorectal anastomosis. J Anim Physiol Anim Nutr 66(5):278–286
Köhler T, Mosenthin R, Verstegen MWA, Huisman J, den Hartog LA, Wensing T (1992a) Effect of ileo-rectal anastomosis and post-valve T-caecum cannulation on growing pigs. (2) Blood variables and mineral balances. Br J Nutr 68:305–315

Köhler T, Mosenthin R, Verstegen MWA, Huisman J, den Hartog LA, Ahrens F (1992b) Effect of ileo-rectal anastomosis and post-valve T-caecum cannulation on growing pigs. 1. Growth performances, N-balance, and intestinal adaptation. Br J Nutr 68:293–303
Kowalik B, Kowalczyk J, Pajak JJ, Zebrowska T (2004) Digestion of starch and crude fibre in segments of the digestive tract of sheep fed different types and amounts of starch. J Anim Feed Sci 13(Suppl 1):123–126
Laplace JP, Darcy-Vrillon B, Picard M (1985) Evaluation de la disponibilité des acides aminés: choix raisonné d'une méthode. Journées de la recherche porcine en France 17:353–370
Laplace JP, Souffrant WB, Hennig U, Chabeauti E, Fevrier C (1994) Measurement of precaecal dietary protein and plant cell digestion in pigs; comparison of four surgical procedures for ileo-rectal anastomosis. Livest Prod Sci 40:313–328
Lindberg JE (1997) A comparison of the total tract digestibility in intact and PVTC-cannulated pigs. In: Laplace JP, Fevrier C, Barbeau A (eds) Digestive physiology in pigs, pp 404–407
Livingstone RM, Fowler VR, White F, Wenham G (1977a) Annealed glass cannulae for use in digestion studies with pigs. Vet Rec 101(18):368
Livingstone RM, Fowler VR, Atkinson T, Baird B (1977b) Diurnal patterns of digestion in pigs fitted with annealed glass cannulae. Anim Prod 24(FEB):156
MacRae JC, Smith JS, White F (1982) Effects of gastrointestinal cannulation and jugular vein catheterization on the metabolism in sheep. Br J Nutr 47:637–644
Marinho MC, Pinho MA, Mascarenhas RD, Silva FC, Lordelo MM, Cunha LF et al (2007) Effect of prebiotic or probiotic supplementation and ileo rectal anastomosis on intestinal morphology of weaned piglets. Livest Sci 108(1–3):240–243
Moughan PJ (2003) Amino acid availability: aspects of chemical analysis and bioassay methodology. Nutr Res Rev 16:127–141
Moughan PJ (2005) Absorption of chemically unmodified lysine from proteins in foods that have sustained damage during processing or storage. J AOAC Int 88:949–954
Moughan PJ, Rutherfurd SM (1996) A new method for determining digestible reactive lysine in foods. J Agric Food Chem 44:2202–2209
Moughan PJ, Rutherfurd SM (2008) Available lysine in foods—a brief historical overview. J AOAC Int 91:901–906
Moughan PJ, Smith WC (1985) Determination and assessment of apparent ileal amino acid digestibility coefficients for the growing pig. N Z J Agric Res 28:365–370
Moughan PJ, Smith WC (1987) A note on the effect of cannulation of the terminal ileum of the growing pig on the apparent ileal digestibility of amino acids in ground barley. Anim Prod 44: 319–321
Moughan PJ, Smith WC, James KAC (1984) Preliminary observations on the use of the rat as a model for the pig in the determination of apparent digestibility of dietary protein. N Z J Agric Res 27(4):509–512
Moughan PJ, Pedraza M, Smith WC, Williams M, Wilson MN (1990) An evaluation with piglets of bovine milk, hydrolysed bovine milk and isolated soybean proteins included in infant milk formulas. I. Effect on organ development, digestive enzyme activities, and amino acid digestibility. Journal of Pediatric Gastroenterology and. Nutrition 10(3):385–394
Moughan PJ, Smith WC, Pearson G, James KAC (1991) Assessment of apparent ileal lysine digestibility for use in diet formulation for the growing pig. Anim Feed Sci Technol 34:95–109
Moughan PJ, Souffrant WG, Hodgkinson SM (1998) Physiological approaches to determining gut endogenous amino acid flows in the mammal. Arch Anim Nutr 51:237–252
Mroz Z, Bakker GCM, Dekker J, Jongbloed AW, Jongbloed R (1994) Application of the steered ileo-cecal valve cannulation (SICV) technique to measure digesta kinetics and ileal digestibility in pigs fed high fibrous diets. Paper presented at the VIth International Symposium on Digestive Physiology in Pigs. Dummersdorf, Germany
Mroz Z, Bakker GCM, Jongbloed AW, Dekker RA, Jongbloed R, vanBeers A (1996) Apparent digestibility of nutrients in diets with different energy density, as estimated by direct and marker methods for pigs with or without ileo-cecal cannulas. J Anim Sci 74(2):403–412

Payne WL, Combs GF, Kifer RR, Synder DG (1968) Investigation of protein quality—ileal recovery of amino acids. Fed Proc 27:1199–1203

Phillipson AT (1952) The passage of digesta from the abomasum of sheep. J Physiol 116(1):84–97

Picard M, Bertrand S, Genin F, Maillard RA (1984) Digestibilité des acides aminés: intérêt de la technique du shunt iléorectal chez le porc. Journées de la recherche porcine en France 16:355–360

Potkins ZV, Lawrence TLJ, Thomlinson JR (1991) Effects of structural and nonstructural polysaccharides in the diet of the growing pig on gastric-emptying rate and rate of passage of digesta to the terminal ileum and through the total gastrointestinal-tract. Br J Nutr 65(3): 391–413

Presto MH, Lyberg K, Lindberg JE (2011) Digestibility of amino acids in organically cultivated white-flowering faba bean and cake from cold-pressed rapeseed, linseed and hemp seed in growing pigs. Arch Anim Nutr 65(1):21–33

Radcliffe JS, Rice JP, Pleasant RS, Apgar GA (2005) Technical note: improved technique for fitting pigs with steered ileocecal valve cannulas. J Anim Sci 83(7):1563–1567

Rapp C, Lantzsch HJ, Drochner W (2001) Hydrolysis of phytic acid by intrinsic plant or supplemented microbial phytase (Aspergillus niger) in the stomach and small intestine of minipigs fitted with re-entrant cannulas. 1. Passage of dry matter and total phosphorus. J Anim Physiol Anim Nutr 85(11/12):406–413

Redman DR, King NB, Henderickx HK, Teague HS (1964) Cecal fistulation of pig using 2 forms of indwelling cannulas. J Anim Sci 23(4):1032–1035

Rutherfurd SM, Moughan PJ (1990) Guanidination of lysine in selected dietary proteins. J Agric Food Chem 38(1):209–211

Rutherfurd SM, Moughan PJ (1998) The digestible amino acid composition of several mile proteins: application of a new bioassay. J Dairy Sci 81(4):909–917

Rutherfurd SM, Moughan PJ, Morel PCH (1997a) Assessment of the true ileal digestibility of reactive lysine as a predictor of lysine uptake from the small intestine of the growing pig. J Agric Food Chem 45:4378–4383

Rutherfurd SM, Moughan PJ, van Osch L (1997b) Digestible reactive lysine in processed feedstuffs: application of a new bioassay. J Agric Food Chem 45:1189–1194

Salgado P, Martins JM, Carvalho F, Abreu M, Freire JPB, Toullec R, Lalles JP, Bento O (2002) Component digestibility of lupin (Lupinus angustifolius) and pea (Pisum sativum) seeds and effects on the small intestine and body organs in anastomosed and intact growing pigs. Anim Feed Sci Technol 98(3–4):187–201

Sauer W, Ozimek L (1986) Digestibility of amino acids in swine: results and their practical application. A review. Livest Prod Sci 15:367–369

Sauer WC, Dugan MER, de Lange CFM, Imbeah M, Mosenthin R (1989) Considerations in methodology for the determination of amino acid digestibilities in feedstuffs for pigs. In: Friedmann M (ed) Absorption and utilization of amino acids, vol III. CRC Press, Boca Raton, FL, pp 217–230

Schröder H (1988) Untersuchungen zur scheinbaren Verdaulichkeit von N-Verbindungen in differenzierten Abschnitten des Intestinaltraktes am wachsenden Schwein, Kiel

Skilton GA, Moughan PJ, Smith WC (1988) Determination of endogenous amino acid flow at the terminal ileum of the rat. J Sci Food Agric 44(3):227–235

Stein HH, Seve B, Fuller MF, Moughan PJ, de Lange CFM (2007a) Invited review: amino acid bioavailability and digestibility in pig feed ingredients: terminology and application. J Anim Sci 85:172–180

Stein HH, Fuller MF, Moughan PJ, Sève B, Mosenthin R, Jansman AJM, Fernández JA, de Lange CFM (2007b) Definition of apparent, true and standardised ileal digestibility of amino acids in pigs. Livest Sci 109:282–285

Thorpe E, Tomlinson JR (1967) Autolysis and post-mortem bacteriological changes in the alimentary tract of the pig. J Pathol Bacteriol 93:601–610 van Leeuwen P, Huisman J, Verstegen MWA, Baak MJ, van Kleef DJ, Van Weerden EJ, den Hartog LA (1988) A new technique for collection of ileal chyme in pigs. Paper presented at the proceedings of the IVth international seminar on digestive physiology in the pig. Warsaw, Poland

van Leeuwen P, van Kleef DJ, van Kempen GJM, Huisman J, Verstegen MWA (1991) The post valve T-cecum cannulation technique in pigs applicated (sic) to determine the digestibility of amino-acid in maize, groundnut and sunflower meal. J Anim Physiol Anim Nutr 65(3–4): 183–193

Van Wijk HJ, Moughan PJ, Hodgkinson SM, Jansen PP, Pearson G (1998) Variation in apparent and true ileal amino acid digestibility in barley using a rat model. Anim Feed Sci Technol 76: 9–22

Viljoen J, Coetzee SE, Fick JC, Siebrits FK, Hayes JP (2000) Evaluation of in vivo techniques for the determination of apparent ileal amino acid digestibilities in feedstuffs for piglets. S Afr J Anim Sci 30(1):7–15

Yin YL, McCracken KJ (1996) Methodological aspects of in vivo measurement of ileal AA digestibility in pigs—A review. Asian-Australas J Anim Sci 9:495–502

Yin YL, McEvoy DG, McEvoy JDG, Schulze H, McCracken KJ (2000a) Studies on cannulation method and alternative indigestible markers and the effects of food enzyme supplementation in barley-based diets on ileal and overall apparent digestibility in growing pigs. Anim Sci 70: 63–72

Yin YL, McEvoy JDG, Schulze H, Hennig U, Souffrant WB, McCracken KJ (2000b) Apparent digestibility (ileal and overall) of nutrients and endogenous nitrogen losses in growing pigs fed wheat (var. Soissons) or its by-products without or with xylanase supplementation. Livest Prod Sci 62(2):119–132

Zhang YC, Jorgensen H, Fernandez JA, Knudsen KEB (2004) Digestibility of carbohydrates in growing pigs: a comparison between the T-cannula and the steered ileo-caecal valve cannula. Arch Anim Nutr 58(3):219–231

Chapter 18
Measurement of Synthesis and Degradation of Proteins

Dingfu Xiao, Fugui Yin, Yulong Yin, Yangxiao Jiang, Jun Fang, Tiejun Li, Ruilin Huang, and Martin Nyachoti

Contents

D. Xiao • J. Fang
College of Animal Sciences and Technology, Hunan Agricultural University, Changsha, Hunan 410128, People's Republic of China

F. Yin • Y. Yin (✉) • Y. Jiang • T. Li • R. Huang
Observation and Experiment Station of Animal Nutrition and Feed Science in South-Central China, Ministry of Agriculture; Hunan Provincial Engineering Research Center for Healthy Livestock and Poultry Production; Key Laboratory of Agro-Ecological Processes in Subtropical Region, Institute of Subtropical Agriculture, the Chinese Academy of Sciences, Changsha, Hunan, 410125, People's Republic of China
e-mail: yinyulong@isa.ac.cn

M. Nyachoti
Department of Animal science, University of Manitoba, Winnipeg, MB, R3T 2N2, Canada

F. Blachier et al. (eds.), *Nutritional and Physiological Functions of Amino Acids in Pigs*,
DOI 10.1007/978-3-7091-1328-8_18, © Springer-Verlag Wien 2013

18.1 Introduction

Continuous synthesis and breakdown or remodeling of proteins (also called protein turnover) is a principal characteristic of protein metabolism. During animal production, the net differences between synthesis and breakdown represent the actual marketable muscle foods. Because protein synthesis is highly endergonic and protein breakdown is metabolically energy dependent, the efficiency of production can be markedly enhanced by lower muscle protein breakdown rates. Various methodological approaches to study protein synthesis and breakdown, with particular emphasis toward food-producing animals, are presented. These include whole-animal tracer amino acid infusion in vivo, quantifying marker amino acid release from muscle proteins, and in vitro amino acid release-based methodologies. From such methods, protein synthesis rates and protein breakdown rates (mass units/time) may be obtained (Fouillet et al. 2002; Deng et al. 2010; Tan et al. 2010; Kong et al. 2012). The paper briefly reviews the methodology developed over the past 30 years to study protein turnover.

18.2 Marker Methods

18.2.1 Whole-Body Steady-State Tracer Method Application

The flooding dose technique for the measurement of fractional protein synthesis rates (FSR) was first described by Henshaw et al. (1971) and since modified by McNurlan et al. (1979) and Garlick et al. (1980). Stable isotope-labeled tracers have been used in the research of protein (or amino acids), lipid, and carbohydrate metabolism. The principles and practice of stable isotope tracer methodology have been described in detail (Wolfe 1992). Protein synthesis has been measured by using the luminal flooding dose technique (Adegoke et al. 1999). Whole-body tracer kinetic models assume that the amount of amino acids fed in the diet is the same as the amount of amino acids that is entering the metabolic pool from outside the body. Any removal of the intragastric tracer on the first pass through the gut (i.e., PDV) and liver will be reflected by a lower (systemic) plasma enrichment of the intra-gastric tracer compared with that achieved with the intravenous tracer.

The ^{15}N-glycine labeling (Waterlow et al. 1978) in either a continuous infusion or a single oral dose was developed as the end-products method (Fern et al. 1981), followed by the measurement of ^{15}N in urine and blood urea. The use of L-[^{15}N]-Phenylalanine provided an alternative tracer to radiolabeled amino acids for measuring in vivo rates of protein synthesis in fish using the "flooding dose" technique and also allows studies of fish protein metabolism to be carried out in environmentally sensitive (field and laboratory) conditions where the use of radiolabels is restricted (Owen et al. 1999). This method is based on several assumptions, which include the existence of a single metabolic nitrogen pool and the possibility of labeling it through the administration of an amino acid tracer. However, this statement has proven to be inaccurate (Matthews et al. 1981). Although the end-product method has been the subject of methodological criticisms and is little used today, in some situations it represents a practical noninvasive way of estimating nitrogen kinetics, e.g., in frail or sick populations. The ^{15}N-glycine end-product method has gradually been replaced by the ^{13}C-leucine technique for measuring whole-body protein kinetics. The most commonly used isotopic method is the precursor method, based on the continuous infusion of L-[1-^{13}C]-leucine (Matthews et al. 1980). The rate of appearance (R_a) of leucine is measured in terms of tracer dilution when enrichment has reached a plateau. Achievement of the plateau is accelerated by the infusion of a primed dose of the tracer. A better estimate of intracellular leucine enrichment is obtained when its α-ketoacid (ketoisocaproate) is considered. The leucine flux from endogenous protein breakdown is calculated from the difference between R_a and intake, whereas the synthesis flux is indirectly assessed from the difference between the rate of disappearance (R_a in the steady state) and oxidative losses of leucine. Leucine oxidation is measured by monitoring $^{13}CO_2$ excretion. These classical scenario for ^{13}C-based flux analysis relies on labeling enrichments observed when the isotopic labeling has become equilibrated (Blank and Kuepfer 2010; Zamboni 2011). Metabolic flux analysis (MFA) aims at the quantification of intracellular metabolic fluxes in the biochemical networks and has evolved into several model-based approaches to assess in vivo fluxes within cell populations that cannot be measured directly (Nöh and Wiechert 2011). Using the isotopic nonstationary metabolic flux analysis (INST ^{13}C-MFA) method, time profiles of metabolite labeling patterns are measured and the transient label information along with the metabolites steady-state concentrations are used to determine the in vivo fluxes (Hasunuma et al. 2010; Noack et al. 2010). Dai et al. (2011) investigated the metabolism of selected amino acids in bacterial strains (*Streptococcus* sp., *Escherichia coli*, and *Klebsiella* sp.) and mixed bacterial cultures derived from the jejunum and ileum of pigs by using [U-(14)C]-labeled tracers.

L-[ring $^2H_5$]-phenylalanine was used as a tracer by continuous infusion to assess the rate of oxidation by enrichment at the plateau of the product of phenylalanine hydroxylation and its subsequent loss, i.e., L-[$^2H_4$]-tyrosine (Thompson et al. 1989). The total production of tyrosine can concomitantly be monitored by an infusion of L-[$^2H_2$]-tyrosine (Marchini 1993). Adegoke et al. (2003) developed an in situ experimental system that allows controlled exposure of intestinal mucosa to nutrients systemically, luminally, or both and examined the effects of systemic glucose and amino acid infusion in overnight-fasted piglets. Tissue protein synthesis was measured using a flooding dose of L-[4-(3)H]phenylalanine in neonatal pigs (Davis et al. 2002).

Tissue protein synthesis has been measured in vivo using a modification of the flooding dose technique (Fan et al. 2006). At 60 min after feeding, pigs were injected via the jugular vein catheter with 10 ml/kg body weight of a flooding dose of phenylalanine, which provided 1.5 mmol phenylalanine/kg body weight and 1 m Ci of L-[4-^{3}H]phenylalanine/kg body weight. Samples of whole blood were taken 5, 15, and 30 min after the injection of [^{3}H]phenylalanine for measurement of the specific radioactivity of the extracellular free pool of phenylalanine. Immediately after the 30-min, blood samples were taken and pigs were given a lethal injection of sodium pentobarbital (50 mg/kg body weight). Fractional rates of protein synthesis (K_s, percentage of protein mass synthesized in a day) for each tissue were calculated as K_s (% per day) = $[S_b/S_a]\cdot(T/t)\cdot 100$, where S_b (dpm/min) is the specific radioactivity of the protein-bound phenylalanine, S_a (dpm/min) is the specific radioactivity of the tissue-free phenylalanine at the time of tissue collection, T is a constant equal to 1,440 min per day, and t is the time of labeling (in minutes) of the specific tissue. The specific radioactivity of the tissue-free phenylalanine after a flooding dose of phenylalanine is in equilibrium with the aminoacyl-tRNA-specific radioactivity. Hence, the tissue-free phenylalanine reflects the specific radioactivity of the tissue precursor pool (Frank et al. 2007).

Leucine is unique among the amino acids in its ability to promote protein synthesis by activating translation initiation via the mammalian target of rapamycin (mTOR) pathway (Yin et al. 2010; Li et al. 2011). The fractional rate of protein synthesis was measured with a flooding dose of 1.5 mmol/kg body weight of L [4-^{3}H]phenylalanine injected 30 min prior to the end of the infusion, using the denaturing gradient gel electrophoresis analysis (Dai et al. 2010; Escobar et al. 2010; Torrazza et al. 2010; Wilson et al. 2011). A continuously perfused gut loop model and intraluminal flooding dose technique including a 30-min ^{3}H-phenylalanine flooding dose was used to study the acute effects of varying luminal availability of threonine on intestinal protein and mucin synthesis in young and well-nourished pigs. This technique was shown to be acutely sensitive to luminal threonine concentration (Nichols and Bertolo 2008). Fractional rates of protein synthesis were measured in muscle and liver using the [(3)H]phenylalanine flooding-dose technique, and the results suggested that dietary arginine supplementation increased the mammalian target of rapamycin (mTOR) signaling activity in skeletal muscle, providing a molecular mechanism for explaining the previous observation that increased circulating arginine stimulated muscle protein synthesis and promoted weight gain in neonatal pigs (Yao et al. 2008; Hou et al. 2011).

Routes of tracer delivery, including comparison between the intragastric (i.g.) and the intravenous (i.v.) route, are an important methodological consideration for application of the flooding dose technique for the measurement of FSR (Davis and Reeds 2001; Fan et al. 2006). Measurements of FSR are widely carried out by both an i.p. route (Martinez 1987; Danicke 2001) and an i.v. route (Burrin et al. 1999; Nyachoti et al. 2000; Davis et al. 2002; Hunter et al. 2004). Time courses of isotopic enrichments of free and bound tracer pools and corresponding FSR in plasma, visceral organs, cardiac muscle, and skeletal muscle after an i.p. route of injection of a flooding dose of containing L-[$^2H_5$]Phenylalanine was investigated between 15 and 75 min of post-tracer injection in studies with pigs (Bregendahl 2004). A "flooding dose" of the stable isotope l-[(2)H(5)]-phenylalanine (125 mg/kg live weight) was given and frequent blood samples (permanent catheter) were collected over a 60-min period, and the results obtained showed that the flooding dose technique appeared to be suitable for distinguishing deoxynivalenol (DON)-related effects on the protein synthesis (Goyarts et al. 2006). Dänicke et al. (2006) measured porcine tissue protein synthesis employing the so-called flooding dose technique using [(2)H(5)]-phenylalanine as tracer in pig. Bregendahl et al. (2008) concluded that FSR can be measured by the flooding dose through an i.p. or an i.v. route and that the i.p. route might underestimate FSR by the flooding dose for plasma, cardiac muscle, and skeletal muscles. This concern might be addressed by a fast regimen of sampling to be completed within 12–20 min after an i.p. route of tracer injection.

Protein synthesis was determined using the [^{2}H] phenylalanine flooding-dose technique (Yin et al. 2010). Three days after the catheter insertion and at 1 h after the last meal, all piglets were injected, through the jugular catheter, with a flooding dose of L-phenylalanine (1.50 mmol/kg body weight) containing L-[ring-^{2}H5] phenylalanine at 40 mol% (0.60 mmol/kg BW) in sterile saline. The intravenous administration of phenylalanine was completed in 5–10 s. The exact time (min) of tracer labeling was recorded from the end of the intravenous injection of phenylalanine to the time of placing tissue samples in liquid nitrogen (Deng et al. 2009). The isotopic enrichment of L-[^{2}H5]phenylalanine in the tissue-free pool was measured according to the procedures of Wang et al. (2007). Ions with mass-to-charge ratios of 91 and 96 were monitored and converted to percentage of molar enrichment (mol%) using calibration curves. The isotopic enrichment of L-[^{2}H5]phenylalanine in the free pool of tissue is in equilibrium with that in the aminoacyl-tRNA pool and is therefore an appropriate measure of fractional synthesis rate (Bregendahl et al. 2004).

Protein turnover, the result of protein synthesis and degradation, represents a dynamic process, which is of equal importance to understand physiological processes. Methods employing isotopic tracers have been developed to measure protein turnover. However, applying these methods to live animals is often complicated by the fact that an assessment of precursor pool relative to isotope abundance is required. Also, data analysis becomes difficult in case of low label incorporation, which results in a complex convolution of labeled and unlabeled peptide mass spectrometry signals. Zhang et al. (2011) presented a protein turnover analysis method that circumvents this problem using a ^{15}N-labeled diet as an isotopic tracer and developed the ProTurnyzer software that allows the determination of protein fractional synthesis rates without the need of precursor relative isotope abundance information.

Protein translation has even been dissected in single-molecule studies, albeit at lower than physiological concentrations. So what has been lacking is the ability to study the timing of translation at micromolar concentrations of components, relates partly to the conditions in cells that enable this fast and efficient process. Joseph Puglisi perfected a technology for single-molecule, real-time studies in tiny (10–21 l) reaction chambers (Kaganman 2010). The enzyme is immobilized in chambers, and the system is called zero-mode waveguides.

However, use of classical approaches to assess protein synthesis and breakdown in food animals is expensive (e.g., reagents, isotopes, and animals) and require suitable experimental animal facilities and skilled technical human resources. Thus, any of the classical isotope and marker amino acid-based approaches to follow protein breakdown are not applicable to studies involving large numbers of animals required to assess (a) genotype variations in protein metabolism efficiency for selection purposes and (b) genotype by performance enhancer interactions in growth trials.

18.2.2 Biomarker Methods

To date, none of the methods is completely satisfactory to generate comprehensive data on amino acid requirements of animals or humans. Because of many influencing factors, amino acid requirements remain a complex and controversial issue in nutrition that warrants further investigations. Benefits from the rapid advances in the emerging omics technologies and bioinformatics/biomarker discovery show great potential for obtaining in-depth understanding of regulatory networks in protein metabolism (Lin et al. 2011). The amino acid signaling pathway leading to the stimulation of protein synthesis has not been fully elucidated. Among the amino acids, leucine is considered to be a principal anabolic agent that regulates protein synthesis. Biomarker methods were used to study the role of amino acids in enhancing muscle protein synthesis in vivo. To examine the effects of enteral protein and carbohydrate on protein synthesis, biomarkers of mRNA translation were also assessed (Frank et al. 2006).

18.3 Arteriovenous Balance Methods

The arteriovenous catheterization of regions (e.g., splanchnic bed, forearm, and leg) or organs (e.g., gut and liver) allows study of their net balance in total amino acids (i.e., dietary plus endogenous amino acids) under various nutritional conditions. Such studies emphasize the importance of substrate supply to tissue metabolism, and the specialized roles by various organs in amino acid homeostasis, as well as the possible roles of specific amino acids in regional protein metabolism (e.g., Tessari et al. 2000). Organ-balance studies have shown that, in the postabsorptive state, the skeletal muscle exhibits a net negative α-amino nitrogen balance, releasing principally alanine and glutamine that are taken up by the tissues of the splanchnic bed, which has a net positive α-amino nitrogen balance. After a protein meal, the splanchnic area switches from basal net amino acid uptake to net amino acid release. The splanchnic bed extracts alanine and glutamine but releases to the periphery many other amino acids, such as BCAA, which account for the majority of muscle uptake and play a key role in postprandial muscle protein repletion. The exchange of most amino acids, such as leucine, reverts from the basal net negative balance of the postabsorptive state to a net uptake after ingestion of a mixed meal, despite a continuous net release of alanine and glutamine.

18.4 Methods Based on the Incorporation of a Tracer

Tracer incorporation studies have been widely used to measure protein synthesis in tissues and specific proteins. During such studies, isotopic tracers in the form of labeled amino acid are administered either by constant infusion or using a flooding dose method, and their incorporation into a protein pool is then quantified

(Balagopal et al. 1998). The constant infusion method requires steady-state conditions and is particularly well suited to measure the synthesis of proteins with slow turnover rates, such as those in the muscles. This method involves the infusion of a labeled tracer amino acid at a constant rate until steady-state labeling of the precursor pool for protein synthesis is achieved. However, due to the technical difficulties of measuring enrichment in the true precursor pool, i.e., the intracellular amino acyl-tRNA pool, other alternatives have been proposed, which are largely discussed elsewhere. For instance, one alternative to directly measuring the labeling of aminoacyl-tRNA is to measure the labeling of proteins with very rapid turnover in condition of isotopic enrichment quite similar to that of the synthetic precursor pool. This approach has been applied to the liver from the enrichment of apolipoprotein B-100 (Stoll et al. 1998, 1999) and to the mucosa by using the labeling of the precursor forms of sucrase and lactase (Dudley et al. 1998). A variant of this method was proposed to determine the albumin fractional synthesis rate under non-steady-state conditions by adjusting the doses of the tracer infused to produce steady state enrichment of the precursor pool (de Meer et al. 2000). *For the determination of protein synthesis rate and level, incorporation of [^{14}C]-leucine into translational products in a cell-free* system was measured on isolated pig hearts (Kassauskas et al. 2004). Stable isotope-labeled amino acid also enable the determination of the metabolic fate of individual amino acid (Stoll et al. 2006).

The flooding dose method, which involves injection of a large dose of the trace amino acid together with the tracer amino acid, may avoid the problem of sampling the true precursor pool, because it minimizes the differences in isotopic enrichment between the extracellular, intracellular, and amino acyl-tRNA pools (Davis et al. 1999). Furthermore, the flooding dose method enables rapid measurements because it rapidly increases the labeling of the intracellular amino acid pool (Davis et al. 2001), thus allowing the determination of tissue protein synthesis under acute, non-steady-state conditions of feeding (Davis et al. 1996) and hormone infusion (Bark et al. 1998). However, the large dose of the trace amino acid that needs to be injected with this method is thought to alter protein synthesis, as extensively discussed elsewhere (Rennie et al. 1994). Despite the fact that protein synthesis rates of liver-exported plasma proteins have frequently been studied in humans using either the constant infusion method or the flooding-dose technique (Caso et al. 2000), the protein synthesis of splanchnic constitutive proteins has been the subject of little investigation, particularly in the fed state (Barle et al. 1999). The studies that have been performed revealed the broad differences between protein synthesis rates in different tissues and specific proteins, which are slower in skeletal muscle and much more rapid in the gut and liver. They also emphasized the variability in the response of the different proteins pools to the ingestion of a mixed meal. Overall, quantitative aspects of protein breakdown and synthesis can be determined with isotope-tracer kinetics methodologies. Whole body protein turnover has been of interest to relate to amino acid needs, whereas in animal agriculture applications, the focus is more often on synthesis and degradation in tissues, particularly in skeletal muscles containing desirable cuts of meat (Bergen et al. 2007). Eden et al. (2011) developed a method called "bleach-chase" that was applicable to fluorescently tagged proteins to measure protein degradation and proposed that this approach opened a way to understand proteome half-life dynamics in living cells.

18.5 Combined Methods

18.5.1 Arteriovenous Catheterization with Isotopic Tracer

The organ-balance tracer technique (sometimes applied in conjunction with sampling of the tissue under study for direct measurements of amino acid labeling) has been applied to the study of protein turnover in the intestine and liver of animals (Stoll et al. 1997; Deutz et al. 1998). The organ-balance tracer technique, which involves arteriovenous catheterization with the concomitant continuous infusion of an isotopic tracer, enables a distinction between amino acid uptake and release in a net amino acid balance; as well as the measurement of the protein turnover under steady-state conditions. It is achieved by combining repeated measurements of the tracee and tracer balances across the tissues under investigation; the first being proportional to protein synthesis whereas the second measures the difference between protein synthesis and degradation. Compared with the tracer incorporation method, the organ-balance tracer approach has the advantage of providing information on the relative rates of both protein synthesis and degradation in specific organs and tissues. Another advantage of the tracer balance technique is its dynamic nature, because it allows measurement of the changes in amino acid uptake and protein turnover over relatively short periods of time (Reeds and Davis 1999). This is a highly significant advantage to the design of studies focusing on acute nutritional or hormonal effects (Biolo et al. 1997; Volpi et al. 1999). However, the invasive tracer balance approach shares the same problem when determining the isotopic enrichment of the protein synthetic precursor pool as the continuous infusion technique. Moreover, some technical problems have been encountered using the tracer balance approach. For instance, although the accurate measurement of blood flow is of crucial importance for determining the net balance of both the tracee and the tracer, there is no consensus in the literature as to whether plasma or whole-blood measurements are preferable. In addition, the siting of the venous catheter is particularly critical when measuring muscle protein turnover by catheterization of the forearm or leg (Reeds and Davis 1999).

Compartmental analysis of organ-balance tracer data has provided new estimates of protein and amino acid kinetics across the muscle under steady-state conditions. Thus, a six-compartment model for the intracellular muscle kinetics of leucine and KIC in the human forearm was developed (Tessari et al. 1995) and used to evaluate muscle protein synthesis and degradation under different nutritional conditions (Zanetti et al. 1999). These models have also enabled calculations of the rate of inward (from artery to muscle) and outward (from muscle to vein) amino acid transport and have demonstrated that infused or ingested amino acids are capable of stimulating muscle protein anabolism by enhancing intracellular amino acid transport and protein synthesis. They have also been able to take account of the reutilization of amino acids derived from proteolysis for protein synthesis, thus overcoming a classical limitation to the organ-balance tracer technique (i.e., recycling of the tracer).

18.5.2 *Modified ELISA*

Bioinformatics has been incorporated into a competitive enzyme-linked immunosorbent assay (ELISA) and was shown to be able to detect the heat-processed parent protein after digestion with either pepsin or trypsin (Reece et al. 2009). Standard indirect ELISA screenings (with immobilization of hapten–protein conjugate and use of enzyme-labeled anti-mouse IgG as tracer) was compared with direct ELISA (with antibody immobilization and use of a hapten–enzyme conjugate as tracer) for the detection of high-affinity hapten-specific monoclonal antibodies (Cervino et al. 2008). The antibodies raised against recombinant Ail protein could specifically identify pathogenic *Y. enterocolitica* strains both by indirect plate ELISA and Western blot immunoassay (Balakrishna et al. 2010). These methods have not yet been widely applied to protein breakdown and synthesis in animal nutrition studies.

18.6 Contemporary Genomic Regulation Strategies

Biochemical mechanisms and molecular regulation of protein biosynthesis and protein breakdown have been extensively documented. Proteolysis is dependent in part on the extent of expression of genes for components of cellular proteolytic machinery during skeletal muscle atrophy. It was proposed that high throughput methods, based on emerging understanding about protein breakdown, may be useful in enhancing production efficiency.

On the contrary, during skeletal muscle atrophy and sepsis, expression of genes for the components of the intracellular protein breakdown mechanism is elevated (Lecker et al. 2004; Sacheck et al. 2007). Upon induction of sepsis or muscle atrophy, a parallel increase in gene regulation of inflammation-related transcription factors and expression of several genes associated with proteolytic mechanisms has been noted, but a direct relationship between upregulation of these genes and increased proteolytic activity has not been demonstrated (Hasselgren et al. 2005). Genomic tools are becoming more available and specific swine and beef cDNA microarrays have been developed and validated (Collier and Largaespada 2007). Although transcriptional profiling is experimentally well defined, changes in mRNA abundance for any gene do not necessarily imply regulation at the level of gene expression but reflect potential capacity for synthesis of a given protein (Carey 2000). Gene expression profiles of skeletal muscle of nursing pigs was investigated using Affymetrix Porcine Gene Chip (Jiang et al. 2010).

Application of proteomics may also be useful in studying proteolytic mechanisms in breeding animals. Assessing products of protein degradation (i.e., the degradome) is another proteomic approach (Schilling 2007) that may be fruitful in quantifying degradation rates in breeding animals. Proteomic approaches have not been widely used or validated to evaluate protein turnover in skeletal muscle. Clearly, new data and new information on regulatory and quantitative aspects of protein breakdown may be expected in the future. The combination of muscle biopsies and laboratory assessment of protein synthesis and breakdown status will be a strategy to explore many unanswered questions in breeding animal protein metabolism.

18.7 Metabonomics

In the post-genomic era, biology of systems is central to the biological sciences. Functional genomics such as transcriptomics and proteomics can simultaneously determine marked gene or protein expression changes following amino acid intervention. Metabonomics and its many pseudonyms (metabolomics, metabolic profiling, etc.) has exploded onto the scientific scene in the past few years. Metabonomics, "the quantitative measurement of the dynamic multiparametric metabolic response of living systems to pathophysiological stimuli or genetic modifications," combines metabolic profiling and multivariate data analysis to facilitate the high-throughput analysis of metabolites in biological samples, providing a system approach to understand global metabolic regulations of organisms (Rezzi et al. 2007). As a sensitive and powerful method, metabonomics can quantitatively measure subtle dynamic perturbations of metabolic pathways in organisms due to changes in nutritional and epigenetic states. Therefore, metabonomics holds great promise to enhance our understanding of the complex relationship between amino acids and metabolism to define the roles for dietary amino acids in maintaining health and the development of disease. Such a technique also aids in the studies of functions, metabolic regulation, safety, and individualized requirements of amino acids (He et al. 2011). Metabonomics is a rapidly growing research area and a system approach for comprehensive and quantitative analysis of the global metabolites in a biological matrix. Analytical chemistry approach is necessary for the development of comprehensive metabonomics investigations. Fundamentally, there are two types of metabonomics approaches: mass-spectrometry based and nuclear magnetic resonance methodologies. Metabonomics measurements provide a wealth of data information and interpretation of these data relies mainly on chemometrics approaches to perform large-scale data analysis and data visualization, such as principal and independent component analysis, multidimensional scaling, a variety of clustering techniques, and discriminant function analysis, among many others (Li et al. 2008).

18.8 Conclusion

Most of the methods currently used to determine protein metabolism in different dietary scenarios are tracer based. Steady-state whole-body studies in humans and animals have provided important data, but these data have been somewhat limited in interpretation. Regional approaches to protein metabolism have enabled a clearer understanding of the different responses of splanchnic and peripheral protein metabolism to nutrient ingestion. The splanchnic zone, which undergoes very rapid turnover, seems to respond to almost all nutritional perturbations, whereas the peripheral muscle protein mass, which is slowly renewed, appears to be only slightly affected by acute dietary factors. However, these pictures become more complicated when dietary and recycled sources of amino acids or nitrogen are taken into account (Tomé et al. 2000). The techniques used at present need to be improved to address directly the still unresolved questions. Organ-balance studies allow determination of the net organ balance of amino acids, without discrimination between dietary and endogenous amino acid metabolism, whereas multiple tracer methods address the splanchnic extraction of dietary amino acid without making a distinction between the proportion used for protein synthesis and that utilized (deaminated) by the splanchnic bed (Dangin et al. 2001). The difficulties encountered in establishing the underlying mechanisms of sarcopenia illustrate well the limitations of the current approaches in the context of understanding regional protein metabolism. The widespread use of leucine in whole-body studies to quantify protein kinetics in response to dietary modulations might be called into question in view of the limited metabolism of this amino acid in the splanchnic zone. This latter metabolism is of crucial importance for postprandial protein accretion. The measurement of fluxes at the whole-body, organ or tissue levels is probably insufficiently specific to accurately detect any chronic slight dysfunction in the normal pattern of daily protein depletion–repletion cycling.

Newer, well-developed methods based on modifications of traditional approaches have emerged during the last 10 years. For example, the fluorescein isothiocyanate-labeled peptide technique is able to detect the synthesis of porcine reproductive and respiratory syndrome virus (PRRSV) (Xu et al. 2011). However, most new technologies have not been widely applied to protein breakdown and synthesis in farm animal research. Because protein breakdown and synthesis are energy-dependent processes, these processes are directly related to the efficiency of animal production. High-throughput methods are needed to study molecular regulation of protein turnover, and such understanding should be applied to regulate protein turnover during animal production.

References

Adegoke OAJ, McBurney MI, Baracos VE (1999) Jejunal mucosal protein synthesis: validation of luminal flooding dose method and effect of luminal osmolarity. Am J Physiol Gastrointest Liver Physiol 276:G14–G20

Adegoke OA, McBurney MI, Samuels SE, Baracos VE (2003) Modulation of intestinal protein synthesis and protease mRNA by luminal and systemic nutrients. Am J Physiol Gastrointest Liver Physiol 284(6):G1017–G1026

Balagopal P (1998) In-vivo measurement of protein synthesis in humans. Curr Opin Clin Nutr Metab Care 1(5):467–473

Balakrishna K, Murali HS, Batra HV (2010) Cloning, expression and characterization of attachment-invasion locus protein (Ail) of Yersinia enterocolitica and its utilization in rapid detection by immunoassays. Lett Appl Microbiol 50(2):131–137

Barle H, Nyberg B, Ramel S, Essén P, McNurlan MA, Wernerman J, Garlick PJ (1999) Inhibition of liver protein synthesis during laparoscopic surgery. Am J Physiol 277(4 Pt 1):E591–E596

Bark TH, McNurlan MA, Lang CH, Garlick PJ (1998) Increased protein synthesis after acute IGF-I or insulin infusion is localized to muscle in mice. Am J Physiol 275(1 Pt 1):E118–E123

Bergen BK, Lindsay S, Matlock T, Narayanan S (2007) Spatial and linguistic aspects of visual imagery in sentence comprehension. Cogn Sci 31(5):733–764

Biolo G, Chinkes D, Zhang XJ, Wolfe RR (1997) Harry M. Vars Research Award. A new model to determine in vivo the relationship between amino acid transmembrane transport and protein kinetics in muscle. J Parenter Enteral Nutr 16(4):305–315

Blank LM, Kuepfer L (2010) Metabolic flux distributions: genetic information, computational predictions, and experimental validation. Appl Microbiol Biotechnol 86(5):1243–1255

Bregendahl K, Liu L, Cant JP, Bayley HS, McBride BW, Milligan LP, Yen JT, Fan MZ (2004) Fractional protein synthesis rates measured by an intraperitoneal injection of a flooding dose of L-[ring-$^{2}H_{5}$] phenylalanine in pigs. J Nutr 134(10):2722–2728

Bregendahl K, Yang X, Liu L, Yen JT, Rideout TC, Shen Y, Werchola G, Fan MZ (2008) Fractional protein synthesis rates are similar when measured by intraperitoneal or intravenous flooding doses of L-[ring-$^{2}H_{5}$] phenylalanine in combination with a rapid regimen of sampling in Piglets. J Nutr 138(10):1976–1981

Burrin DG, Wester TJ, Davis TA, Fiorotto ML, Chang X (1999) Dexamethasone inhibits small intestinal growth via increased protein catabolism in neonatal pigs. Am J Physiol 276(2 Pt 1): E269–E277

Carey MM (2000) Managing conflict to gets your needs met. Can Oper Room Nurs J 18(3):23–26

Caso G, Scalfi L, Marra M, Covino A, Muscaritoli M, McNurlan MA, Garlick PJ, Contaldo F (2000) Albumin synthesis is diminished in men consuming a predominantly vegetarian diet. J Nutr 130(3):528–533

Cervino C, Asam S, Knopp D, Rychlik M, Niessner R (2008) Use of isotope-labeled aflatoxins for LC-MS/MS stable isotope dilution analysis of foods. J Agric Food Chem 56(6):1873–1879

Collier LS, Largaespada DA (2007) Transposable elements and the dynamic somatic genome. Genome Biol 8(Suppl 1):S5

Dai ZL, Zhang J, Wu G, Zhu WY (2010) Utilization of amino acids by bacteria from the pig small intestine. Amino Acids 39(5):1201–1215

Dai ZL, Zhu WY, Wu G (2011) Amino acid metabolism in intestinal bacteria: links between gut ecology and host health. Front Biosci 16:1768–1786

Dangin M, Boirie Y, Garcia-Rodenas C, Gachon P, Fauquant J, Callier P, Ballevre O, Beaufrere B (2001) The digestion rate of protein is an independent regulating factor of postprandial protein retention. Am J Physiol 280(2):E340–E348

Danicke S, Bottcher W, Simon O, Jeroch H (2001) The measurement of muscle protein synthesis in broilers with a flooding dose technique: use of ^{15}N-labelled phenylalanine, GC-MS and GC-C-IRMS. Isotopes Environ Health Stud 37(3):213–225

Davis TA, Fiorotto ML, Burrin DG, Reeds PJ, Nguyen HV, Beckett PR, Vann RC, O'Connor PMJ (2002) Stimulation of protein synthesis by both insulin and amino acids is unique to skeletal muscle in neonatal pigs. Am J Physiol Endocrinol Metab 282(4):E880–E890

Davis TA, Reeds PJ (2001) Of flux and flooding: the advantages and problems of different isotopic methods for quantifying protein turnover in vivo. II. Methods based on the incorporation of a tracer. Curr Opin Clin Nutr Metab Care 4(1):51–56

Davis TA, Fiorotto ML, Nguyen HV, Burrin DG (1999) Aminoacyl-tRNA and tissue free amino acid pools are equilibrated after a flooding dose of phenylalanine. Am J Physiol 277(1 Pt 1): E103–E109

Davis TA, Burrin DG, Fiorotto ML, Nguyen HV (1996) Protein synthesis in skeletal muscle and jejunum is more responsive to feeding in 7- than in 26-day-old pigs. Am J Physiol 270(5 Pt 1): E802–E809 de Meer K, Smolders HC, Meesterburrie J, de Sain-van der Velden M, Voorbij HA, Okken A, Reijngoud DJ, Kulik W (2000) A single food bolus stimulates albumin synthesis in growing piglets. J Pediatr Gastroenterol Nutr 31(3):251–257

Deng D, Yin YL, Li T, Huang R, Yin Y, Liu Z, Zhang J, Wu G (2009) Impaired translation initiation activation and reduced protein synthesis in weaned piglets fed a low-protein diet. J Nutr Biochem 20(7):544–552

Deng J, Wu X, Bin S, Li TJ, Huang R, Liu Z, Liu Y, Ruan Z, Deng Z, Hou Y, Yin YL (2010) Dietary amylose and amylopectin ratio and resistant starch content affects plasma glucose, lactic acid, hormone levels and protein synthesis in splanchnic tissues. J Anim Physiol Anim Nutr (Berl) 94(2):220–226

Deutz NE, Bruins MJ, Soeters PB (1998) Infusion of soy and casein protein meals affects interorgan amino acid metabolism and urea kinetics differently in pigs. J Nutr 128(12):2435–2445

Dudley MA, Burrin DG, Wykes LJ, Toffolo G, Cobelli C, Nichols BL, Rosenberger J, Jahoor F, Reeds PJ (1998) Protein kinetics determined in vivo with a multiple-tracer, single-sample protocol: application to lactase synthesis. Am J Physiol 274(3 Pt 1):G591–G598

Eden E, Geva-Zatorsky N, Issaeva I, Cohen A, Dekel E, Danon T, Cohen L, Mayo A, Alon U (2011) Proteome half-life dynamics in living human cells. Science 331(6018):764–768

Escobar J, Frank JW, Suryawan A, Nguyen HV, Van Horn CG, Hutson SM, Davis TA (2010) Leucine and α-ketoisocaproic acid, but not norleucine, stimulate skeletal muscle protein synthesis in neonatal pigs. J Nutr 140(8):1418–1424

Fan MZ, Chiba LI, Matzat PD, Yang X, Yin YL, Mine Y, Stein HH (2006) Measuring synthesis rates of nitrogen-containing polymers by using stable isotope tracers. J Anim Sci 84(Suppl): E79–E93

Fern EB, Garlick PJ, McNurlan MA, Waterlow JC (1981) The excretion of isotope in urea and ammonia for estimating protein turnover in man with [15N] glycine. Clin Sci 61(2):217–228

Frank JW, Escobar J, Nguyen HV, Jobgen SC, Jobgen WS, Davis TA, Wu G (2007) Oral N-carbamylglutamate supplementation increases protein synthesis in skeletal muscle of piglets. J Nutri 137(2):315–319

Frank JW, Escobar J, Suryawan A, Nguyen HV, Kimball SR, Jefferson LS, Davis TA (2006) Dietary protein and lactose increase translation initiation factor activation and tissue protein synthesis in neonatal pigs. Am J Physiol Endocrinol Metab 290(2):E225–E233

Fouillet H, Bos C, Gaudichon C, Tomé D (2002) Approaches to quantifying protein metabolism in response to nutrient ingestion. J Nutr 132(10):3208S–3218S

Garlick PJ, McNurlan MA, Preedy VR (1980) A rapid and convenient technique for measuring the rate of protein synthesis in tissues by injection of [^{3}H] phenylalanine. Biochem J 192(2):719–723

Goyarts T, Grove N, Dänicke S (2006) Effects of the Fusarium toxin deoxynivalenol from naturally contaminated heat given subchronically or as one single dose on the in vivo protein synthesis of peripheral blood lymphocytes and plasma proteins in the pig. Food Chem Toxicol 44(12):1953–1965

Hasselgren M, Gustafsson D, Ställberg B, Lisspers K, Johansson G (2005) Management, asthma control and quality of life in Swedish adolescents with asthma. Acta Paediatr 94(6):682–688

Hasunuma T, Harada K, Miyazawa SI, Kondo A, Fukusaki E, Miyake C (2010) Metabolic turnover analysis by a combination of in vivo ^{13}C-labelling from $^{13}CO_2$ and metabolic profiling with CE-MS/MS reveals rate-limiting steps of the C3 photosynthetic pathway in Nicotiana tabacum leaves. J Exp Bot 61(4):1041–1051

He Q, Yin Y, Zhao F, Kong X, Wu G, Ren P (2011) Metabonomics and its role in amino acid nutrition research. Front Biosci 16:2451–2460

Henshaw EC, Hirsch CA, Morton BE, Hiatt HH (1971) Control of protein synthesis in mammalian tissues through changes in ribosome activity. J Biol Chem 246(2):436–446

Hou Y, Wang L, Ding B, Liu Y, Zhu H, Liu J, Li Y, Kang P, Yin Y, Wu G (2011) Alpha-Ketoglutarate and intestinal function. Front Biosci 16:1186–1196

Hunter RJ, Patel VB, Baker AJ, Preedy VR (2004) Liver dysfunction induced by bile duct ligation and galactosamine injection alters cardiac protein synthesis. Metabolism 53(8):964–968

Jiang C, Shi P, Li S, Dong R, Tian J, Wei J, Luo S (2010) Gene expression profiling of skeletal muscle of nursing piglets. Int J Biol Sci 6(7):627–638

Kaganman I (2010) A glow on protein synthesis. Nat Methods 7(6):422–423

Kassauskas A, Viezeliene D (2004) Effect of Polyscias filicifolia Bailey biomass on protein synthesis process in isolated pig heart. Medicina (Kaunas) 40(10):991–996

Kong X, Tan B, Yin Y, Gao H, Li X, Jaeger LA, Bazer FW, Wu G (2012) Arginine stimulates the mTOR signaling pathway and protein synthesis in porcine trophectoderm cells. J Nutr Biochem 23(9):1178–1183

Lecker SH, Jagoe RT, Gilbert A, Gomes M, Baracos V, Bailey J, Price SR, Mitch WE, Goldberg AL (2004) Multiple types of skeletal muscle atrophy involve a common program of changes in gene expression. FASEB J 18(1):39–51

Li F, Yin Y, Tan B, Kong X, Wu G (2011) Leucine nutrition in animals and humans: mTOR signaling and beyond. Amino Acids 41(5):1185–1193

Li H, Jiang Y, He FC (2008) Recent development of metabonomics and its applications in clinical research. Yi Chuan 30(4):389–399

Lin G, Liu C, Wang T, Wu G, Qiao S, Li D, Wang J (2011) Biomarkers for optimal requirements of amino acids by animals and humans. Front Biosci (Schol Ed) 3:1298–1307

Marchini JS, Castillo L, Chapman TE, Vogt JA, Ajami A, Young VR (1993) Phenylalanine conversion to tyrosine: comparative determination with L-[ring-$^{2}H_5$]phenylalanine and L-[1-^{13}C]phenylalanine as tracers in man. Metabolism 42(10):1316–1322

Martinez JA (1987) Validation of a fast, simple and reliable method to assess protein synthesis in individual tissues by intraperitoneal injection of a flooding dose of [^{3}H] phenylalanine. J Biochem Biophys Methods 14(6):349–354

Matthews DE, Bier DM, Rennie MJ, Edwards RHT, Halliday D, Millward DJ, Clugston GA (1981) Regulation of leucine metabolism in man: a stable isotope study. Science 214 (4525):1129–1131

Matthews DE, Motil KJ, Rohrbaugh DK, Burke JF, Young VR, Bier DM (1980) Measurement of leucine metabolism in man from a primed, continuous infusion of L-[^{13}C]leucine. Am J Physiol 238(5):E473–E479

McNurlan MA, Tomkins AM, Garlick PJ (1979) The effect of starvation on the rate of protein synthesis in rat liver and small intestine. Biochem J 178(2):373–379

Torrazza RM, Suryawan A, Gazzaneo MC, Orellana RA, Frank JW, Nguyen HV, Fiorotto ML, El-Kadi S, Davis TA (2010) Leucine supplementation of a low-protein meal increases skeletal muscle and visceral tissue protein synthesis in neonatal pigs by stimulating mTOR-dependent translation initiation. J Nutr 140(12):2145–2152

Nichols NL, Bertolo RF (2008) Luminal threonine concentration acutely affects intestinal mucosal protein and mucin synthesis in piglets. J Nutr 138(7):1298–1303

Noack S, Nöh K, Moch M, Oldiges M, Wiechert W (2010) Stationary versus non-stationary ^{13}C-MFA: a comparison using a consistent dataset. J Biotechnol 154(2–3):179–190

Nöh K, Wiechert W (2011) The benefits of being transient: isotope-based metabolic flux analysis at the short time scale. Appl Microbiol Biotechnol 91(5):1247–1265

Nyachoti CM, de Lange CFM, McBride BW, Leeson S, Gabert VM (2000) Endogenous gut nitrogen losses in growing pigs are not caused by increased protein synthesis rates in the small intestine. J Nutr 130(3):566–572

Owen SF, McCarthy ID, Watt PW, Ladero V, Sanchez JA, Houlihan DF, Rennie MJ (1999) In vivo rates of protein synthesis in Atlantic salmon (Salmo salar L.) smolts determined using a stable isotope flooding dose technique. Fish Physiol Biochem 20(1):87–94

Reece P, Bremer M, Stones R, Danks C, Baumgartner S, Tomkies V, Hemetsberger C, Smits N, Lubbe W (2009) A bioinformatics approach to the development of immunoassays for specified risk material in canned meat products. Anal Bioanal Chem 394(7):1845–1851

Reeds PJ, Davis TA (1999) Of flux and flooding: the advantages and problems of different isotopic methods for quantifying protein turnover in vivo. I. Methods based on the dilution of a tracer. Curr Opin Clin Nutr Metab Care 2(1):23–28

Rennie MJ, Bowtell JL, Millward DJ (1994) Physical activity and protein metabolism. In: Bouchard C, Shepherd RJ, Stephens T (eds) Physical activity, fitness and health. Human Kinetics, Champaign, IL, pp 432–450
Rezzi S, Ramadan Z, Fay LB, Kochhar S (2007) Nutritional metabonomics: applications and perspectives. J Proteome Res 6(2):513–525
Sacheck JM, Hyatt JP, Raffaello A, Jagoe RT, Roy RR, Edgerton VR, Lecker SH, Goldberg AL (2007) Rapid disuse and denervation atrophy involve transcriptional changes similar to those of muscle wasting during systemic diseases. FASEB J 21(1):140–155
Schilling G (2007) Dark riddles. Sci Am 297(5):32–33
Stoll B, Burrin DG, Henry J, Jahoor F, Reeds PJ (1997) Phenylalanine utilization by the gut and liver measured with intravenous and intragastric tracers in pigs. Am J Physiol 273(6 Pt 1): G1208–G1217
Stoll B, Burrin DG, Henry J, Yu H, Jahoor F, Reeds PJ (1998) Dietary amino acids are the preferential source of hepatic protein synthesis in piglets. J Nutr 128(9):1517–1524
Stoll B, Burrin DG, Henry JF, Jahoor F, Reeds PJ (1999) Dietary and systemic phenylalanine utilization for mucosal and hepatic constitutive protein synthesis in pigs. Am J Physiol 276(1 Pt 1):G49–G57
Stoll B, Price PT, Reeds PJ, Chang X, Henry JF, van Goudoever JB, Holst JJ, Burrin DG (2006) Feeding an elemental diet vs a milk-based formula does not decrease intestinal mucosal growth in infant pigs. JPEN J Parenter Enteral Nutr 30(1):323–329
Tan BE, Yin YL, Kong XF, Li P, Li XL, Gao HJ, Li XG, Huang RL, Wu GY (2010) L-Arginine stimulates proliferation and prevents endotoxin-induced death of intestinal cells. Amino Acids 38(4):1227–1235
Tessari P, Garibotto G (2000) Interorgan amino acid exchange. Curr Opin Clin Nutr Metab Care 3(1):51–57
Tessari P, Inchiostro S, Zanetti M, Barazzoni R (1995) A model of skeletal muscle leucine kinetics measured across the human forearm. Am J Physiol 269(1 Pt 1):E127–E136
Thompson GN, Pacy PJ, Merritt H, Ford GC, Read MA, Cheng KN, Halliday D (1989) Rapid measurement of whole body and forearm protein turnover using a [$^2H_5$] phenylalanine model. Am J Physiol 256(5 Pt 1):E631–E639
Tomé D, Bos C (2000) Dietary protein and nitrogen utilization. J Nutr 130(7):1868S–1873S
Volpi E, Mittendorfer B, Wolf SE, Wolfe RR (1999) Oral amino acids stimulate muscle protein anabolism in the elderly despite higher first-pass splanchnic extraction. Am J Physiol 277(3 Pt 1):E513–E520
Waterlow JC, Golden MHN, Garlick PJ (1978) Protein turnover in man measured with 15N: comparison of end products and dose regimes. Am J Physiol 235(2):E165–E174
Wang X, Qiao SY, Yin YL, Yue L, Wang Z, Wu G (2007) A deficiency or excess of dietary threonine reduces protein synthesis in jejunum and skeletal muscle of young pigs. J Nutr 137(6):1442–1446
Wilson FA, Suryawan A, Orellana RA, Gazzaneo MC, Nguyen HV, Davis TA (2011) Differential effects of long-term leucine infusion on tissue protein synthesis in neonatal pigs. Amino Acids 40(1):157–165
Wolfe RR (1992) Radioactive and stable isotope tracers in biomedicine: principles and practice of kinetic analysis. Wiley, New York
Xu H, Wang J, Yu Z, Lv F, Hou J (2011) Screening of polypeptides binding to porcine reproductive and respiratory syndrome virus by phage display library. Wei Sheng Wu Xue Bao 51(1):127–133
Yao K, Yin YL, Chu W, Liu Z, Deng D, Li T, Huang R, Zhang J, Tan B, Wang W, Wu G (2008) Dietary arginine supplementation increases mTOR signaling activity in skeletal muscle of neonatal pigs. J Nutr 138(5):867–872
Yin Y, Yao K, Liu Z, Gong M, Ruan Z, Deng D, Tan B, Liu Z, Wu G (2010) Supplementing L-leucine to a low-protein diet increases tissue protein synthesis in weanling pigs. Amino Acids 39(5):1477–1486

Zamboni N (2011) ^{13}C metabolic flux analysis in complex systems. Curr Opin Biotechnol 22(1):103–108

Zanetti M, Barazzoni R, Kiwanuka E, Tessari P (1999) Effects of branched-chain-enriched amino acids and insulin on forearm leucine kinetics. Clin Sci (Lond) 97(4):437–448

Zhang Y, Reckow S, Webhofer C, Boehme M, Gormanns P, Egge-Jacobsen WM, Turck CW (2011) Proteome scale turnover analysis in live animals using stable isotope metabolic labeling. Anal Chem 83(5):1665–1672

Chapter 19
Methods for Measuring Amino Acids of Endogenous Origin in Intestines

Yulong Yin, Zhang Yuzhe, Feng Zemeng, Ziwei Xu, Jie Feng, Bo Deng, Zhou Chuanshe, and Ruin Huang

Contents

Abbreviations

AA	Amino acids
CaR	Ca^{2+}-sensing receptor
ENL	Endogenous nitrogen loss

Y. Yin (✉) • Z. Yuzhe • F. Zemeng • Z. Chuanshe • R. Huang
Chinese Academy of Sciences, Institute of Subtropical Agriculture, Research Center for Healthy Breeding of Livestock and Poultry, Hunan Engineering and Research Center of Animal and Poultry Science, Key Laboratory for Agro-ecological Processes in Subtropical Region, Scientific Observing and Experimental Station of Animal Nutrition and Feed Science in South-Central, Ministry of Agriculture, Changsha, Hunan 410125, Peoples Republic of China
e-mail: yinyulong@isa.ac.cn

Z. Xu • B. Deng
Institute of Animal Husbandry and Veterinary Science, Zhejiang Academy of Agricultural Sciences, Hang Zhou, Zhe Jiang Province 310021, Peoples Republic of China

J. Feng
The Key Laboratory of Molecular Animal Nutrition of Ministry of Education, College of Animal Science, Zhejiang University, Hangzhou, Peoples Republic of China

F. Blachier et al. (eds.), *Nutritional and Physiological Functions of Amino Acids in Pigs*, DOI 10.1007/978-3-7091-1328-8_19, © Springer-Verlag Wien 2013

GPRC6A	G-protein-coupled receptor family C member 6A
HA	Homoarginine
mGlu	Metabotropic glutamate
TCA	Trichloroacetic acid
VFT	Venus Fly Trap

19.1 Introduction

Evaluation of the true ileal digestibility of dietary proteins and amino acids (AA) in pigs is essential to improve growth and reduce production costs and nitrogen excretion in the environment. Several methods have been used to quantify endogenous nitrogen flow at the distal ileum of the pig. Earlier studies made use of conventional methods. These include feeding protein-free diets, feeding diets containing protein sources with an assumed 100 % digestibility, and mathematical regression techniques. More recently, isotope dilution, homoarginine, and peptide alimentation ultrafiltration techniques have been used. They all have their difficulties and limitations, as well be demonstrated in the following sections.

19.2 Feeding Protein-Free Diets Method

The classical method for measuring endogenous loss along the gastrointestinal tract in to measure fecal N excretion after feeding with N-free diet. When protein-free diets are given, all nitrogen-containing compounds found in the ileal digesta are assumed to be of endogenous origin. The main criticism of this method is its nonphysiological nature (Low 1980) that may affect the normal body protein metabolism and in turn may reduce secretion of nitrogenous compounds into the gut lumen and affect the efficiency of reabsorption (Darragh et al. 1990; Schulze 1994). The fact that animals are in a negative protein balance appears to have minor effects on endogenous loss of essential amino acids at the distal ileum of the pig (de Lange et al. 1989). In this study, only the recovery of endogenous proline, glycine, and total N was affected being reduced when amino acids rather than saline (control) were administered intravenously to animals fed a protein-free diet. As some research indicated, a protein-free diet may lack the stimulatory effect on endogenous gut protein secretions (Butts et al. 1993; Donkoh et al. 1995). This may lead to an underestimation of endogenous nitrogen loss (ENL) at the distal ileum and in feces. In addition, dietary constituents, such as fibers and other compounds, that are associated with dietary proteins, may enhance endogenous nitrogen loss. A further limitation of this approach is that it only yields information under the specific feeding conditions applied. Thus, for instance the effects of specific feed ingredients on endogenous losses, like proteins, cannot be measured.

19.3 Homoarginine Method

This method is based on the transformation of dietary lysine to homoarginine (HA) through a guanidination process (a reaction with *O*-methylisourea), which was first suggested by Hagemeister and Erbersdobler in 1985 and allows a differentiation between dietary lysine (converted to HA) and endogenous lysine (Hagemeister and Erbersdobler 1985). Homoarginine does not occur naturally in protein because it is not a substrate for protein synthesis and hence a return into the intestine of absorbed label in endogenous protein can be excluded a priori (Libao-Mercado et al. 2006).

Prerequisites for an effective use of this method include (1) uniform, or almost complete, guanidination of lysine in test protein, (2) guanidination must not affect protein digestibility, (3) HA must not be hydrolyzed to urea and lysine in the gut, (4) HA must be absorbed and metabolized like other amino acids, (5) HA must not re-enter the digestive tract, i.e., must not be used for protein synthesis, and (6) the presence of HA in the plasma or in the gut, per se, must not affect endogenous nitrogen loss (Nyachoti et al. 1997).

The suitability of this method for direct estimation of endogenous lysine flow and true lysine digestibility has been demonstrated in a number of studies (Rutherfurd and Moughan 1990) (Schmitz et al. 1991; Marty et al. 1994). However, since homoarginine is not used for protein synthesis, this technique may only be used for a short time to avoid lysine deficiency. Tews and Harper in a study in rats showed that the addition of an amino acid analogue (e.g., homoarginine) to a diet that is limiting in a natural amino acid and that has similar characteristics to that analogue (e.g., lysine) can induce an amino acid imbalance (Tews and Harper 1986). However, if HA would be rapidly hydrolyzed by arginase in the liver to lysine and urea and then if HA would serve as an effective source of lysine to the animal, then this concern would be unjustified. The fact that optimum guanidination conditions may vary for individual proteins and that near complete guanidination is not always achieved may impose some limitations on this method (Rutherfurd and Moughan 1990; De Vrese et al. 1994). So far this method has largely been used with diets having just one, purified protein source. A key requirement, however, for any method is that it has to be applied to a wide variety of feedstuffs. It needs to be determined whether this method can be equally employed to measure endogenous nitrogen flow when feeding a practical diet containing more than one protein source, and in particularly when plant proteins are evaluated. With this method, the true digestibility of amino acids other than lysine is not directly determined. Instead, they are estimated based on an assumed constant relationship between lysine and other amino acids in endogenous nitrogen (De Lange et al. 1990; Marty et al. 1994).

The main advantage with the method is that it allows the endogenous ileal excretion of lysine to be determined with the animal being fed a near-normal diet. A disadvantage is that it provides direct information about the endogenous flow of lysine only, and the flow of other amino acid are estimated by assuming a constant ratio of lysine to the other amino acids in the endogenous protein. In addition, another disadvantage of this technique is that homoarginine, although partially transformed to lysine, accumulates in the liver over time and has toxic effect on the animal, thus restricting the duration of experiments, especially when the direct method is used. Whereas the direct method does not rely on all of the assumptions that are inherent with the indirect method, there are often difficulties in achieving 100 % homoarginine of dietary origin as required for the direct method.

Overall, the homoarginine method appears to offer considerable promise for the determination of endogenous ileal amino acid flows. Endogenous flows can be determined in animals that are in a physiologically normal state, after ingestion of a wide range of dietary protein sources. Further studies are required to fully establish the validity of some of the assumptions inherent in the method. Comparisons of endogenous ileal lysine flows determined using the homoarginine and protein-free diet approaches have shown a lower lysine flow with the protein-free diet method (Moughan et al. 1990).

19.4 Enzymatically Hydrolyzed Casein Method

In this method (also referred to as Peptide Alimentation Ultrafiltration Method), first suggested by Moughan and Rutherfurd, animals are fed a diet in which the sole nitrogen source is enzymatically hydrolyzed casein (Moughan et al. 1990). Hydrolyzed casein contains approximately 56 % free amino acids and 41 % di- and tri-peptides with molecular weights of less than 5,000 Da (Leterme et al. 1996; Yin et al. 2004, 2008).

It is assumed that all endogenous nitrogen is present in fractions with molecular weights greater that 10,000 Da. Therefore, by separating low (<5,000 Da) and high (>10,000 Da) molecular weight nitrogen compounds in the ileal digesta, a distinction is made between any unabsorbed exogenous nitrogen and nitrogen of endogenous origin (Moughan et al. 1990).

In order to achieve this separation, protein-precipitating agents like perchloric acid and trichloroacetic acid have been tested but found to be unsuitable (Darragh et al. 1990).

As an alternative, Moughan et al. suggested separation of protein in ileal digesta immediately after collection by centrifugation and ultrafiltration in which low molecular fractions (amino acids and peptides) are removed with the supernatant. Any free amino acids and small peptides of endogenous origin present in ileal digesta are discarded as well (Moughan et al. 1990).

However, according to Moughan and Schuttert free and peptide-bound amino acids are only a small portion (11 %) of ENL (Moughan and Schuttert 1991). This suggests that any endogenous nitrogen loss in the supernatant is likely to result in slight underestimations of ENL. However, the exact proportion of these components needs to be clearly determined to enhance the reliability of this method. Some research reported values of 21 % and 9.2 %, respectively, suggesting the presence of considerable variation in the contribution of free and peptide-bound amino acids to ENL (Butts et al. 1992; Marty et al. 1994). Loss of material during sample preparation (centrifugation and ultrafiltration) may also lead to lower estimates with this method (Butts et al. 1993). A major restriction of the peptide alimentation ultrafiltration method is the fact that it cannot be used to estimate endogenous protein secretion since there is a direct association between protein and dietary factors that are known to influence endogenous secretions and ENL (e.g., in diets containing practical ingredients and/or with high levels of ANFs). Donkoh indicated that this method can be applied to feeds containing animal protein sources such as meat and bone meal (Donkoh et al. 1995).

A limitation of the method is that low molecular weight fractions of endogenous or endogenous origin, like peptides, free amino acids, urea, and ammonia (which account for 10–20 % of total digesta) are discarded (Butts et al. 1992, 1993).

19.5 Stable Isotopes Labeled Endogenous Nitrogen Methods

The use of stable isotopes allows the determination of endogenous protein flows while the animal is fed a diet that contains protein. The use of ^{15}N and ^{13}C seems most appropriate for this purpose. What can be used is the labeling of the dietary (Leterme et al. 1994) or animal nitrogen pool (Schulze 1994) with distinction between endogenous nitrogen and the proportion of endogenous nitrogen in the digesta and calculation from dilution of the isotope.

The disadvantage of labeling the dietary compounds is that contamination of endogenous secretions starts almost immediately after feeding (Schulze et al. 1995), though sequential slaughtering of animals shortly after the introduction of the labeled dietary is possible in rats (Moughan et al. 1992). However, with farm animals this approach may not be applicable because of the expensive cost and thus difficulty to repeat measurements. So labeling the animal or at least the precursor pool for endogenous secretions is commonly used.

Labeling the animal's precursor pool usually involves continuous intravenous infusion of labeled amino acids and then measurement of the amount of labeled amino acids present in the ileal digesta relative to the labeling in the precursor pool for the measurement of the endogenous gut protein synthesis. The deproteinized fraction of blood plasma is generally considered to be the precursor pool (Moughan et al. 1992). ^{15}N-labeled leucine is the most commonly used amino acid as it is presumed to be readily metabolized resulting in labeling of other amino acid by transamination. However, ^{15}N-labeled leucine is not transferred to lysine or threonine, and the isotopic enrichment of other amino acids is less than that of the labeled leucine (Lien et al. 1997). Alternatively, the use of mixtures of ^{15}N-labeled amino acids may solve the problem, but the cost is very expensive (Yin et al. 2000).

It is very important to choose an appropriate precursor pool (Moughan et al. 1992). The TCA-soluble blood plasma fraction is commonly used to indicate ^{15}N enrichment. It is assumed that the free amino acid pool in the TCA-soluble fraction is used for the synthesis of body protein and thus for most of the endogenous secretions into the digestive tract. The assumption is not completely true, however, for small intestinal mucosa secretions, in which case absorbed dietary amino acids are used directly for protein synthesis (Leterme et al. 1994). Therefore, plasma enrichment underestimates the enrichment of endogenous nitrogen secreted into the intestinal lumen which leads to an overestimation of endogenous protein excretion (Lien et al. 1997). More work is required to isolate the actual precursor pool for endogenous secretion.

A limitation with the ^{15}N method is that only endogenous nitrogen is directly measured, not the amino acid. Thus to calculate the endogenous amino acid flows, it must be assumed that there is a constant ratio in the endogenous material between nitrogen and the amino acid.

The ^{15}N-isotope dilution technique is a valuable method for studying factors that influence the recovery of endogenous protein in digesta, such as antinutritional factors and plant fiber, but requires modification to improve accuracy.

The stable carbon isotope ^{13}C has also been investigated for the determination of endogenous protein flows. When infused intravenously or given orally to the animals, ^{13}C is incorporated into body proteins (Minson et al. 1975; Kong et al. 2012).

Atmospheric carbon dioxide contains approximately 1.1 % of the heavier isotope ^{13}C and 98.9 % of the lighter ^{12}C. Plants discriminate between ^{12}C and ^{13}C during photosynthesis in ways that reflect plant metabolism (O'Leary 1981). Plants that fix carbon dioxide via the dicarboxylic acid pathway (C4 pathway) show less isotopic discrimination than plants that fix carbon dioxide via the Calvin cycle(Minson et al. 1975). In plant physiology, ^{13}C abundances are commonly expressed as a delta ^{13}C value in parts per thousand or parts per mil (‰). Absolute values are difficult to obtain, and for most purposes, it is adequate to give delta ^{13}C values relative to carbon from a standard. Therefore, feedstuffs that are normally used to formulate diets differ in the degree of labeling with ^{13}C, so that diets can be prepared with different abundances of ^{13}C. The endogenous ileal protein flows in pigs by feeding ^{13}C-labeled dietaries has been studied and the results obtained demonstrate that the endogenous flows are similar to those previously obtained using ^{15}N (Arentson and Zimmerman 1995). Leterme et al. compared endogenous protein flows determined following intravenous infusion of ^{13}C-leucine with infusion of ^{15}N-leucine in the pig and found lower values for endogenous ileal nitrogen flows with the ^{13}C method i.e. 1.4 vs. 5.7 mg/g dry matter intake for the ^{13}C and ^{15}N method, respectively. This difference in endogenous flows between the two labels was suggested to be due to an overestimation of endogenous N with ^{15}N due to underestimation of the enrichment of the endogenous N which occurs when the TCA-soluble blood plasma fraction is used as the precursor pool, as described above. The use of ^{13}C for the determination of endogenous protein flows has the same disadvantages as ^{15}N in terms of recycling of the label and the determination of an appropriate precursor pool, but the advantage of using ^{13}C over ^{15}N is the lower cost of labeling the experimental diets. When ^{13}C-leucine is intravenously administered to the animal, however, the ^{13}C cannot be transferred to other amino acids, so that only the endogenous flow of leucine can be determined (Leterme et al. 1998).

19.6 Model of Association Between Amino Acid-Sensing Genes and the Amino Acids in Plasma

A group of extracellular amino acid sensors has recently been identified within class 3 of the G-protein-coupled receptor superfamily (Feng et al. 2012). These sensors control G-protein-regulated signaling pathways and are widely distributed. They seem to be designed for the detection of subgroups of amino acids or even specific amino acids (including the acidic amino acid, L-glutamate and the metabotropic glutamate (mGlu) receptors). They also include several broad-spectrum amino acid sensors; the extracellular Ca^{2+}-sensing receptor (CaR), which responds to aromatic, aliphatic, and polar amino acids (but not to positively charged or branched-chain amino acids) (Conigrave et al. 2000); a heterodimeric taste receptor composed of the T1R1 and T1R3 receptors (which respond to aliphatic, polar, charged, and branched-chain amino acids but not to aromatic amino acids) (Nelson et al. 2002); and finally the goldfish 5.24 receptor (Speca et al. 1999; Kuang et al. 2003) and its putative mammalian ortholog, G-protein-coupled receptor family Cmember 6A (GPRC6A) (Wellendorph and Bräuner-Osborne 2004; Kuang et al. 2005; Wellendorph et al. 2005), which respond to basic, aliphatic, and polar amino acids.

The binding of amino acids to these receptors, including CaR, GPRC6A, and the T1R taste receptors provide mechanisms by which changes in the intestinal and plasma levels of amino acids regulate digestion, absorption, growth and metabolism; and mechanisms by which oral amino acids mediate the sensation of umami taste. Amino acid binding depends on highly conserved residues in the N-terminal VFT domains that bind the a-amino and a-carboxylate functional groups.

The VFT domains mediate amino acid binding in the class 3 GPCRs. These bilobed structural motifs have an ancient lineage providing the basis of nutrient sensing by the periplasmic binding proteins of Gram-negative bacteria (Felder et al. 1999).

The CaR shares pluripotency along with 3 GPCRs that mediate taste (Conigrave and Brown 2006). With various signals to process and numerous signaling pathways to access, the CaR acts as a multichannel switch box, providing ligand selective control of physiological responses. The CaR's response to amino acids is dependent not only on variations in amino acid concentration to which the receptor is exposed but also on the prevailing Ca^{2+}concentration. Indeed, the different amino acids exhibit differential Ca^{2+} concentration thresholds for receptor activation.

The CaR is widely expressed in mammalian tissues including cells with recognized roles in amino acid sensing. These include acid-, enzyme-, and hormone-secreting cells of the stomach and small intestine, the liver, exocrine and endocrine cells of the pancreas, and endocrine cells of the anterior pituitary and parathyroid (for a review, see Hofer and Brown 2003). It is also expressed in glia and neurons in the brain (Ruat et al. 1995; Chattopadhyay et al. 1997, 1998).

CaR is selective for aromatic amino acids because these amino acids are its most potent activators (Conigrave et al. 2000). However, when the plasma concentrations of different amino acids are taken into consideration (aliphatic and polar amino acids > aromatic amino acids), CaR can be recognized as a broad-spectrum amino acid sensor that is "tuned" to respond to comparable relative changes in 6–10 of the genetically encoded amino acids including L-Phe, L-Trp, L-Tyr, L-His, L-Thr, L-Ser, L-Ala, L-Gln, L-Asn, and Gly (Conigrave et al. 2004).

Although CaR is a broad-spectrum amino acid sensor, it is not a universal amino acid sensor. Thus, it is significantly less sensitive to the acidic amino acids aspartate and glutamate and is essentially unresponsive to basic and branched-chain amino acids. This means that universal amino acid sensing would require coexpression of the CaR with one or more additional amino acid sensors such as GPRC6A or the amino acid-sensing T1R1/T1R3 taste receptor (Nelson et al. 2002).

Activated CaR markedly stimulate gastric acid production (Busque et al. 2005). It seems likely that amino acid dependent activation of CaR also modulates the secretion of hormones involved in the control of calcium metabolism, growth, nutrient disposition, and intestinal digestion and absorption of nutrients. GPRC6A was classified as an orphan class 3 GPCR until the recent demonstration that both the mouse and human orthologs are activated by amino acids (Kuang et al. 2005; Wellendorph et al. 2005).

CaR can integrate signals arising from distinct classes of nutrients: mineral ions and amino acids.

The extracellular CaR is activated by a variety of divalent and polyvalent cations that act as receptor agonists (Hofer and Brown 2003). In addition, recent work indicates that the CaR is allosterically activated by L-amino acids (Conigrave et al. 2000). These effects are stereoselective for natural (i.e., L-) isomers of CaR activated by amino acids including aromatic amino acids such as Phe, Trp, and His as well as aliphatic and polar amino acids. Branched-chain amino acids are almost inactive, and charged amino acids, including arginine and lysine, are much less effective than aromatic and other amino acids (Conigrave et al. 2000). The CaR Venus flytrap domain, but none of its other domains, is required for L-amino acid sensing (Mun et al. 2004). Ca^{2+} and L-amino acid-sensing function of CaR can be dissociated (Mun et al. 2005). Mutational analysis of conserved binding residues in the CaR Venus flytrap domain has led to the identification of two mutants, T145A and S170T, which selectively impair L-amino acid sensing, leaving extracellular Ca^{2+} sensing intact. Furthermore, the double mutant T145A/S170T selectively disables amino acid sensing (Mun et al. 2005).

The CaR is, however, the only member of this receptor family identified thus far to select for aromatic L-amino acids and to respond more generally to a large number of L-amino acids.

The CaR is widely expressed in mammalian tissues and is expressed at high levels in endocrine cells such as parathyroid chief cells, thyroid C cells, and anterior pituitary cells (Brown and MacLeod 2001). In the brain, the CaR is expressed in the ionic strength-sensing subfornical organ that provides inputs to hypothalamic centers that control antidiuretic hormone secretion, various other organs including the hippocampus and myelin-producing oligodendrocytes.

In the GI tract, the CaR is expressed at high levels on the basolateral membranes of gastric parietal cells (Cheng et al. 1999). It provides an explanation for the recognized phenomena of Ca^{2+}-dependent and amino acid-dependent activation of acid secretion (Busque et al. 2005), and the CaR's selectivity for aromatic amino acids provides an explanation for the recognized selectivity of gastric acid secretion for aromatic amino acids. The CaR is also expressed in epithelial cells and enteroendocrine cells of the small intestine (Ray et al. 1997). It seems plausible that the CaR mediates the known effects of aromatic amino acids, on cholecystokinin release and thus bile flow and pancreatic enzyme secretion (Conigrave and Brown 2006).

GPRC6A shows the highest degree of amino acid sequence identity to the goldfish 5.24 chemosensory receptor (41 %), which has a broad spectrum of amino acid sensing and an apparent preference for the basic amino acids L-Arg and L-Lys (Speca et al. 1999). The 5.24 receptor is expressed in the olfactory epithelium and is thought to have a role in feeding-related navigation by sensing gradients of amino acids in water.

GPRC6A is widely expressed in mammalian tissues including brain, skeletal muscles, heart, lung, spleen, kidney, and liver (Kuang et al. 2005), although its roles in these tissues are currently unknown.

Three alternatively spliced forms of GPRC6A are present in both the mouse and the human; two of these are short forms that have inframe deletions of all, or parts, of exons 3 and 4, which code for sequences in the Venus Fly Trap (VFT) domain. Surprisingly, whereas the mouse long protein isoform is readily expressed on the cell surface of transfected HEK293 cells, the corresponding human isoform is expressed at much lower levels (Kuang et al. 2005; Wellendorph et al. 2005).

Both GPRC6A and the goldfish 5.24 chemosensory receptor have broad-spectrum amino acid sensing with greatest sensitivities to the basic amino acids L-Arg and L-Lys.

GPRC6A also responds sensitively to polar and aliphatic, but not other, amino acids (Kuang et al. 2005; Wellendorph et al. 2005). These findings seem to indicate that the 5.24 receptor is more tolerant of variations in the size, shape, and even charge of the amino acid side chain than is GPRC6A.

Much less is known about the expression and function in enteroendocrine cells of heterodimers of T1R1 and T1R3, thought to act as broadly tuned amino acid sensors in the mouse (Nelson et al. 2002; Wellendorph et al. 2009).

References

Arentson R, Zimmerman D (1995) True digestibility of amino acids and protein in pigs with 13C as a label to determine endogenous amino acid excretion. J Anim Sci 73(4):1077–1085

Brown EM, MacLeod RJ (2001) Extracellular calcium sensing and extracellular calcium signaling. Physiol Rev 81(1):239–297

Busque SM, Kerstetter JE et al (2005) L-type amino acids stimulate gastric acid secretion by activation of the calcium-sensing receptor in parietal cells. Am J Physiol Gastrointest Liver Physiol 289(4):G664–G669

Butts CA, Moughan PJ et al (1992) Protein nitrogen, peptide nitrogen and free amino acid nitrogen in endogenous digesta nitrogen at the terminal ileum of the rat. J Sci Food Agric 59(3):291–298

Butts CA, Moughan PJ et al (1993) Endogenous lysine and other amino acid flows at the terminal ileum of the growing pig (20 kg bodyweight): The effect of protein-free, synthetic amino acid, peptide and protein alimentation. J Sci Food Agric 61(1):31–40

Chattopadhyay N, Légrádi G et al (1997) Calcium-sensing receptor in the rat hippocampus: a developmental study. Brain Res Dev Brain Res 100(1):13–21

Chattopadhyay N, Ye CP et al (1998) Extracellular calcium-sensing receptor in rat oligodendrocytes: expression and potential role in regulation of cellular proliferation and an outward K+ channel. Glia 24(4):449–458

Cheng I, Qureshi I et al (1999) Expression of an extracellular calcium-sensing receptor in rat stomach. Gastroenterology 116(1):118–126

Conigrave AD, Brown EM (2006) Taste receptors in the gastrointestinal tract II: L-Amino acid sensing by calcium-sensing receptors: implications for GI physiology. Am J Physiol Gastrointest Liver Physiol 291(5):G753–G761

Conigrave AD, Quinn SJ et al (2000) L-amino acid sensing by the extracellular Ca2+−sensing receptor. Proc Natl Acad Sci 97(9):4814

Conigrave AD, Mun HC et al (2004) L-amino acids regulate parathyroid hormone secretion. J Biol Chem 279(37):38151–38159

Darragh AJ, Moughan PJ et al (1990) The effect of amino acid and peptide alimentation on the determination of endogenous amino acid flow at the terminal ileum of the rat. J Sci Food Agric 51(1):47–56

de Lange C, Sauer W et al (1989) The effect of protein status of the pig on the recovery and amino acid composition of endogenous protein in digesta collected from the distal ileum. J Anim Sci 67(3):755

De Lange C, Souffrant W et al (1990) Real ileal protein and amino acid digestibilities in feedstuffs for growing pigs as determined with the 15N-isotope dilution technique. J Anim Sci 68(2):409–418

De Vrese M, Middendorf K et al (1994) Prevention of amino acid racemization during guanidination—a prerequisite for measurement of protein digestibility by homoarginine labeling. Z Ernahrungswiss 33(4):310–312

Donkoh A, Moughan PJ et al (1995) Comparison of methods to determine the endogenous amino acid flow at the terminal ileum of the growing rat. J Sci Food Agric 67(3):359–366

Felder CB, Graul RC et al (1999) The Venus flytrap of periplasmic binding proteins: an ancient protein module present in multiple drug receptors. AAPS J 1(2):7–26

Feng Z-M, Guo J-P, Kong X-F, Wang W-C, Li F-N, Nyachoti M, Yin Y-L (2012) Molecular cloning and expression profiling of G protein coupled receptor 120 in Landrace pig and different Chinese indigenous pig breeds. J Food Agric Environ 10(3&4):809–814

Hagemeister H, Erbersdobler H (1985) Chemical labeling of dietary protein by transformation of lysine to homoarginine: a new technique to follow intestinal digestion and absorption. Proc Nutr Soc 44:133A

Hofer AM, Brown EM (2003) Extracellular calcium sensing and signalling. Nat Rev Mol Cell Biol 4(7):530–538

Kong XF, Tan BE, Yin YL, Li XL, Jaeger LA, Bazer FW, Wu GY (2012) Arginine stimulates the mTOR signaling pathway and protein synthesis in porcine trophectoderm cells. J Nutr Biochem 23(9):1178–1183

Kuang D, Yao Y et al (2003) Molecular similarities in the ligand binding pockets of an odorant receptor and the metabotropic glutamate receptors. J Biol Chem 278(43):42551–42559

Kuang D, Yao Y et al (2005) Cloning and characterization of a Family C orphan G-protein coupled receptor. J Neurochem 93(2):383–391

Leterme P, van Leeuwen P, Théwis A (1994) Effect of the carbohydrate fractions of pea seeds on the endogenous amino acid flow at the terminal ileum of the pig. In: Souffrant WB, Hagemeister H (eds) Digestive physiology in pigs. EAAP Publ. No. 80. Forschungsinstitut für die Biologie landwirtschaftlicher Nutztiere, Dummerstorf, p 67

Leterme P, Monmart T et al (1996) Effect of oral and parenteral N nutrition vs N-free nutrition on the endogenous amino acid flow at the ileum of the pig. J Sci Food Agric 71(2):265–271

Leterme P, Sève B et al (1998) The current 15N-leucine infusion technique is not suitable for quantitative measurements of ileal endogenous amino acid flows in pigs. J Nutr 128 (11):1961–1968
Libao-Mercado AJ, Yin Y, Van Eys J, De Lange CFM (2006) True ileal amino acid digestibility and endogenous ileal amino acid losses in growing pigs fed wheat shorts- or casein-based diets. J Anim Sci 84:1351–1361
Lien K, Sauer W et al (1997) Evaluation of the 15N-isotope dilution technique for determining the recovery of endogenous protein in ileal digestion of pigs: effect of dilution in the precursor pool for endogenous nitrogen secretion. J Anim Sci 75(1):148
Low AG (1980) Nutrient absorption in pigs. J Sci Food Agric 31(11):1087–1130
Marty B, Chavez E et al (1994) Recovery of amino acids at the distal ileum for determining apparent and true ileal amino acid digestibilities in growing pigs fed various heat-processed full-fat soybean products. J Anim Sci 72(8):2029–2037
Minson DJ, Ludlow MM et al (1975) Differences in natural carbon isotope ratios of milk and hair from cattle grazing tropical and temperate pastures. Nature 256(5518):602
Moughan PJ, Schuttert G (1991) Composition of nitrogen-containing fractions in digesta from the distal ileum of pigs fed a protein-free diet. J Nutr 121(10):1570
Moughan PJ, Darragh AJ et al (1990) Perchloric and trichloroacetic acids as precipitants of protein in endogenous ileal digesta from the rat. J Sci Food Agric 52(1):13–21
Moughan PJ, Buttery PJ et al (1992) Evaluation of the isotope dilution technique for determining ileal endogenous nitrogen excretion in the rat. J Sci Food Agric 58(2):165–172
Mun HC, Franks AH et al (2004) The Venus Fly Trap domain of the extracellular Ca2+−sensing receptor is required for L-amino acid sensing. J Biol Chem 279(50):51739–51744
Mun HC, Culverston EL et al (2005) A double mutation in the extracellular Ca2+-sensing receptor's Venus flytrap domain that selectively disables L-amino acid sensing. J Biol Chem 280(32):29067–29072
Nelson G, Chandrashekar J et al (2002) An amino-acid taste receptor. Nature 416(6877):199–202
Nyachoti C, De Lange C et al (1997) Significance of endogenous gut nitrogen losses in the nutrition of growing pigs: a review. Can J Anim Sci 77(1):149–163
O'Leary MH (1981) Carbon isotope fractionation in plants. Phytochemistry 20(4):553–567
Ray JM, Squires PE et al (1997) Expression of the calcium-sensing receptor on human antral gastrin cells in culture. J Clin Invest 99(10):2328
Ruat M, Molliver ME et al (1995) Calcium sensing receptor: molecular cloning in rat and localization to nerve terminals. Proc Natl Acad Sci 92(8):3161
Rutherfurd SM, Moughan PJ (1990) Guanidination of lysine in selected dietary proteins. J Agric Food Chem 38(1):209–211
Schmitz M, Hagemeister H et al (1991) Homoarginine labeling is suitable for determination of protein absorption in miniature pigs. J Nutr 121(10):1575
Schulze H (1994) Endogenous ileal nitrogen losses in pigs: dietary factors[D]. Department of Animal Nutrition, Wageningen Agricultural University, Wageningen
Schulze H, Butts CA et al (1995) The 15N-isotope dilution method for determining ileal endogenous nitrogen excretion in the young (10 kg liveweight) pig. J Sci Food Agric 69(1):41–50
Speca DJ, Lin DM, Sorensen PW, Isacoff EY, Ngai J, Dittman AH (1999) Functional identification of a goldfish odorant receptor. Neuron 23:487–498
Tews JK, Harper AE (1986) Induction in rats of lysine imbalance by dietary homoarginine. J Nutr 116(10):1910
Wellendorph P, Bräuner-Osborne H (2004) Molecular cloning, expression, and sequence analysis of GPRC6A, a novel family C G-protein-coupled receptor. Gene 335:37–46
Wellendorph P, Hansen KB, Balsgaard A, Greenwood JR, Egebjerg J, Bräuner-Osborne H (2005) Deorphanization of GPRC6A: a promiscuous L-α-amino acid receptor with preference for basic amino acids. Mol Pharmacol 67(3):589–597
Wellendorph P, Johansen LD et al (2009) Molecular pharmacology of promiscuous seven transmembrane receptors sensing organic nutrients. Mol Pharmacol 76(3):453–465

Yin YL, McEvoy J, Souffrant WB, Schulze H, McCracken KJ (2000) Apparent digestibility (ileal and overall) of nutrients and endogenous nitrogen losses in growing pigs fed wheat or wheat by-products without or with xylanase supplementation. Livest Prod Sci 62:119–132

Yin YL, Huang RL, Jeaurond E, Rademacher M, de Lange CFM (2004) Effect of including purified Jack Bean lectin in a casein based diet on apparent and true ileal amino acid digestibility in growing pigs. Anim Sci 79:283–291

Yin YL, Li TJ, Huang RL (2008) Evaluating standardized ileal digestibility of amino acids in growing pigs. Anim Feed Sci Technol 140:385–401

Index

DOI 10.1007/978-3-7091-1328-8,

N

O

P

MIX
Papier aus verantwortungsvollen Quellen
Paper from responsible sources
FSC® C105338

If you have any concerns about our products,
you can contact us on
ProductSafety@springernature.com

In case Publisher is established outside the EU,
the EU authorized representative is:
Springer Nature Customer Service Center GmbH
Europaplatz 3, 69115 Heidelberg, Germany

Printed by Libri Plureos GmbH
in Hamburg, Germany